How To Build Solid-State Audio Circuits

By Mannie Horowitz

TAB BOOKS
Blue Ridge Summit, Pa. 17214

FIRST EDITION

FIRST PRINTING—NOVEMBER 1972

Printed in the United States
of America

Hardbound Edition: International Standard Book No. 0-8306-2606-9

Paperbound Edition: International Standard Book No. 0-8306-1606-3

Library of Congress Card Number: 72-80560

Preface

Scientists perform experiments and then generate "theories" in an attempt to explain their observations. Some of these theories are supported by evidence gathered over the years; others are simply disproved and discarded as so much nonsense.

This is what happened in the engineering field regarding transistors and transistor circuits. Many engineers have tried to explain the various phenomena they observed, especially in the case of audio power amplifiers. Many of these explanations held up well while others had to be disposed of as invalid. To one degree or another every engineer was guilty of this weakness. I was no exception. In this book I made every attempt to avoid theory unless it was required to do actual practical design work. It is my goal here to provide the tools necessary to design transistor audio circuits successfully. As you proceed through the book, you will find some sections covering elementary topics. I am certain that most readers know the information covered. But the basic facts are included primarily to help anyone who is not too familiar with the topic discussed and secondarily to present a complete discussion of organized knowledge.

Circuit design is a science based primarily on fact, and to an even greater degree, on trial and error. Before reaching the trial-and-error stage, a circuit should be designed on paper. Design requires mathematics—mostly simple arithmetic. Some basic knowledge of algebra is assumed, as is the ability to convert to different units such as from amperes to milliamperes to microamperes, or from ohms to kilohms to megohms, and so on. This information is readily available in various text books. Also assumed is a fundamental knowledge of electric circuits and a familiarity with the basic "hardware" such as capacitors, resistors and inductors. Everyone reading this is probably as familiar with voltage, current, impedance and reactance, as he is with his postal zip code number. However, much of this will be reiterated throughout the book just for the sake of clarifying a more involved point.

The technical discussion must be initiated by describing two basic conventions concerned with the direction of current flow. In the heyday of vacuum tubes, it was taken for granted that current flows from the negative terminal of a battery through the load and back to the positive terminal. This is the actual direction in which electrons flow. With tubes it is significant only because electrons flow from the heated cathode in the electron tube to the plate.

When talking about circuits involving semiconductors, it is much easier to use conventional current flow as the standard. Here, current starts to flow from the positive terminal of the power supply, through the load, and back to the negative terminal. This direction is just the reverse of electron flow. The arrow in diode and transistor symbols indicates the direction of current—not electron-flow. It is a great help when describing transistor circuits to remember that current flow is in the direction of the arrow drawn in the symbol of the particular semiconductor.

I would like to thank the editor of Radio-Electronics, Larry Steckler, Gernsback Publications and M. Harvey Gernsback, the editor and publisher, for permission to use the material previously published in the magazine as the basis of this book.

Mannie Horowitz

Contents

Fundamentals of Semiconductor Devices

Chapter 1

Every experimenter is familiar with electrical conductors. Ordinary copper hookup wire used to connect two components in a piece of electronic equipment is a common example of a conductor. The chassis used to support the components of an audio amplifier is frequently used as a conductor.

Insulators are just as important in audio work as are the conductors. Air resists the flow of electricity at the most used voltage levels, and is thus an excellent insulator. The plastic, cotton, or enamel covering over hookup wire are important insulators. These insulators are frequently rated by the amount of voltage they can safely withstand before they break down and start conducting.

Semiconductors, as the name implies, are materials that fall somewhere between conductors and insulators in their ability to conduct electricity. Although germanium and silicon are fair insulators, in their pure state, the addition of impurities changes all this. Impure germanium and silicon are the basic materials of most semiconductors currently in use.

The characteristics of the semiconductor materials are determined by the impurities added. While one type of impurity will cause an excess of electrons to float around in the semiconductor material, a second type will provide a material lacking in electrons or positive in nature. These materials are assigned names: N-type material contains an excess of electrons while P-type material is lacking in electrons.

JUNCTION DIODES

An independent piece of N-type or P-type material is of little value by itself. But if an N-type semiconductor is joined to a P-type semiconductor, the practical value of the resulting device is without bounds. This is the geometry of the P-N junction diode. Depending upon the application and specific design, it may be called a silicon rectifier, a zener diode, a noise diode, a signal diode, a switching diode, as well as by many other designations.

The symbol of the junction diode is shown in Fig. 1-1A. The terminal with the arrow is made of P-type material and is

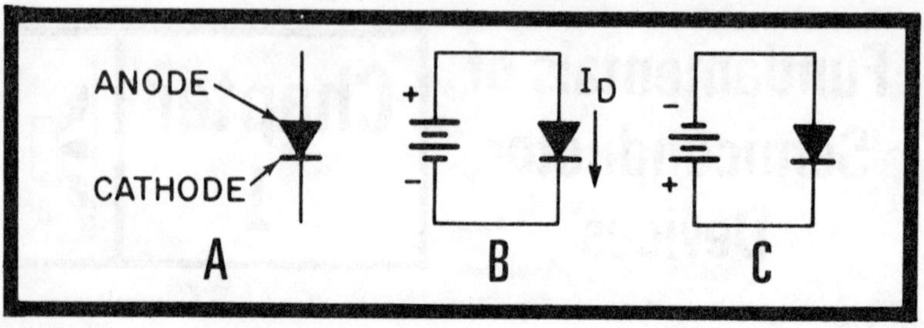

Fig. 1-1. Junction diode symbol (A). A diode can conduct when forward biased (B) and cannot conduct when reverse biased (C).

referred to as the anode. The straight line is the cathode and is made of an N-type slab. Figs. 1-1B and 1-1C indicates what happens if a battery is connected across the diode. (Do not try this experimentally unless you include a resistor in series with the circuit to limit diode current to a safe value.)

A diode will conduct (Fig. 1-1B) when the voltage supplied at the anode is positive with respect to that at the cathode. In other words, the P-type material must be positive with respect to the N-type material if the diode is to conduct. This is an important fact to remember. It is also important to notice the direction in which the electron current will flow. It is indicated by the arrow next to the diode symbol. Current flows from the negative battery terminal, through the diode, to the positive terminal. In Fig. 1-1C, the battery is connected in the reverse direction. Theoretically, no current will flow. This last statement will be modified to some degree later.

Now let us turn our attention to the forward-biased diode, the arrangement where it will conduct current. There is no intention of implying that a diode will conduct if a supply voltage is connected as shown in Fig. 1-1B; other conditions could prevent conduction. However, the voltages must be applied with the polarity shown if the diode is to conduct.

The amount of current a diode conducts depends on the circuit, the applied voltage and the characteristics of the particular diode. A typical forward-biased diode curve is shown in Fig. 1-2A. The solid line indicates the relationship between the current the diode passes and the voltage across the diode.

It is simple to determine from the curve just how much current will flow when a specific voltage is placed across the diode. Just draw a vertical line from the voltage axis (horizontal axis) in Fig. 1-2B, starting at a point equalling the voltage (V_1) across the diode. Extend the line until it touches the solid line diode characteristic curve. Now draw a horizontal line from the point where the diode curve and the vertical line just drawn intersect. Extend this line to the vertical current axis and read the current at the point where this line touches the vertical axis. Current I_1 flows through the diode when voltage V_1 is present across the diode.

Returning to Fig. 1-2A, you will notice that at low voltages, very little or no current flows. As the voltage is increased, the amount of current that flows increases rapidly. This can be seen in a more instructive way in Fig. 1-3. Here, the diode is in a circuit with a resistor and variable voltage power supply. Assume that E is increased to very large values. The current that can flow will be limited by resistor R. Regardless of the current flowing, the voltage across the diode will rise very slowly after voltage V has been exceeded. The bulk of voltage E will be developed across the load resistor R. The voltage across the diode will increase only somewhat above V in Fig. 1-2A.

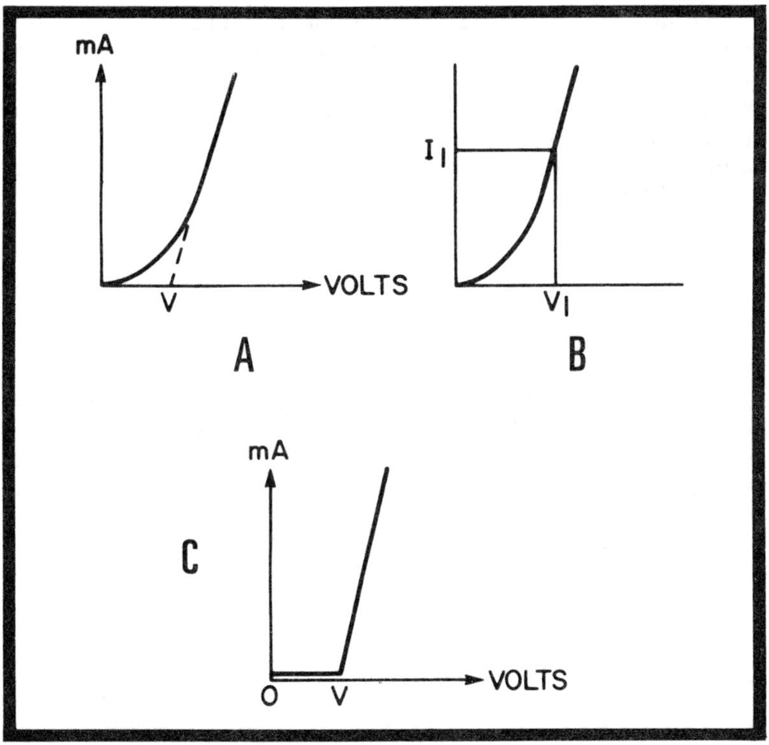

Fig. 1-2. Forward-biased diode curves. Curve A represents a forward-biased diode. Curve B shows the method of determining diode current. Curve C approximates a diode's characteristics.

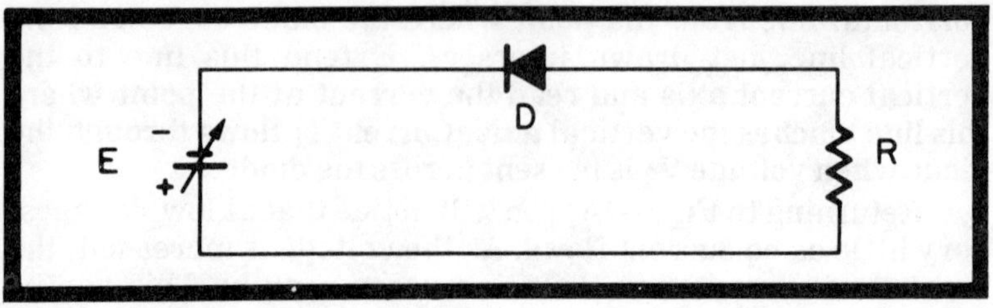

Fig. 1-3. This diode is forward biased by a variable voltage source. Resistor R limits the diode current to a safe level.

It is interesting to stop here for a moment to see just what is happening. Input voltage E rises rapidly and the current through the circuit increases rapidly. Yet the voltage across the diode remains at about V and rises only slightly above V. The voltage across a load in parallel with the diode will remain quite constant at V despite supply voltage variations. This is the basis of a voltage regulator.

Point V in Fig. 1-2A is quite important: This is basic to the approximate characteristic curve of the diode. V is determined by assuming that the straight portion of the curve is extended to the zero current axis. The total approximate curve for the forward biased diode, is shown in Fig. 1-2C. From 0 volts to V volts, no current flows. Therefore, it can be assumed that no current flows when the voltage across the diode is less than V. As soon as the voltage across the diode is increased even slightly above V, current will flow and the voltage across the diode will remain at about V volts or rise slightly. The amount of current flowing through the circuit in Fig. 1-3 will be relatively independent of the diode; instead, it will depend primarily upon E and R.

Two important numbers should be remembered. For germanium diodes, V ranges from 0.2 to 0.4 volt. For silicon devices, it is about 0.6 or 0.7 volt. We use these approximations throughout the book. It should also be noted that these voltages are relatively accurate at room temperatures, 25 degrees C. They decrease at about 2.5 mv (.0025 volt) for each degree Celsius rise in temperature. The effect of this change in temperature moves the entire curve in Fig. 1-2A slightly to the left.

The diode is also characterized by its forward resistance. Fig. 1-4 is a drawing of curves representing the forward characteristics of two different diodes. Curve A could be that of a silicon power rectifier while curve B may be the curve of a point-contact germanium signal diode. The DC resistance of the diode is r_{DC} equals V divided by I at any point on the curve. It is different at different points on each curve. As an example, determine the DC resistance at 0.5 volt for both rectifiers. For power rectifier A, the current flowing is 0.05 ampere and the DC resistance is 0.5 volt divided by 0.05 ampere or 10 ohms. For diode B, the current flowing is 0.1 ampere and the DC resistance is 0.5 volt divided by 0.1 ampere or 5 ohms. In this region, the power rectifier has a higher DC resistance than the signal diode.

Now consider the DC resistance at 1 volt. The current flowing through the signal diode is 0.3 ampere and that flowing through the power rectifier is 1.5 amperes. The DC resistance of the signal diode is then 1 volt divided by 0.3 ampere or 3.33 ohms, and the DC resistance of the power rectifier is 1 volt divided by 1.5 amperes or 0.67 ohms. This is the usual situation that exists under normal operating conditions. Power rectifiers have a lower resistance than point-contact germanium signal diodes. Notice that point-contact signal diodes have few, if any, audio applications; therefore, they are not discussed here. Point-contact diodes are used primarily in RF work. The structure does not consist of a junction of dissimilar semiconductors. It is introduced here only because of the distinct difference of the curve when compared with junction diode curves. It will also be found that the resistance of junc-

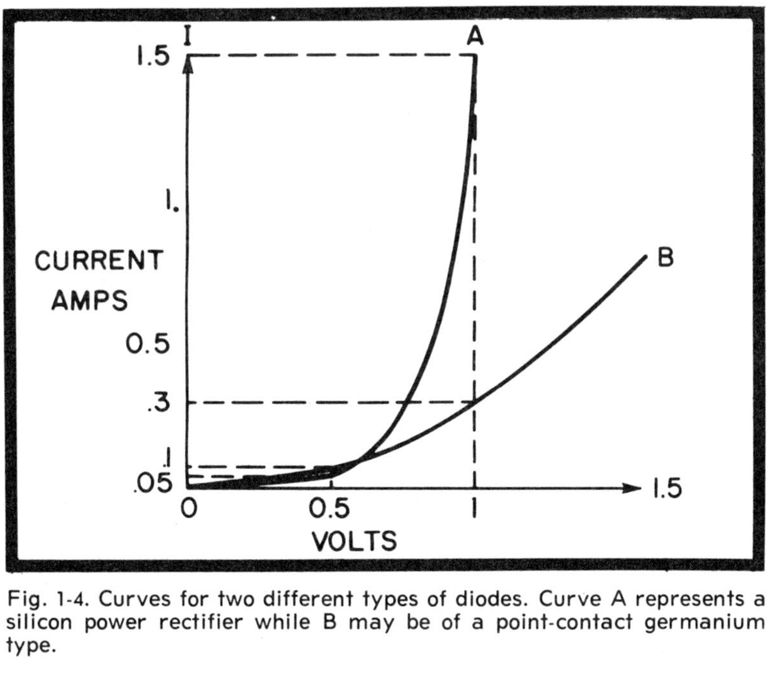

Fig. 1-4. Curves for two different types of diodes. Curve A represents a silicon power rectifier while B may be of a point-contact germanium type.

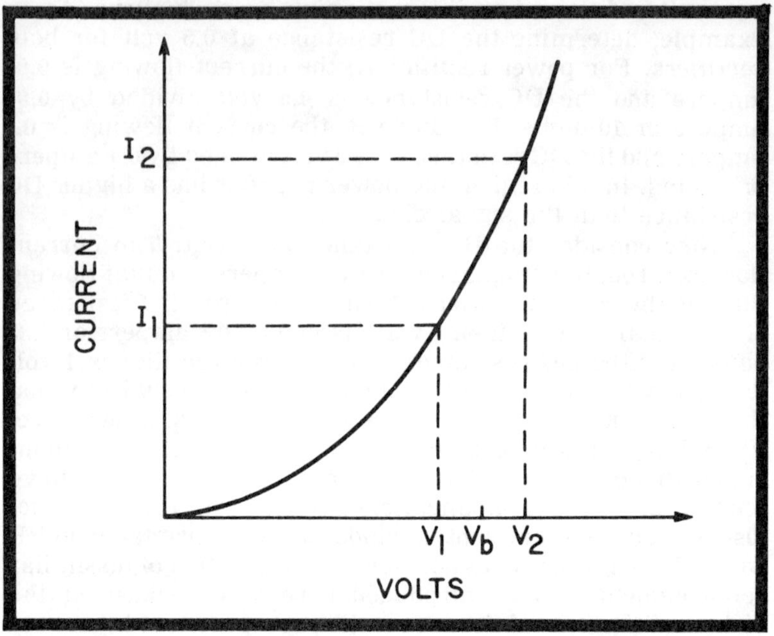

Fig. 1-5. Curves useful in calculating the AC resistance of a diode.

tion signal diodes is lower than that of point-contact devices, but is usually larger than that of power rectifiers.

A more important diode characteristic is the AC resistance, r_{AC}. In the discussion above, it was assumed that only direct current flows through the diode. In the more usual case, the current is AC. The AC resistance can be readily and accurately determined from the diode curve. The procedure is described with the help of Fig. 1-5.

As you know, an AC waveshape is basically sinusoidal, varying over the 360-degree cycle. Assume an AC voltage is applied to the diode. When the AC signal is at a minimum, the voltage across the diode is V_1; while at the maximum, it is V_2. At these instants in the cycle, the currents flowing through the diode are I_1 and I_2, respectively. The AC resistance, determined from the slope of the curve, is $(V_2\text{-}V_1)$ divided by $(I_2\text{-}I_1)$. This is the actual resistance the sine wave "sees." It will be found that the relative AC resistances of the various types of diodes are like the DC resistances in that power rectifiers exhibit the lowest resistances. Point-contact diodes have the highest resistances and silicon junction signal devices fall somewhere between the two, but close to the magnitude of the AC resistance of power rectifiers.

A convenient formula to remember for all junction diodes (not point-contact types) is:

$$r_{AC} = 26/I_F \qquad (1\text{-}1)$$

where I_F is the average or DC current flowing through the diode when this current is expressed in milliamperes (one-thousandth of an ampere). I_F is frequently midway between I_1 and I_2. At room temperature, this formula will provide AC resistances close in value to those determined from the curves.

The complete equivalent circuit of a forward-biased diode can now be drawn as in Fig. 1-6. The arrow-line symbol represents the ideal diode showing that current will flow only if the anode (arrow) is positive with respect to the cathode. For the ideal diode, V in Fig. 1-2A is zero. However, V is not forgotten in the equivalent circuit. The battery V represents this cut-in voltage encountered in the description of Fig. 1-2A. It is assumed that a positive voltage greater than V must be applied at the terminal marked (+) in the equivalent circuit (or at the anode of a real diode) before the diode will conduct. r_{AC} in the equivalent circuit is the AC diode resistance discussed above and is drawn as a resistor in the equivalent circuit.

We now turn our attention to the reverse bias characteristic. It is drawn graphically in Fig. 1-7. It was stated earlier, with reservations, that current will not flow through the diode when it is reverse biased. Actually, current will flow. When the reverse voltage is about 0.1 volt, a "saturation" current flows. This current increases only with temperature. It doubles for every 10-degree C rise in temperature when the diode is made of germanium material. For silicon devices, the current is assumed to double for every 6-degree C rise in conservative designs. In actual devices, this number can be anywhere between 6 degrees C and 16 degrees C, depending upon the diode used.

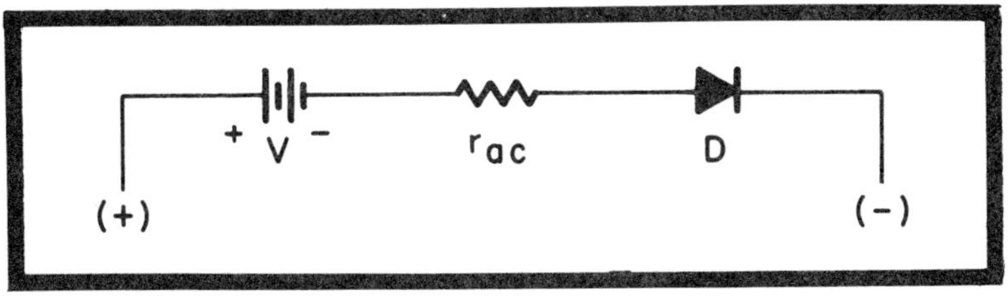

Fig. 1-6. Equivalent circuit of a forward-biased diode. The diode conducts when the applied voltage exceeds V.

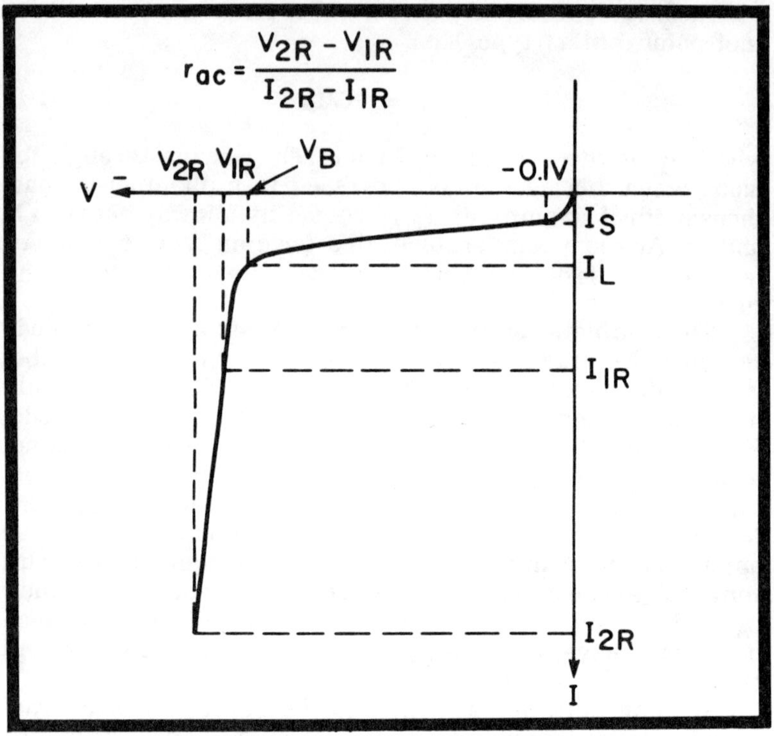

Fig. 1-7. Characteristics of reverse-biased diode.

In addition to the undesirable reverse saturation current, impurities associated with the diode add to the unwanted current, producing a leakage current. The size of this factor depends upon the reverse voltage.

As the reverse voltage is increased, a breakdown voltage is reached, V_B, above which the current increases rapidly. Any slight increase in voltage results in a large increase in the amount of reverse current flowing through the diode. The reverse AC resistance can be determined for this or for other sections of the curve, using the methods applied to Fig. 1-5. The sole difference here is that the arithmetic and construction are now accomplished in the third rather than in the first graphical quadrant. (See the mathematics and construction in Fig. 1-7.) Notice that Equation 1-1 does **not** apply to the reverse characteristic.

If you have a scope, a 12-volt (about 17 volt peak) RMS supply, a 9-volt zener diode, and a 1500-ohm resistor, you can perform an interesting and instructive experiment. First it must be understood that the zener diode is merely a silicon diode with breakdown voltage V_B controlled. In this case, it is equal to 9 volts. The forward characteristics are identical to those of any silicon device. It should also be noted that the 1500-ohm resistor is necessary to limit the diode current to safe values.

Now wire the circuit in Fig. 1-8. Place your scope input leads across the 12-volt supply. If the sweep frequency and phase controls on your scope are properly adjusted, you will see a sine wave as shown at the top of Fig. 1-9.

Now connect your scope across the zener diode. The trace should resemble the second curve in Fig. 1-9 approaching the shape of a square wave. Should your scope have a DC vertical amplifier, you will find that the voltage increases sinusoidally above the zero DC level (it is below the zero level if the connections to your scope are reversed) until the 9-volt reverse breakdown voltage of the diode is reached. The voltage across the diode will remain flat at 9 volts despite the fact that the peak voltage applied to the circuit rises to 17 volts. Below the 9-volt breakdown, the curve is identical in shape to the upper sine-wave curve as applied to the circuit. This proves that the curve in Fig. 1-7 is proper, for up to V_B the diode is an open circuit and the entire input voltage is across the diode; above V_B the voltage remains relatively constant at V_B (9 volts in this case). However, the reverse diode current increases rapidly above V_B.

The bottom and second half of the middle curve shows the effect of the forward characteristic of Fig. 1-2. Here, the voltage across the diode increases sinusoidally until V is reached. It remains at V (about 0.6 volts for silicon) until the applied sine-wave signal once again falls below V. In the forward direction this diode is an open circuit below V volts and is a relative short circuit when the voltage applied to the circuit is above V.

The current flowing through the diode is identical to that flowing through the 1500-ohm resistor. Since the waveshape of

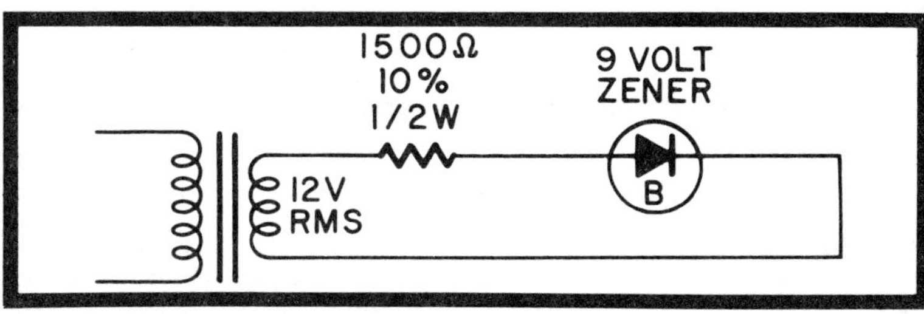

Fig. 1-8. Setup for an experiment to determine the characteristics of a zener diode.

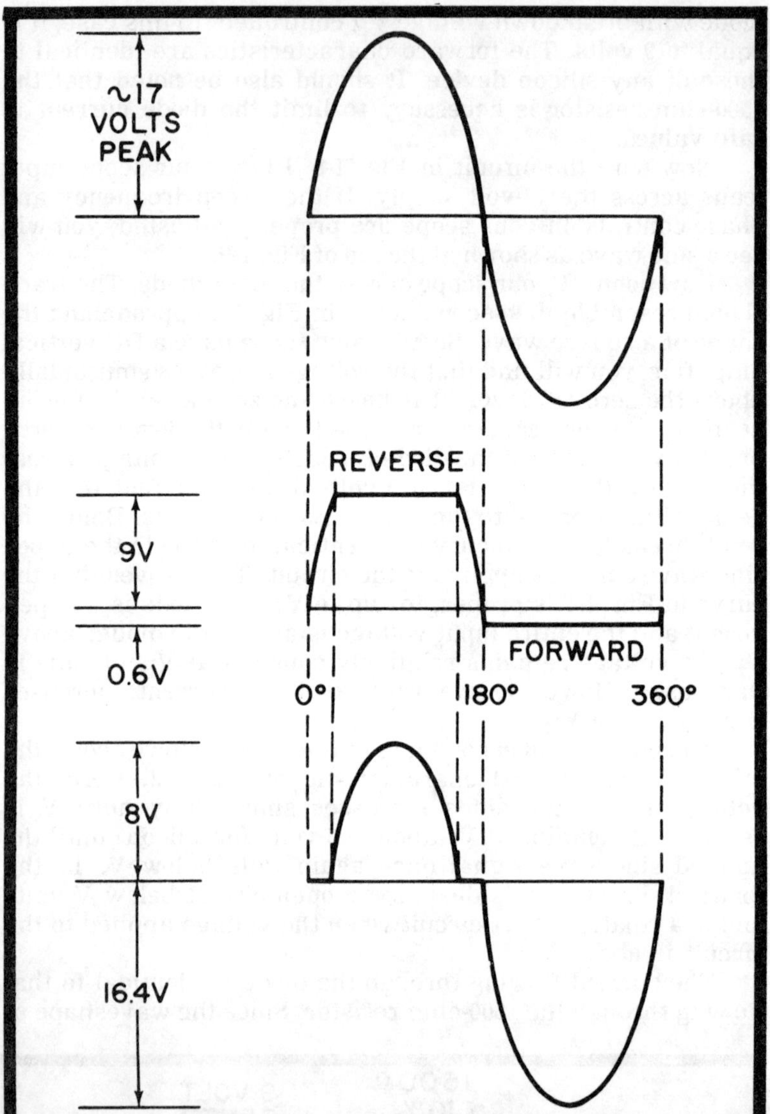

Fig. 1-9. Patterns observed on a scope connected across the components in Fig. 1-8. The top pattern is a sine wave across the secondary winding of a transformer. The middle curve shows the voltage across the zener diode. In the reverse-biased direction it is clamped to the zener breakdown voltage, while when forward biased it is at about 0.6 volts. The bottom pattern shows the current and voltage across the resistor. The sum of the voltage across the zener and the voltage across the resistor equals the voltage at the secondary of the transformer.

the voltage and current across a resistor are identical, a scope across the resistor will show a display of the current curve. If the scope is properly connected, you should see a curve similar to the bottom curve in Fig. 1-9. Notice that in the first half of the curve, when the diode is reverse biased, no current flows until the breakdown voltage of the diode is exceeded. The current flowing through this portion of the cycle increases with the voltage applied to the circuit, due to cyclic variations. But notice once again that in the middle drawing the voltage across the diode remains at 9 volts despite the increase in current during this portion of the cycle.

In the forward direction, current flows practically throughout the entire cycle. (See the 180-degree to 360-degree section of the lower drawing.) This is because conduction starts at a low forward diode voltage (V). Once again, notice in the middle drawing that the voltage across the diode will remain constant at V, regardless of the amount of current the diode is conducting.

BIPOLAR TRANSISTORS

If a slab of P-type semiconductor is placed in intimate contact with a piece of N-type material, the resulting device is a junction diode. Now, add a second N-type slab in close contact with the other side of the P-type material. The new configuration is an NPN bipolar transistor. Fig. 1-10 shows both a physical representation and a schematic drawing of the transistor. The three semiconductor slabs are assigned names—emitter, base and collector.

Notice that the base and emitter form one diode and the base and collector a second diode. In audio applications, the

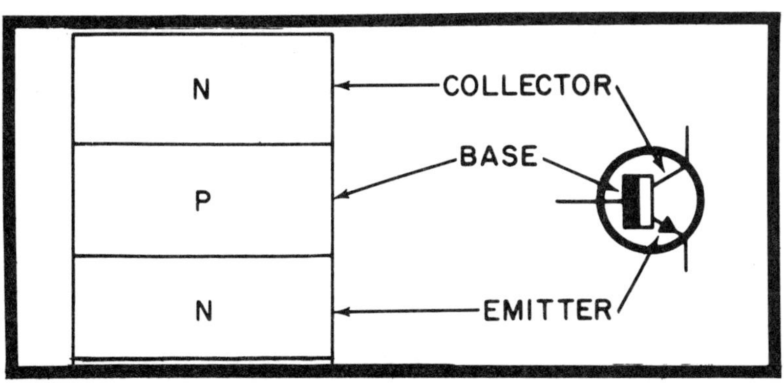

Fig. 1-10. Physical and schematic representation of an NPN bipolar transistor.

base-emitter junction is forward biased and the base-collector is reverse biased. In the case of the NPN transistor, the base is made positive with respect to the emitter and negative with respect to the collector. As is the case with the diode, electron current flows opposite to the direction of the arrow in the schematic. In the NPN transistor shown, it flows from the emitter to the base and collector.

Emitter current I_E consists of the sum of two currents—collector current I_C and the base current I_B. Stated mathematically, I_E equals $I_C + I_B$.

One important relationship between the currents is the ratio of DC collector current to the DC emitter current. This ratio is assigned the Greek letter alpha (α). The direct current gain of the transistor, DC, is:

$$\alpha_{dc} = I_C / I_E \qquad (1\text{-}2)$$

As I_C is less than I_E, α_{DC} is always less than one. In most transistors, it is so close to this limiting value that it is indistinguishable from 1.

To establish a second relationship between α_{DC} and the various currents, assume for the moment that the emitter current is equal to 1 ampere. The collector current is α_{DC} while the current flowing through the base is I_B equals $(1 - \alpha_{DC})$ amperes.

Another important ratio to be remembered is the DC collector current relationship to the DC base current. The Greek beta (β) is the letter representing this ratio. The direct current gain from the base to the collector is β_{DC}.

$$\beta_{dc} = I_C/I_B \qquad (1\text{-}3)$$

I_C is usually much greater than I_B. β_{DC} normally ranges from 2 to 1000.

A bit of arithmetic manipulation can be used to establish the relationship between α and β.

$$\alpha = \beta/(\beta + 1) \qquad (1\text{-}4)$$

$$\beta = \alpha/(1 - \alpha) \qquad (1\text{-}5)$$

A Bipolar Circuit

α_{DC} and β_{DC} are actually the current gain of a transistor. If a DC current, I_B, is fed into the base, the collector current should be, from Equation 1-3, about $\beta_{DC} I_B$. (The "about" in this statement stems from leakage factors to be discussed below.)

Put a transistor into the circuit shown in Fig. 1-11. The transistor used is the 2N3393, specified as having a beta somewhere between 100 and 300. We will assume a nominal beta of 200. We will also assume zero DC resistance and zero volts across the base-emitter diode. Set up the circuit and see if the results you find through measurements are as predicted in the calculations below.

The base current is 3 volts divided by 300,000 ohms or 10^{-5} ampere. The collector current should be about β x 10^{-5} equals 200 x 10^{-5} or 2 x 10^{-3} ampere. If this is true, then the voltage across the 2200-ohm resistor is (2200 ohms) (2 x 10^{-3} ampere) or 4.4 volts. This voltage can be checked accurately with an ordinary VOM if the resistance of the meter range used is much higher than 2200 ohms. Otherwise, use a TVM or VTVM.

Variations will occur. The beta of the 2N3393 can be anywhere between 100 and 300. Assume that the beta is at one extreme—100. Then I_C is 10^{-5} x 100 equals 10^{-3} ampere and the voltage across the 2200-ohm resistor is 2200 x 10^{-3} or 2.2 volts. Should beta be at the upper extreme of the range at 300, the collector current is 10^{-5} x 300 equals 3 x 10^{-3} ampere and the voltage across the 2200-ohm resistor is 2200 x 3 x 10^{-3} or 6.6 volts. The voltage across the resistor can vary at least from 2.2 volts to 6.6 volts, depending upon the particular 2N3393 transistor used in the circuit. Methods have been devised to reduce the size of this variation.

PNPs

Only one form of transisor is shown in Fig. 1-10. Should N-type material be sandwiched between two slabs of P-type

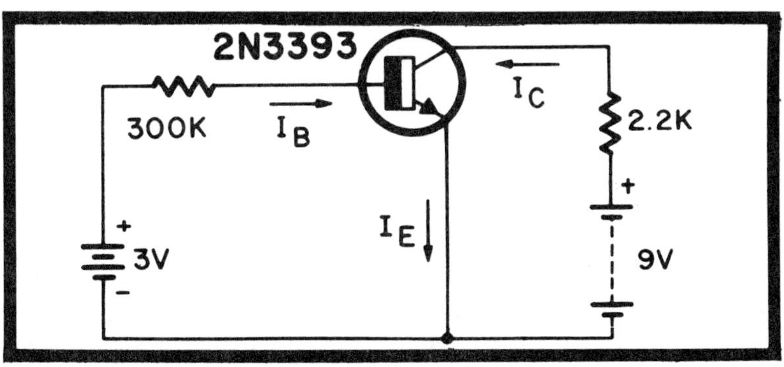

Fig. 1-11. Transistor in an experimental circuit.

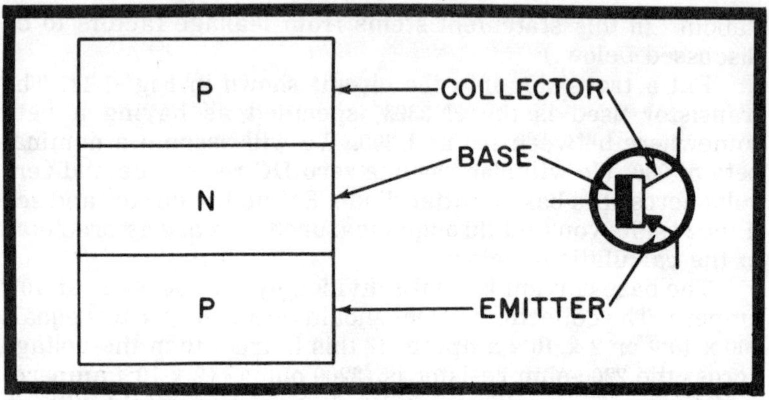

Fig. 1-12. Physical and schematic representations of a PNP bipolar transistor.

material, as shown in Fig. 1-12, a second form of the bipolar transistor is formed. This is the PNP transistor. Everything that has been written here about the NPN device applies equally to the PNP transistor, except the polarities of the power supplies must be reversed. For the base-emitter junction to be forward biased, the base must be negative with respect to the emitter. Following this logic, the base must be positive with respect to the collector if there is to be no conduction through this junction.

Although PNP and NPN devices are made from either germanium or silicon semiconductor materials, silicon transistors are primarily NPN and germanium are mostly PNP. Since silicon transistors currently dominate the field, our discussion generally considers NPN devices. However, all facts are applicable to PNP transistors as well. Only the polarity of the power supplies must be reversed to accommodate these devices.

AC Alpha and Beta

α_{DC} and β_{DC} define the current gain for the transistor at one point of operation. In the example, when 10^{-5} ampere is fed to the base, a current of 2×10^{-3} ampere flows through the collector circuit, assuming β_{DC} equals 200.

Now, feed an AC current into the base with a peak-to-peak value of 5×10^{-6} ampere. Assume this current is riding on top of the 10^{-5} DC ampere current. The situation is shown in Fig. 1-13. Since the peak AC current is 0.25×10^{-5} ampere, the current fed to the base at this instant is $10^{-5}A + 0.25 \times 10^{-5}A$ equals $1.25 \times 10^{-5}A$. At the crest of the signal, the current fed to the base is

$10^{-5}A - 0.25 \times 10^{-5}A$ equals $0.75 \times 10^{-5}A$. Considering β_{DC} as 200, the collector current at the instant of peak base current is $200\ (1.25 \times 10^{-5})$ equals $2.5 \times 10^{-3}A$. At the moment the current fed to the base is at a minimum, the collector current is 200 (0.75×10^{-5}) equals $1.5 \times 10^{-3}A$. The peak-to-peak output current is $2.5 \times 10^{-3} - 1.5 \times 10^{-3}$ equals $10^{-3}A$. Thus, with an AC input to the base of $5 \times 10^{-6}A$, the output at the collector is $10^{-3}A$, and the AC current gain in this example is 10^{-3} divided by 5×10^{-6} or 200.

In this very approximate example, the AC current gain is identical in value to the DC beta. AC beta or current gain, β_{AC}, is usually very close in value to β_{DC}. β_{AC} is defined as the ratio of the difference of two DC current levels:

$$\beta_{ac} = \frac{I_{C2} - I_{C1}}{I_{B2} - I_{B1}} = \frac{\Delta I_C}{\Delta I_B} \qquad (1\text{-}6)$$

Similarly,

$$\alpha_{ac} = \frac{I_{C2} - I_{C1}}{I_{E2} - I_{E1}} = \frac{\Delta I_C}{\Delta I_E} \qquad (1\text{-}7)$$

where I_{C2} is the collector current flowing when I_{B2} or I_{E2} is fed to the base or emitter, respectively. I_{C1} is the collector current flowing when I_{B1} or I_{E1} is fed to the base or emitter, respectively. Δ is a symbol indicating a change of value, such as ΔI_C is a change of collector current, ΔI_B is a change in base current, etc.

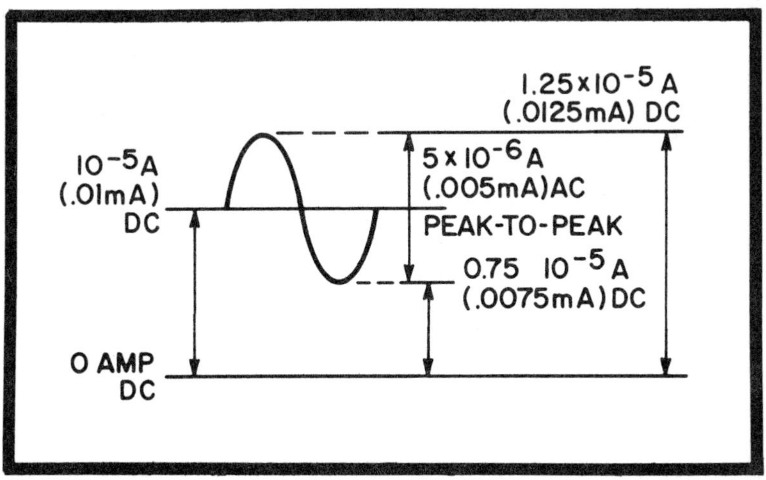

Fig. 1-13. AC input signal superimposed on the DC base bias current.

FIELD-EFFECT TRANSISTORS

The input impedance of a bipolar transistor is relatively low due to the forward bias requirement of the input base-emitter diode. An important comparative characteristic of the field-effect transistor is the higher input impedance. The input circuit is reverse biased.

FETs can be divided into two basic groups. One type is abbreviated JFET for Junction Field-Effect Transistor and the other is known by the letters IGFET for Insulated Gate Field-Effect Transistor, or MOS and MOST for Metal Oxide Field-Effect Transistor. The IGFET is identical to the MOS and MOST. The structure of the JFET differs from that of the IGFET. The former is currently more applicable to audio circuits due to its inherent lower noise characteristic.

JFETs

The physical representation and schematic of the N-channel JFET are shown in Fig. 1-14. An N-type semiconductor channel connects two terminals known as the drain and

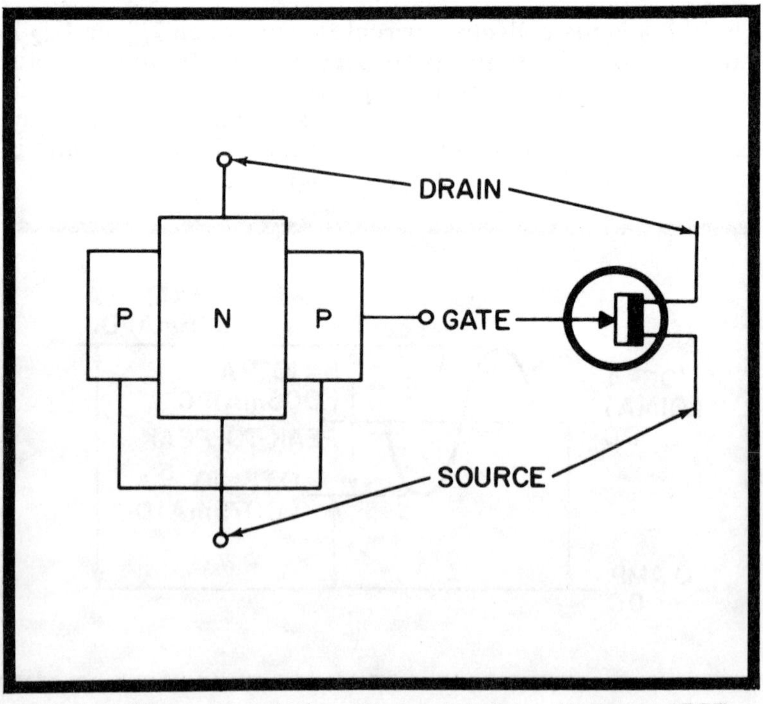

Fig. 1-14. Structural and schematic symbols for an N-channel JFET.

the source. Current flows through this channel. A third terminal, the gate, is connected to the P-slabs. These P-slabs form a junction with the channel. The relative voltage on the gate with respect to the source determines the amount of current flowing through the channel. In the usual bias arrangement, the drain is made positive with respect to the source while the gate is made negative.

The resistance from the source to the drain is merely the resistance of the N slab. It is usually only several hundred ohms. Current flows through this slab from the negative source to the relatively positive drain.

The gate and source are normally the input terminals of the JFET. Should the gate-source junction be forward biased, in which case the gate is made positive with respect to the channel (and more positive than the +0.6 volts cut-in voltage discussed with respect to forward-biased diodes), there would be conduction between the gate and the channel. The input impedance would be that of a forward-biased diode. But this junction is normally reverse biased. The negative voltage controls the quantity of current flowing from the source to the drain. An increase in negative voltage reduces the source-drain current.

Should the channel be made of a P-type semiconductor and the gate of N slabs, the resulting configuration is a P-channel transistor. The schematic change is but a reversal of the arrow representing the gate. The operation is exactly as described for the N-channel device. However, here the bias must be reversed to make the drain negative with respect to the source and the gate positive with respect to this latter lead.

Since the input impedance is high, the FET is a voltage-activated device. Whereas β_{AC} is the ratio of currents, ΔI_C divided by ΔI_B, the AC gain here is the ratio of the change of drain current, ΔI_D, to the change of gate voltage, ΔV_G. This ratio is known as transconductance. Using the symbol g_m for this factor:

$$g_m = \frac{\Delta I_D}{\Delta V_G} \tag{1-8}$$

Assume for the moment that you have a transistor with a transconductance of 5000 micromhos (equal to 0.005 mho). Should 0.2 volt AC be applied between the gate and source, the output drain current is (0.2) (0.005) equals 0.001 ampere AC or 1 ma AC. The AC voltage developed across a load resistor that may be in the drain circuit is the drain current flowing through that resistor multiplied by 1 ma.

IGFETs

As in the case of the JFET, the input impedance of the IGFET is extremely high. The magnitude of this impedance is emphasized by the fact that a static charge can be and

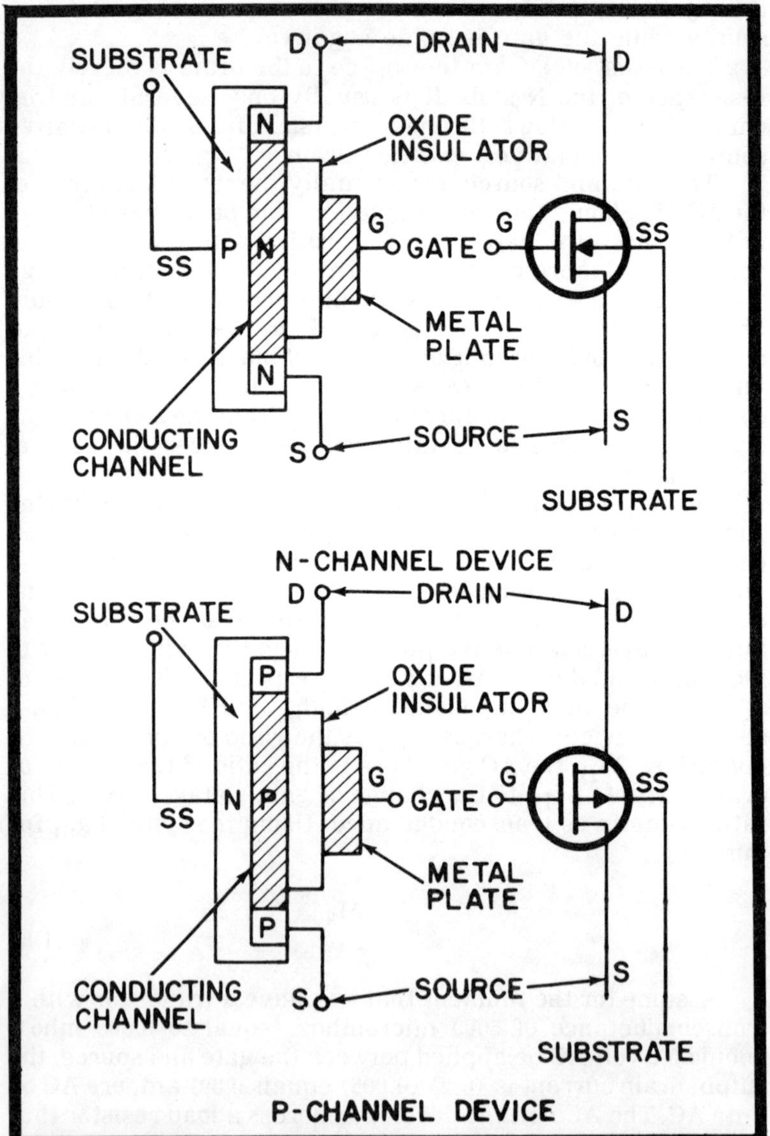

Fig. 1-15. Structural and schematic symbols for a depletion-type IGFET.

frequently is induced at the various elements of the device. (Large amounts of static charge can be generated across extremely high impedances.) This static voltage can be so high that it requires special precautions to be taken when handling an IGFET. To avoid damage to the IGFET due to the static charge, all leads must be shorted together until after the IGFET has been soldered into its circuit.

Drawings of the physical structures of the N- and P-channel depletion-type IGFETs, along with the schematic representation of each, are shown in Fig. 1-15. The following discussion centers around the N-channel device. Once again, everything refers to the P-channel transistor as well, except the types of material and the polarities of the applied voltages are reversed.

Two N slabs of the N-channel IGFET are diffused (joined) to, or put in intimate contact with, a highly resistive P semiconductor foundation known as a substrate. A medium resistance N-type material connects the two N slabs. One of the two N slabs serves as the source and the other N slab is the drain. The gate is made out of metallic material and is insulated from the semiconductors by an oxide. Should the drain be made positive with respect to the source, current will flow between the two semiconductors. A negative voltage on the gate will affect the amount of source-to-drain current flowing. The current will be reduced as the gate is made more negative with respect to the source. This is referred to as the **depletion** mode of operation.

An additional feature of the IGFET over the junction type of transistor is that more source-to-drain current will flow as the gate is made more positive. The IGFET gate can actually be made positive with respect to the source. This positive bias will not reduce the input impedance because the gate material does not form a junction with the channel, but is physically insulated from it by the oxide. When biased positively, the transistor is said to be operating in the **enhancement** mode.

Some IGFETs are strictly enhancement types. A drawing of this type is shown in Fig. 1-16. It is similar to the drawing of the types just discussed, with the exception that there is no N-type material connecting the source to the drain. Current flows between the source and drain only when the positively biased gate induces negative charges into the P substrate. This reduces the resistance between the two N slabs, allowing conduction.

As for gain, the definition of g_m discussed for the JFET applies here as well. The example discussed above does not change for this type of transistor; only the voltages involved at

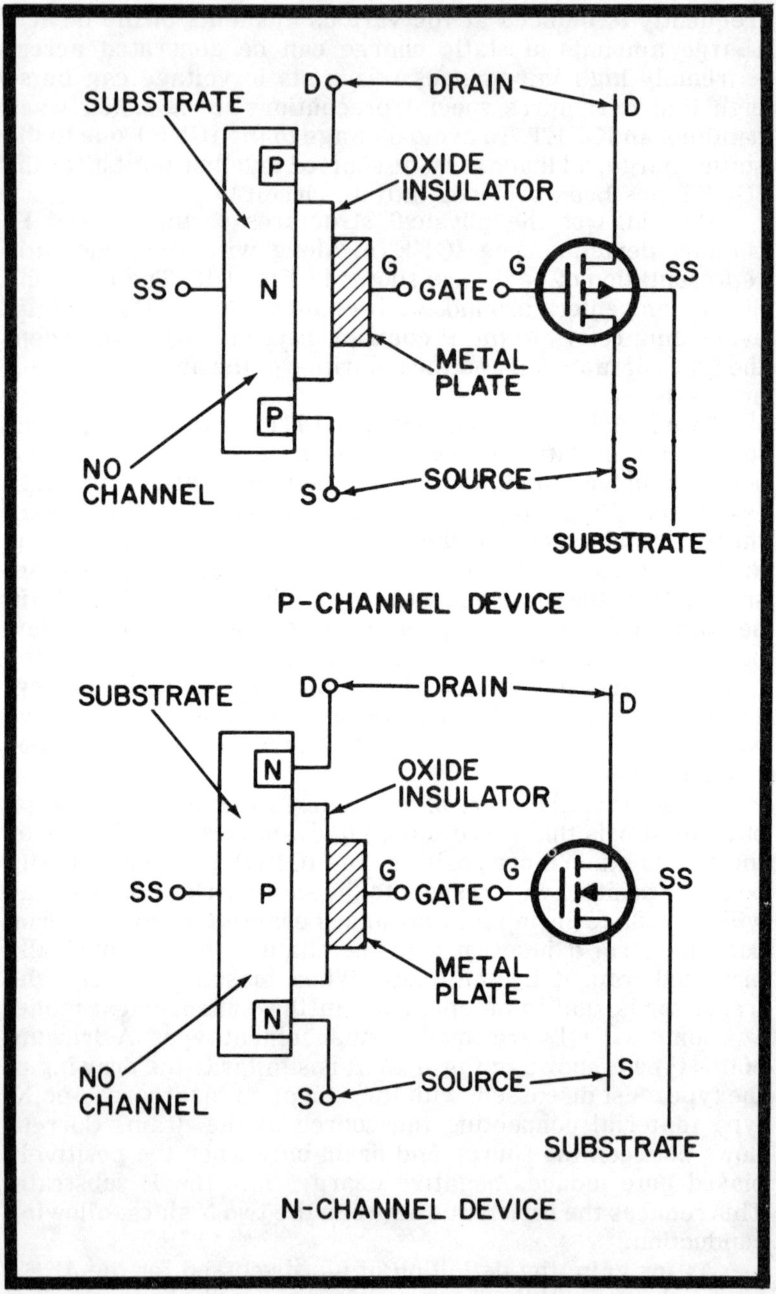

Fig. 1-16. Structural and schematic symbols for an enhancement-type IGFET.

the gate have the additional flexibility of being either positive or negative. The substrate lead is normally tied to the source.

In all types of FETs, there is also a DC transconductance characteristic, written g_{FS}. g_{FS} defines the ratio of DC drain current to the DC gate-to-source voltage.

Chapter 2 Biasing the Bipolar Transistor

When a sine wave is fed to the input side of a transistor audio amplifier stage, the goal may be to recover this sine wave, unaltered except for amplitude, at the output of the device. In other types of amplifier stages, it may be desirable to recover half or even less than half of the sine wave at the output. A major factor in determining the type of signal that will appear at the output is the bias current, or the current flowing through the transistor when it is idling.

BASIC CIRCUITS AND BIAS ARRANGEMENTS

Common-Emitter Circuit

A very rudimentary transistor circuit is shown in Fig. 2-1. It is referred to as the common- or grounded-emitter arrangement. In this type of circuit, the input signal is fed between the base and emitter and the output signal appears across load resistor R_C, which, as far as signal voltage is

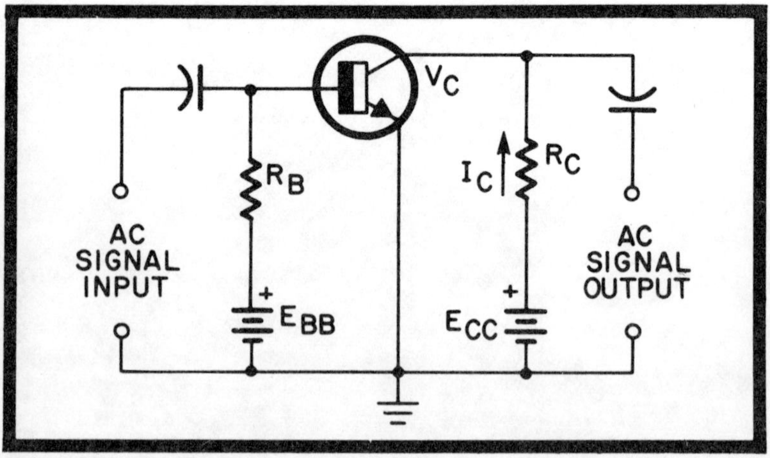

Fig. 2-1. A common-emitter circuit provides voltage, current and power gain.

concerned, is between the collector and emitter. With this type of circuit, it is possible to get voltage, current and power gain.

Now let us start by examining the collector circuit. The maximum DC current that can flow in this circuit is equal to (by Ohm's Law) the supply voltage, E_{CC}, divided by all resistance in the circuit. This resistance consists primarily of R_C, the internal resistance of the battery, and the collector resistance of the transistor. To facilitate matters, the internal resistance of the supply is considered zero and may be entirely ignored. Since the maximum current flows when the voltage drop across the transistor is zero, the collector resistance may likewise be ignored. At times, it is taken into consideration by assuming that there is 1 volt across the transistor when the maximum DC current flows. This 1 volt, opposing the battery voltage, may then be subtracted from E_{CC} when making the calculations.

With little error, it can then be assumed that the maximum current, I_C(max), that will flow in the collector circuit is E_{CC} divided by R_C. The voltage V_C at the collector is then equal to the supply voltage, E_{CC}, less the voltage drop across the collector resistor, I_C(max) x R_C equals V_C, or zero.

The minimum collector current, I_C(min), that can flow is zero, of course. With no current flowing, no voltage is developed across R_C, so the collector voltage is equal to the supply voltage less the zero volts across R_C, or simply the supply voltage. Hence V_C equals E_{CC}.

Let us now assume that a complete sine wave must be reproduced at the output. Collector current must flow at all times throughout the cycle. At the peak of the cycle, the current required to flow must never exceed E_{CC} divided by R_C and the required voltage at the collector must never be less than zero. A circuit which has collector current flowing through the entire cycle is defined as biased in a Class A mode of operation.

The best way of establishing a Class A state is to set the idling or quiescent conditions so that half the maximum collector current flows in the absence of audio signal. In this way the collector current will be capable of swinging equal amounts during both halves of the cycle—from zero to $\frac{1}{2}I_C$(max) during half of the cycle and from $\frac{1}{2}I_C$(max) to I_C(max) during the second half of the cycle. The identical condition is established if the collector voltage is set at $\frac{1}{2}E_{CC}$ so that V_C can then swing from zero to $\frac{1}{2}E_{CC}$ and from $\frac{1}{2}E_{CC}$ to E_{CC} on alternate halves of the cycle. Ideal collector current and collector voltage conditions would then exist simultaneously.

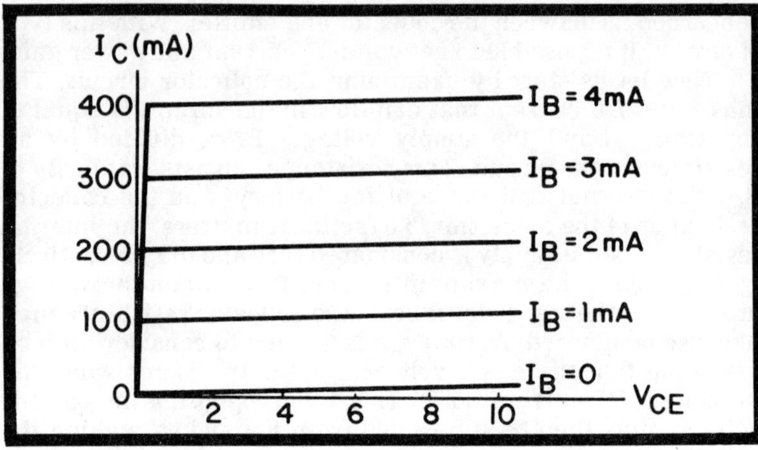

Fig. 2-2. Collector curves of a hypothetical transistor with a beta equal to 100.

A set of curves describing the collector characteristics of a hypothetical transistor is shown in Fig. 2-2. Each curve shows the amount of collector current that flows when specific amounts of current are flowing into the base. Notice that the collector current is theoretically independent of the voltage applied between the collector and emitter of the transistor, since the characteristic curves are perfectly horizontal. For the actual transistor, there is a dependence since the I_B curves do vary from the horizontal and rise somewhat with increasing collector-emitter voltage.

Using the curves, the amount of collector current that flows can be determined from the amount of base current flowing. Two equations must be plotted on one graph to determine the one point (or series of points) relating the output to the input. The first and extremely complex equation describes the transistor curves. No mathematical relationship need be stated, as the plot is already present in Fig. 2-2. The second equation relates components in the collector circuit of Fig. 2-1. It is:

$$E_{CC} = I_C R_C + V_{CE}. \qquad (2\text{-}1)$$

Plot this equation on the graph, as shown in Fig. 2-3. This will be the solution to the simultaneous equation. The combination of these two equations will determine the operating points of the transistor.

From the equation stated, when I_C equals 0, V_{CE} equals E_{CC}. Plot this one point on the graph. It is on the horizontal

V_{CE} axis. A second point can be found from the equation when V_{CE} equals 0. It is I_C equals E_{CC} divided by R_C. Plot this on the vertical I_C axis. Connect these two points. This line is known as the **load line** for the circuit. Circuit operation is defined by the intersection of the load line and the various characteristic curves. For any value of current fed to the base, there is a specific voltage and current flowing, as determined from the various points of intersection. All conditions of operation must lie along this load line.

Try a numerical example. Assume the transistor has a beta of 100. The collector supply is 12 volts and R_C, the load, is 3000 ohms. A sine wave is amplified by the transistor and a sine wave is to appear at the output. First, plot the load line on the curve of the transistor in Fig. 2-4 as follows.

When no collector current flows, I_C equals 0. Since there is no voltage across the 3000-ohm resistor (3000 x 0), the voltage at the collector is equal to the supply voltage. Then I_C equals 0, and V_{CE} equals 12 volts as one point on the graph.

When the maximum current flows, the voltage at the collector is zero. At that instant, I_C equals 12 volts divided by 3000 ohms equals 4×10^{-3} ampere. (3000 ohms is assumed to be the only resistance in the collector circuit.) Plot this point and then connect it with a straight line to the point previously plotted. This is the load line.

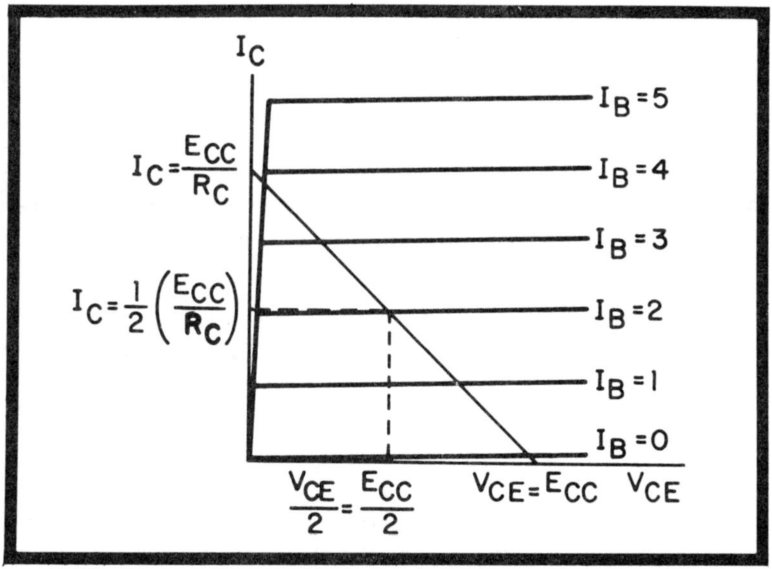

Fig. 2-3. Plot of the collector load line on transistor curves determines the operating locus.

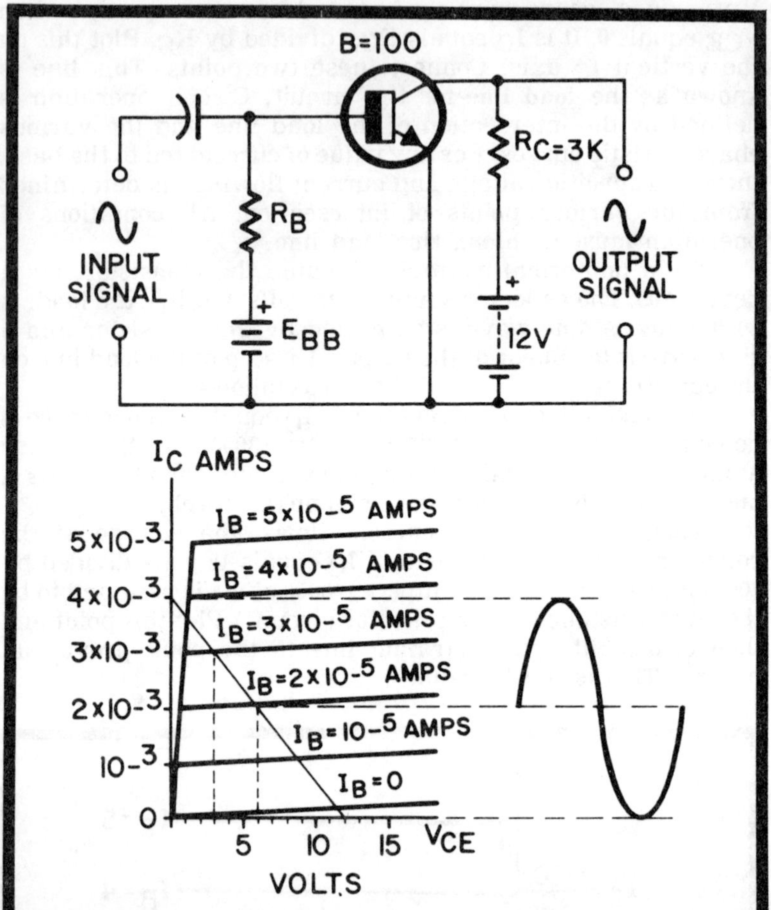

Fig. 2-4. Transistor circuit used in the text example, along with a plot of the load line.

Choose an idling current point midway on the load line. Use the point where V_{CE} equals 6 volts and I_C equals 2 x 10^{-3} ampere. This is the quiescent or bias collector current.

Let the sine wave at the input to the transistor swing the base current up to 3 x 10^{-5} ampere. The base current crosses the load line at a point corresponding to 3 collector volts and 3 x 10^{-3} ampere collector current. Should the peak of the sine wave force the base current further up to 4 x 10^{-5} ampere, the collector current would be 4 x 10^{-3} ampere while the collector voltage is zero. At the crest, if the base current is forced to zero, the collector would swing to 12 volts with zero current flowing. The maximum sinusoidal output is possible only if the collector voltage is set at ½E_{CC} and the collector current is at one half its maximum value.

In the above example, if the input signal drives the base so hard that it will conduct more than 4 x 10^{-5} ampere at the peak, the collector current would be limited to 4 x 10^{-3} ampere. One peak of the signal would be flattened at 4 x 10^{-3} ampere of collector current or at zero collector volts. Likewise, if the base current were forced below zero, a similar flattening of the other end of the curve would occur at the output.

Should the bias be set so that collector current flows for only one half of the cycle, the condition is referred to as Class B operation. In this case, the base current is normally set at I_B equals 0. Only the positive portion of the input signal would cause the transistor to conduct. It would not conduct during the negative portions of the cycle. However, now the swing of the conducting half cycle is not limited to ½E_{CC} as in Class A, but it can swing the collector the entire E_{CC} volts of the supply.

Normally, Class AB is used in audio output stages. In this case, the transistor conducts for slightly more than one-half of the cycle. The quiescent base bias current is easily established once the requirement has been determined. An equivalent of the forward-biased base-emitter circuit is shown in Fig. 2-5. A number of factors must be considered in this circuit.

First, a voltage supply is necessary to establish the forward base current. Then there is the DC resistance in the base material, r_B and the DC resistance in the emitter material, r_E, multiplied by beta. Both of these items may be considered as negligible. The only resistance in the circuit of any

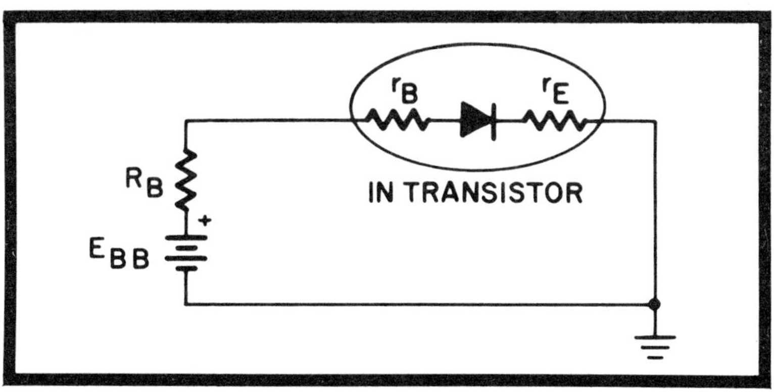

Fig. 2-5. The equivalent circuit seen from the input side of the transistor consists of a diode, a DC base resistance r_B, and a DC emitter resistance multiplied by beta, $\beta\, r_E$.

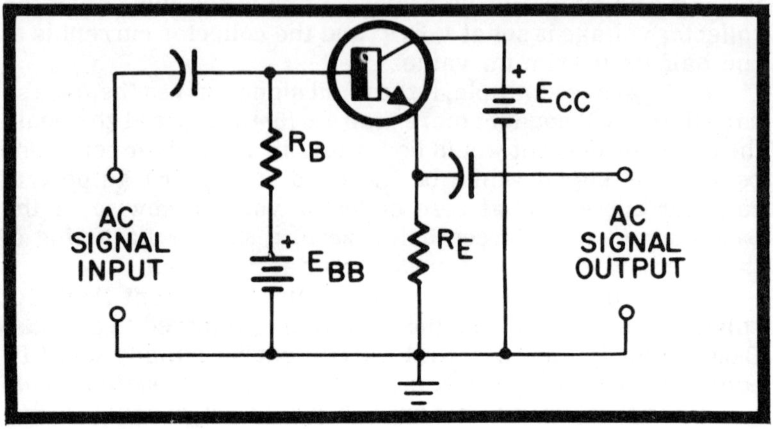

Fig. 2-6. The common-collector circuit provides current and power gain only. Voltage gain is almost 1. The stage has a high input and a low output impedance.

significance is R_B, the external resistor used to establish the base bias current.

Finally, there is the voltage drop across the base-emitter junction. As in the case with the diodes, about 0.2 volt appears across this junction if the transistor is made of germanium semiconductor material and 0.6 volt if it is a silicon device. From Ohm's Law, the base bias current is $(E_{BB} - 0.6)$ divided by R_B for a silicon device.

The effect of r_E is negligible only because r_E is very small. If a sizable resistor, R_E, were placed in the emitter circuit, between the emitter lead and the common ground, the effect of the resistor would be far from negligible. The emitter resistor appears to the base bias circuit as if it were a resistor in series with the base but multiplied by beta. (It is actually $\beta + 1$, but only a small error is involved if the 1 is omitted from the multiplication operation.) For example, if the resistor in the emitter lead were 270 ohms, and the beta of the transistor were 100, it would appear as a (270 ohms) (100) equals 27,000-ohm resistor in series with R_B.

Common-Collector Circuit

The common-collector arrangement does not require a resistor in the collector circuit, but does use a load resistor in the emitter. The basic amplifier stage is shown in Fig. 2-6. The input is fed between the base and collector while the output appears between the emitter and collector. This may not be obvious at first, but can be seen clearly when it is recalled that a battery or most other DC voltage supplies have zero resistance. They appear as short circuits to all AC or signal voltages.

The output load resistor is R_E. All output voltages appear across this resistor, and the load line is drawn as before for the common-emitter circuit, using R_E as the load resistor.

In calculating the base current, R_E is a very important factor here due to its size. The base bias current for the silicon transistor in Fig. 2-6 is (E_{BB} - 0.6) divided by ($R_B + \beta R_E$).

Common-Base Circuit

Voltage gain and power gain but no current gain are characteristics of the common-base circuit in Fig. 2-7. The input signal is applied between the emitter and base while the output is developed across R_C between the collector and base. (The supply is a short circuit for all signal frequencies.) A set of characteristic curves for this mode of operation is shown in Fig. 2-8. In the curves drawn, the collector current is plotted against the collector-to-base voltage. Each line indicates how the collector current varies with the collector-base voltage for various magnitudes of emitter current. The load line can be plotted as before, but this time on a different set of axes.

In establishing the bias here, the quiescent emitter current is adjusted to provide the quiescent collector current. The idling emitter current in the circuit in Fig. 2-7, for a silicon transistor, is approximately equal to (E_{EE} - 0.6) divided by R_E. Should there be a resistor in the base circuit, its presence is reflected in the emitter circuit by the ratio of 1 over β .

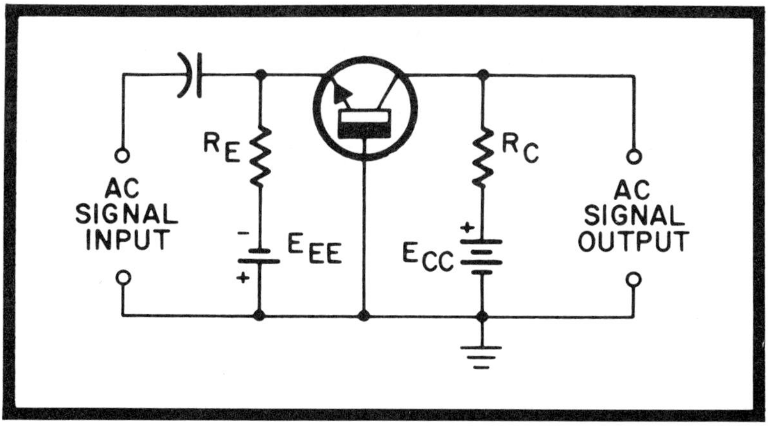

Fig. 2-7. The common-base circuit provides power and voltage gain. Current gain is almost 1.

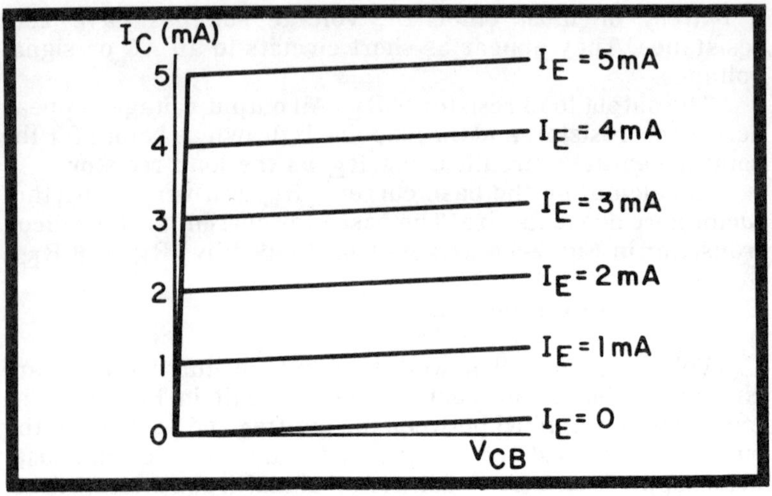

Fig. 2-8. Common-base curves show that the emitter current exceeds the collector current. I_E is slightly greater than I_C.

Bias Circuits

There are numerous bias circuits that can be applied to all the circuit arrangements. Each is detailed below. The reasons for many of these variations may not be evident now, but will become more obvious when stability is discussed. Here we will also divert from the previous notation indicating a power supply in the schematic. Up to this point, a battery with a voltage and polarity marked was our symbol for a voltage source. From here on, the battery symbol will be omitted. Only a notation indicating a supply and the polarity of one end of the voltage source will be marked on the schematic. It should always be understood that the remaining end of the supply is connected to a common or ground return point.

Fig. 2-9 shows the simplicity of the power supply notation we will be using compared with the one used until now. It is also a drawing of the simplest of bias methods, sometimes referred to as fixed bias. This method was discussed above when bias requirements were presented. The design procedure begins with deciding on the idling voltage needed at the collector. Next, determine the quiescent collector current from Ohm's law: I_C equals $(E_{CC} - V_C)$ divided by R_C, where V_C is the voltage at the collector. The base current, I_B, is equal to this divided by beta.

Now look at the base circuit. The base supply voltage in Fig. 2-9 is E_{BB}, although the collector supply, E_{CC}, frequently doubles in this function. The current flowing into the base starts at $+E_{BB}$, flows through R_B, the base-emitter junction, and finally through R_E. Assuming a silicon transistor where the voltage drop across the base-emitter junction is about V_{BB} equals 0.6 volt, the base current is (E_{BB} - 0.6) divided by ($R_B + \beta R_E$). (Note: Through the remainder of the discussion of bias, the 0.6 base-emitter voltage will be ignored. This voltage is

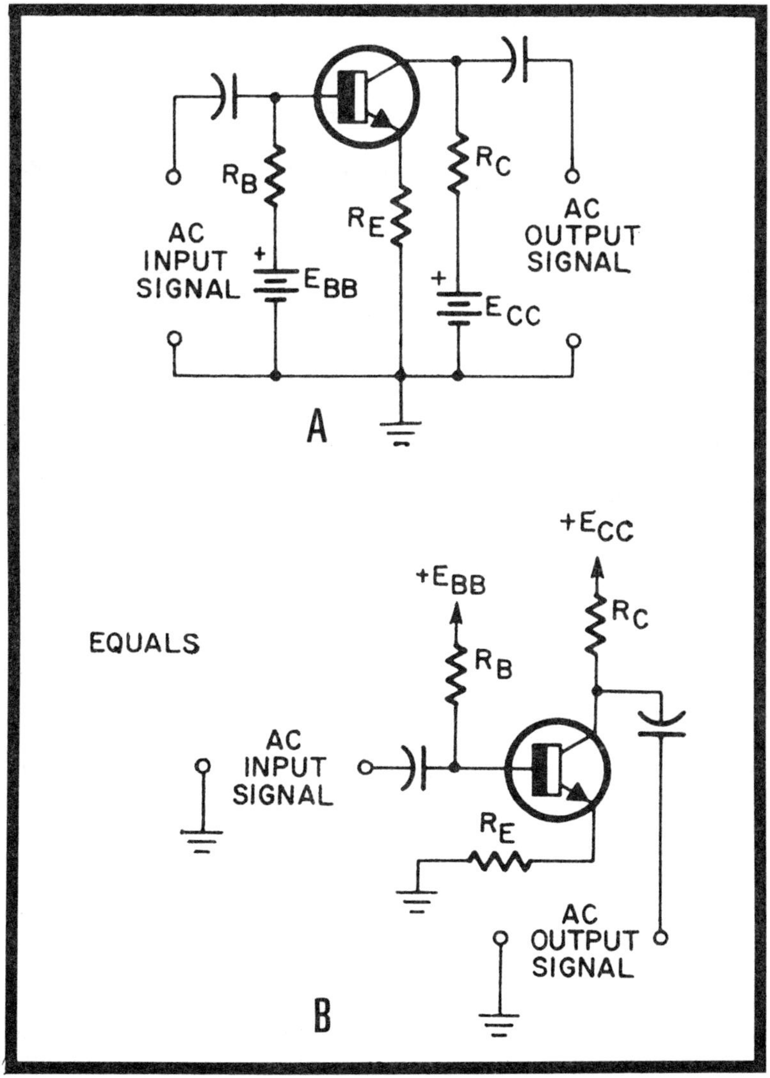

Fig. 2-9. These power supply designations simplify the circuit diagram.

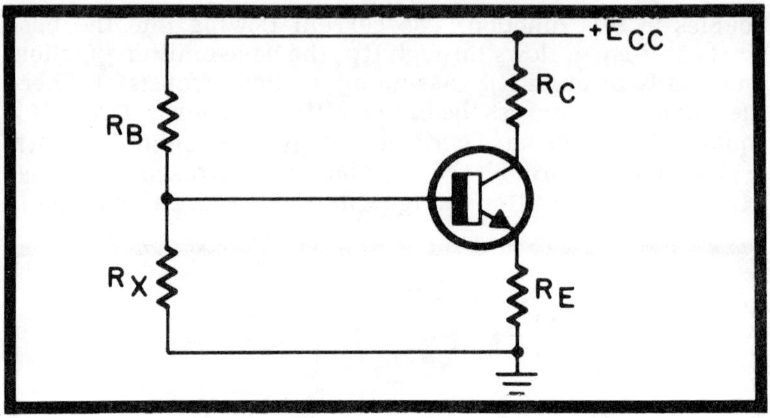

Fig. 2-10. This bias arrangement improves stability.

usually negligible when compared to E_{BB}. If you wish to include it in your calculations, you should merely subtract it from the base voltage supply considered, such as V_{TH} below.)

If we take the circuit in Fig. 2-9 and add a resistor, R_X, from the base to ground, the more complex but more stable circuit of Fig. 2-10 results. For a first approximation, R_X is usually specified at less than ten times the size of R_E.

In order to analyze this circuit, the simple but extremly useful Thevenin's Theorem is most helpful. By applying this theorem, any linear complex circuit consisting of a voltage source and a passive network can be reduced to an equivalent series combination of a voltage source and a series resistor. This can be accomplished in a few easy steps. For an example of the procedure, the circuit in Fig. 2-10 will be "thevenized." Follow the procedure using Fig. 2-11.

The simplification of the circuit begins with the splitting of the power supply for the base and collector circuits. Since both circuits are connected to one supply, E_{CC}, the "splitting" is done by simply indicating two identical supplies in the schematic, one for the collector circuit and the second for the base circuit. Now for the various steps to apply Thevenin's Theorem.

1. Separate the supply circuit from the load circuit. In the case of Fig. 2-11A, this means breaking the circuit at A-B. Redraw the circuit to the left of A-B as shown in Fig. 2-11B.

2. Next find the voltage across the terminals where the load was formerly connected. This is the Thevenin voltage, V_{TH}. In the example, R_X and R_B form a simple voltage

divider, so that the voltage at A-B is V_{TH} equals (E_{CC}) (R_X) divided by (R_X + R_B).

3. The third step is to determine the resistance looking into the open terminals. This is done by simply shorting all voltage sources in the original circuit. In the example, E_{CC} is shorted so that the end of R_B formerly connected to $+E_{CC}$ is now connected to B. R_B is effectively in parallel with R_X. The Thevenin resistance, R_{TH}, is this parallel resistance; R_{TH} equals R_BR_X divided by (R_B + R_X).

4. Draw the complete Thevenin equivalent circuit. For the bias network in Fig. 2-10, see Fig. 2-11C.

Returning to our bias circuit, it can be analyzed by connecting the Thevenin equivalent network back to the base

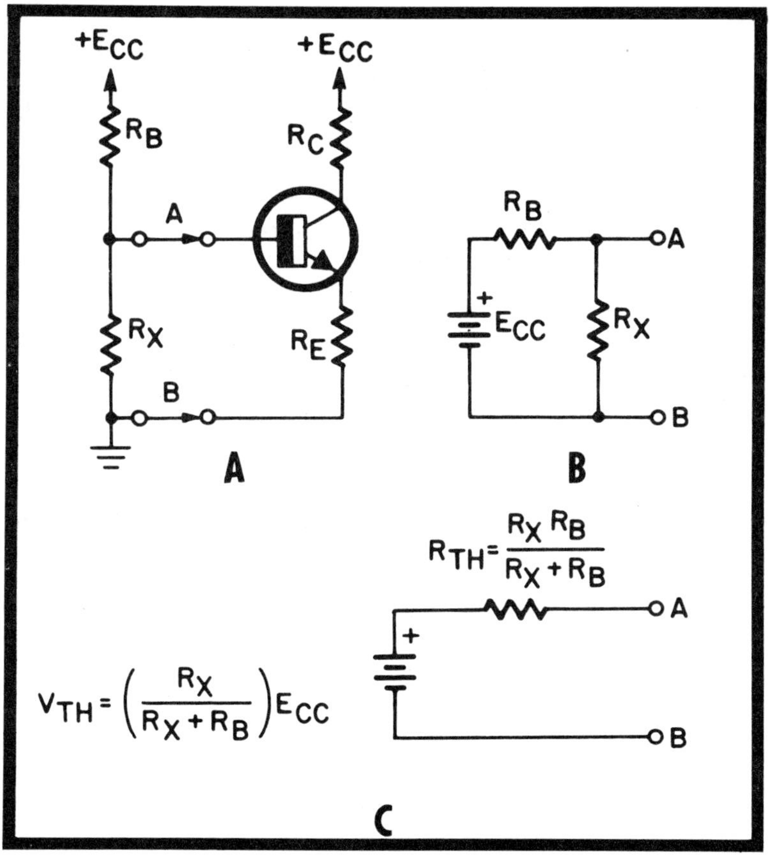

Fig. 2-11. Thevenin equivalent circuit derived for the bias arrangement in Fig. 2-10.

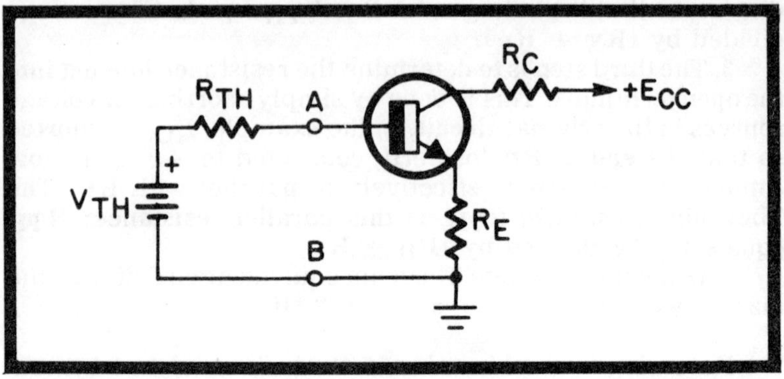

Fig. 2-12. Thevenin equivalent circuit connected to the transistor input, replacing the original circuit. The final arrangement reverts to the form of the simple circuit in Fig. 2-9.

circuit, as shown in Fig. 2-12. By applying this theorem, the circuit reverts to the one previously discussed for Fig. 2-9. Here the bias current is V_{TH} divided by ($R_{TH} + \beta R_E$) while the collector current is once again equal to the base current multiplied by beta.

A common variation of the circuit in Fig. 2-10 involves the addition of a second power supply connected to the bottom of R_X. The circuit is shown in Fig. 2-13. In this case, in addition to the Thevenin equivalent circuit, the superposition principle must be used. This method dictates that when more than one supply is used in a circuit, the effect of each supply upon a circuit factor must be considered individually. Then, the sum of the effects of all supplies are totaled algebraically.

Let us apply this method to the circuit in Fig. 2-13. Use the steps outlined below. Refer to Fig. 2-14 for clarification.

1. First, break the connections between the load and the supply circuit at A-B. This is identical to the procedure used above.

2. Short the negative supply. The voltage at A-B due to $+E_{CC}$ alone is ($+E_{CC}$) (R_X) divided by ($R_B + R_X$).

3. Next, short the $+E_{CC}$ supply. The voltage at A-B due to $-E_{BB}$ is ($-E_{BB}$) (R_B) divided by ($R_B + R_X$). If this is not too obvious, remember that when E_{CC} is shorted, the top of R_B is connected to ground or point B. R_X and R_B thus form a voltage divider.

4. The total Thevenin voltage is the sum of the voltages due to these two sources. It is:

$$V_{TH} = E_{CC}\ R_X/(R_B+R_X) - E_{BB}\ R_B/(R_B+R_X). \qquad (2\text{-}2)$$

5. Short **both** voltage sources. Looking into terminal A-B, R_B is in parallel with R_X. The Thevenin resistance is this parallel combination or:

$$R_{TH} = R_B R_X / (R_B + R_X) \qquad (2\text{-}3)$$

Now, draw the total Thevenized circuit in conjunction with the base circuit of the transistor. This is shown in Fig. 2-14B. Once again, the quiescent base current is:

$$I_B = V_{TH} / (R_{TH} - \beta R_E), \qquad (2\text{-}4)$$

and the idling collector current is beta multiplied by I_B.

Fig. 2-15 shows another method of biasing a transistor: an emitter supply rather than a base voltage source. It should be recalled that a resistor in the emitter is reflected as a resistor in the base circuit, but multiplied by a factor of beta. Thus R_B appears as a resistor, βR_B in series with the base lead.

In a similar manner, any resistor in the base circuit is reflected into the emitter circuit as a resistor divided by beta. R_B appears as a resistor R_B divided by β in series with R_E.

The emitter (and collector) currents flowing through the transistor is:

$$V_{EE} / (R_E + \frac{R_B}{\beta}) \qquad (2\text{-}5)$$

The base current is this value divided by beta.

The final circuit to be considered, sometimes referred to as the self-bias circuit, as shown in Fig. 2-16. It should be noted

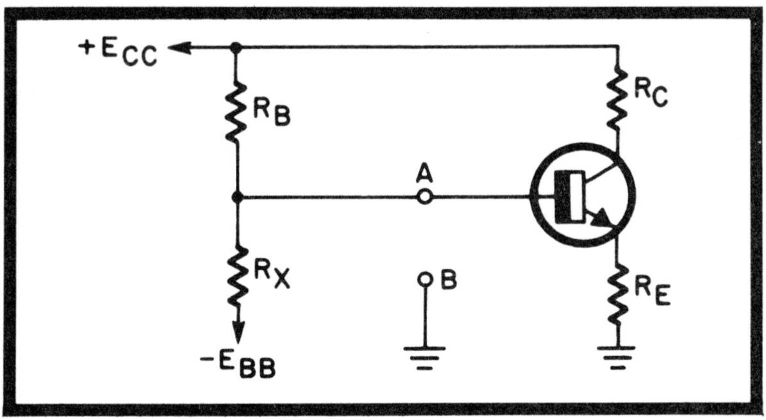

Fig. 2-13. Two voltage sources are used to bias the transistor base circuit.

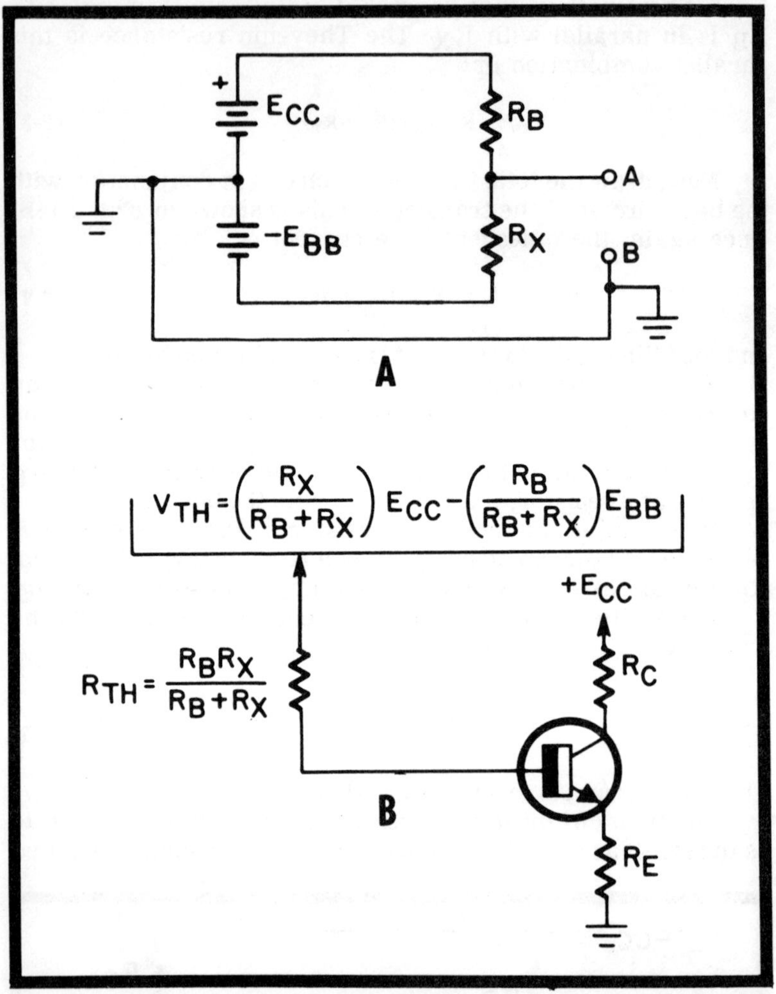

Fig. 2-14. In these drawings, the superposition and Thevenin methods are applied to the circuit in Fig. 2-13.

that the sum of the collector current and base current flows through R_C. The current flowing through the collector resistor is:

$$(E_{CC} - V_C)/R_C = I_B + I_C \tag{2-6}$$

while the base current is:

$$I_B = V_C/(R_B + \beta R_E). \tag{2-7}$$

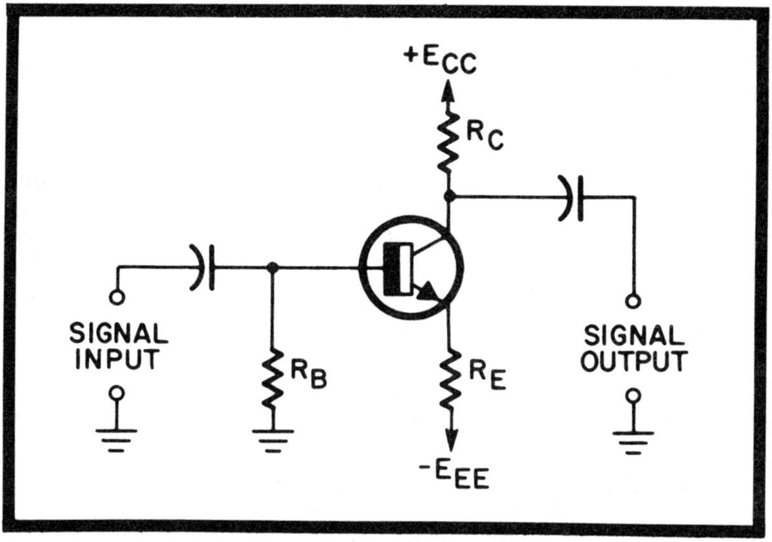

Fig. 2-15. Circuit with the bias supply voltage in the emitter.

STABILIZING THE BIAS

There are two primary factors which lead to instability in transistor bias circuits; parameter variations with collector current, and parameter variations with temperature.

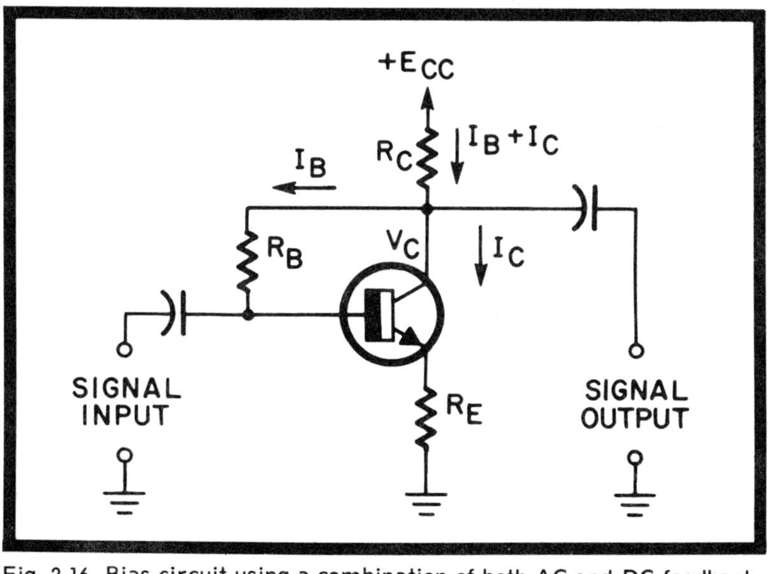

Fig. 2-16. Bias circuit using a combination of both AC and DC feedback.

Parameter Variations with Collector Current

Since the emitter and collector currents are about equal, any increase in collector current, I_C, is followed by a similar increase in emitter current, I_E. Alpha and beta do not necessarily vary with collector current, except that they normally increase rapidly as I_C becomes very large. Consequently, the base current I_B, which is equal to the collector current divided by a relatively constant beta, will also increase at moderate values of I_C.

The AC emitter resistance r_e, since it is equal to 26 divided by I_E, will obviously decrease as the emitter and collector current rise.

An AC resistance, r_c exists between the collector and base of transistors arranged in common-base circuits. r_c is large when the input impedance is low, as is the case for the common-base arrangement. It can range up to 1 megohm.

The common-emitter and common-collector circuits are characterized by high impedances in the input base circuit. The collector-to-emitter resistance is thus low compared to r_c. It is assigned the symbol r_d, which is approximately equal to r_c divided by beta. r_c, and consequently r_d, decrease with any increase in I_C.

Here is an important voltage to watch: V_{BE}. This is the base-to-emitter voltage of the forward-biased transistor junction. It is a primary factor in stability considerations. The base-emitter voltage has been approximated at 0.2 volt for germanium transistors and 0.6 volt for silicon devices. It was implied that these values are constants regardless of the current flowing through this junction. Well, this is almost so. Actually, the voltage does increase with base and collector current.

Parameter Variations with Temperature

V_{BE} also varies with temperature. For a good general approximation, it is assumed to decrease at 2.5 millivolts (0.0025 volts) for every degree Celsius that the temperature rises. This is significant, for as V_{BE} decreases, the voltage bucking the base circuit power supply is reduced. Should the base supply voltage remain constant, the reduced base-emitter voltage will result in an increase of base current and, consequently, will produce an increase in collector current. Under most circumstances, alpha and beta will also increase with temperature.

The third important factor which increases with temperature is leakage current, I_{CBO}. This leakage current is significant when germanium transistors are involved, but it loses its primary importance with silicon devices. This leakage is discussed in the next section. At the moment, it should be remembered that for germanium transistors, I_{CBO} doubles at about every 10 degree C rise in temperature. When designing circuits around silicon transistors, it is best to assume that I_{CBO} doubles every time the temperature increases 6 degrees C. In reality, the doubling of I_{CBO} for silicon devices can vary from 6 degrees C to 16 degrees C. Due to I_{CBO}, the collector current will increase rapidly with any rise in temperature.

Leakage Current

The transistor has been described as consisting of two junction diodes arranged as in Fig. 2-17. In the normal mode of operation for audio amplification purposes, the base-emitter junction is forward biased. For the NPN transistor, the base is made positive with respect to the emitter. At the same time, the base-collector diode is reverse biased and the collector is positive with respect to the base. Theoretically, or more precisely, ideally, the base-collector diode does not conduct any current.

As with all diodes, a leakage current does exist in the reversed-biased base-collector junction. Current flows through the diode even though the cathode end is positive with respect to the anode. To be sure, this leakage current is much smaller than any forward current that would flow had the diode been biased in the opposite direction. The leakage current is frequently of sufficient proportions to significantly affect the transistor collector idling current or bias. I_{CBO} is formally defined as the leakage or reverse current flowing through the collector-base junction while the emitter lead remains open.

Now, what would the leakage current do if the emitter lead were connected to the negative terminal of a voltage source

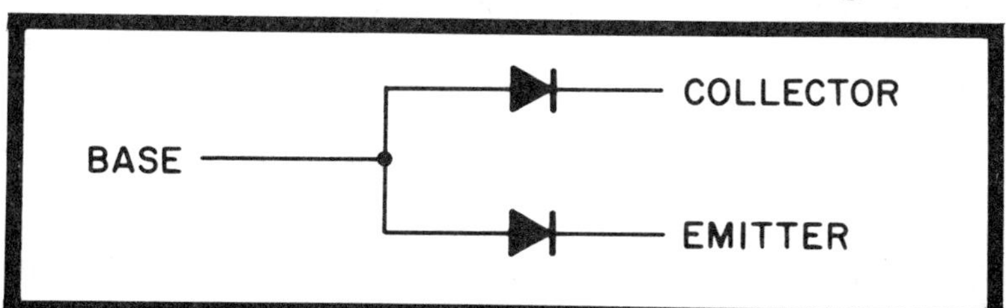

Fig. 2-17. Junction diodes form the equivalent circuit of a bipolar transistor. The NPN version is shown.

while the base lead remained open? The base-collector diode is then reverse biased while the base-emitter diode is biased in the forward direction. Obviously, all the I_{CBO} leakage current will flow into the base and through the base-emitter diode. Since this leakage current flows into the base, the current flowing through the collector due to I_{CBO} would be this base current multiplied by the amplification factor of the transistor, or $\beta_{DC}I_{CBO}$. It is assigned the symbol I_{CEO}—the collector-emitter current that flows when the base lead is open.

In the common-emitter circuit bias equations, the DC collector current has been noted roughly as $\beta_{DC}I_B$, the DC current amplification factor multiplied by the base current. To determine the total collector current flowing, the leakage current must be added to the desirable current fed to the base, so that the **entire** collector current is now:

$$I_C = \beta I_B + \beta I_{CBO} = \beta(I_B + I_{CBO}) \tag{2-8}$$

As for the common-base circuit, the collector current is no longer merely $\alpha_{DC}I_E$, the direct current amplification factor multiplied by the desirable current fed to the emitter, but it is:

$$I_C = \alpha I_E + I_{CBO} \tag{2-9}$$

The emitter-follower also suffers from the effects of leakage current. The emitter current is approximately equal to the collector current modified by a small leakage factor:

$$I_E = I_C - I_{CBO} \tag{2-10}$$

Major Stability Factors

Whatever the established bias current is, and regardless of the circuit used, it is important that that bias point be maintained. Of the various factors that can cause a transistor to drift away from the preset bias condition, the most important ones are:

1. **Leakage current:** The leakage current is a primary factor in establishing the idling current. This current varies, as noted above, with temperature. For germanium transistors, leakage current just about doubles for every 10 degree C rise in temperature, while for silicon devices, conservative designs assume that it doubles for every approximate 6 degree C rise in temperature. However, it is usually a less significant

factor where silicon transistors are used, because in these devices the leakage current at 25 degrees C is nearly non-existent.

2. **Base supply voltage:** Obviously, a variation in the supply voltage which establishes the quiescent base current will drastically affect the bias current. In theoretical drawings, this supply voltage is drawn as a battery. But when practical circuits are used, the voltage is derived from the variable power supply generators at the electric company's plants. Furthermore, the bias supply voltage can and will vary with amplified signals, temperature, and the portion of the cycle in which the applied signal happens to exist. This supply voltage also has a significant influence on V_{BE}, the voltage across the forward-biased base-emitter junction.

3. **Beta:** This factor varies with temperature as well as with collector current. Beta is frequently at a peak at one specific value of collector current. It may also increase with temperature.

Although variations in beta due to temperature and collector current will affect the bias current, it is outshined by beta variations within device categories. Thus, for any transistor type, beta is specified as being somewhere within a range of values. Transistor circuits should be designed to accommodate devices with any beta values specified for a particular type of transistor.

Stability Criteria

The three primary factors determining the stability of a transistor in a particular bias circuit arrangement should be stated mathematically if the merits of the circuit are to be evaluated. In each case, it is desirable to determine the change in collector current, ΔI_C, due to each of the three factors. Once the effect of each individual factor on I_C is established, the total effect can be determined by simply summing the individual ΔI_C values.

One stability factor, S, relates the change in collector current to the change in leakage current, ΔI_{CBO}. Stated mathematically, it is:

$$S = \Delta I_C / \Delta I_{CBO} \qquad (2\text{-}11)$$

In a similar manner, stability factors can be established to determine the change of collector current due to the change in base supply voltage, ΔE_{BB}, and the change in collector current due to the change in beta, $\Delta\beta$. These stability factors are assigned symbols S_E and S_B, respectively. Stated mathematically, they are:

$$S_E = \Delta I_C / \Delta E_{BB} \tag{2-12}$$

$$S_\beta = \Delta I_C / \Delta \beta \tag{2-13}$$

As an example in the use of the stability factors, assume that for a particular bias circuit, S is 5. If I_{CBO} increases from 1 ma at 25 degrees C to 5 ma at an elevated temperature, or if it increases a total of 4 ma, the collector current will increase 4 ma x 5 or 20 ma due to the rise in temperature. This is based on numbers being plugged into Equation 2-11, where ΔI_C equals S times (ΔI_{CBO}) equals 5 (4 ma) or 20 ma.

Similar solutions can be derived if the stability factors S_E and S_β are known for a circuit and ΔE_{BB} and $\Delta\beta$ have been determined. In any problem or circuit, it is desirable to maintain a stability factor as close to its minimum value of 1 as is possible and practical.

The variation of collector current is of immense significance in the design of audio circuits. In order to explore this fact, assume a common-emitter circuit with a 667-ohm load resistor in the collector. Plot the load line in Fig. 2-18A and B.

Now, suppose that at 25 degrees C the transistor were biased at Q1 and that a fixed input signal would swing the collector current from 0.5 to 5.5 ma. The output current swing is drawn alongside the collector current axis in Fig. 2-18A. The quiescent collector current for a perfect sine wave is midway in the swing of the output curve, at 3 ma in this example.

Next, suppose that the temperature rises and I_{CBO} increases, or some other factors cause β and E_{BB} to increase, or different transistors of the same type are used where β and V_{BE} differ from the average device, or any one of a number of other factors causes the quiescent point to shift. All significant factors cause the quiescent collector current to increase. Assume it increases from 3 ma to 4 ma, up to quiescent point Q2. This condition is now drawn in Fig. 2-18B.

With the quiescent current at Q2, the negative portion of the input signal can still cause the negative portion of the collector current to swing the full 2½ ma to a low of 1½ ma. The positive portion of the input signal can swing the collector current only to its maximum value of 6 ma. When the input signal tries to force I_C beyond the 6 ma, only 6 ma will be delivered at the collector. The output current curve will have a

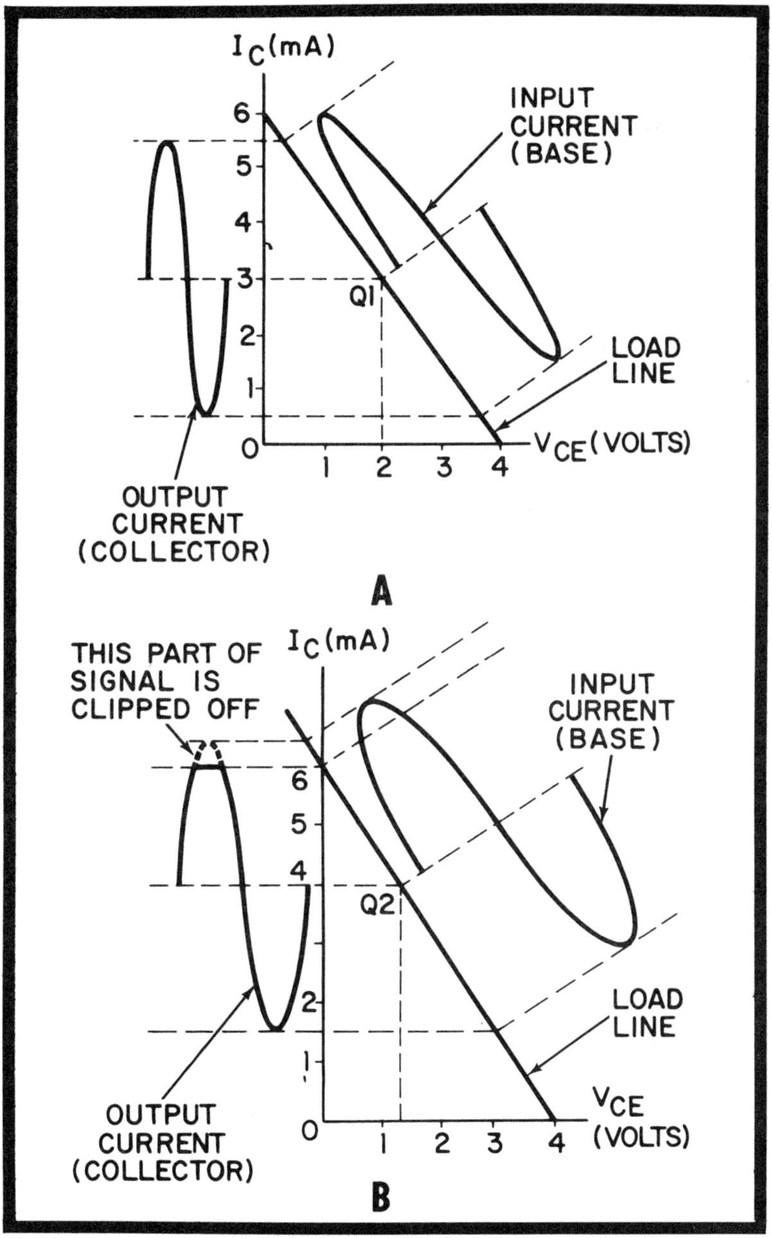

Fig. 2-18. Notice the effect of the bias current shift on the output collector current and the flattened output signal. Transistor characteristic curves are omitted for clarity. All signal currents must travel along the load line. At A the bias current was set at Q1. At B the bias current was shifted up to Q2.

flat top during the period of time when the 6 ma collector current should be exceeded.

The designer should be capable of determining just how much leeway his circuit must have so that a full sine wave can be reproduced at the output at all times. He must realize that the shift in quiescent conditions may be due to factors besides current and temperature variations in any one transistor. Different transistors of identical type vary somewhat in their characteristics. In some types, beta can vary by as much as 5 to 1, although it is frequently held to a 2-to-1 ratio. V_{BE}, a significant factor in determining collector current, is usually specified with a maximum rating, while a 0.5-volt spread between different transistors of the same type is not unusual.

Stability Factors and the Bias Circuits

Three stability factors, S, S_E and S_β, were just defined. Each of these stability factors can be calculated from the bias circuits and the components used in the particular circuit. Then the calculated values of S, S_E and S_β can be substituted into Equations 2-11, 2-12, and 2-13, respectively. The changes in collector current, ΔI_C, for each factor can be calculated by multiplying the specific stability factor by the change involved.

Once S has been determined from the equations stated in the figures, it should be multiplied by the change in leakage current, ΔI_{CBO}, to determine the change in collector current due to this item. In a similar manner, determine S_E from the equations and multiply it by the change in base supply voltage, ΔE_{BB}, to establish the change in collector current due to ΔE_{BB}. Likewise, the product of S_β (as determined from the equations in the figures) and the change of beta, $\Delta\beta$, will provide the data required to find the collector current change due to this variation.

One of the bias circuits which was discussed with respect to Fig. 2-10 is redrawn in Fig. 2-19. The collector current flowing through R_C, as well as the stability factors, can be approximated from the equations in the drawing. (It should be noted here that the equations are not for memorization, but are to be used for simply determining circuit characteristics. They may appear complex, but step-by-step substitution of circuit component values will indicate just how easily they can be utilized.)

As can be seen in Fig. 2-20, the equations for this circuit are simpler than those applying to the single-battery bias circuit in Fig. 2-19. This is due to the absence of R_C. The two sets of equations are identical, but in the latter figure, R_X was made infinite in value. Should the circuit be further simplified to that in Fig. 2-21, where R_E is omitted, the equations in the figure are to be used. The stability here is at its worst for no external circuitry is used to minimize variations.

The bias circuit in Fig. 2-22 is similar to the one in Fig. 2-19, but with the addition of DC feedback. This arrangement provides the best stability of all when using simple resistor circuits for compensation. Once again, the equations are

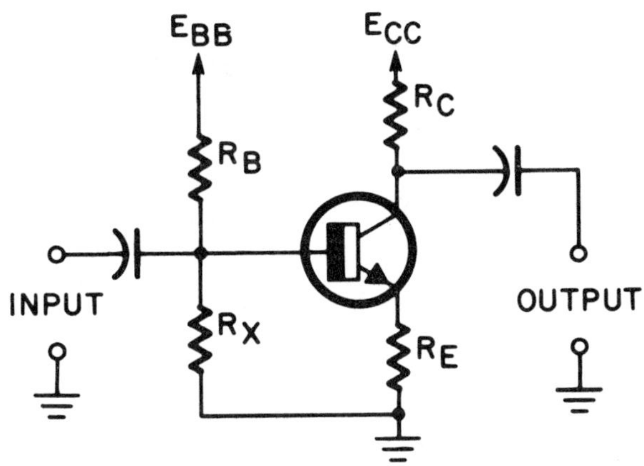

$$I_C = \frac{E_{BB}R_X + (R_E R_B + R_E R_X + R_B R_X) I_{CBO}}{R_E (R_B + R_X) + R_B R_X/\beta}$$

$$S = \frac{R_E R_B + R_E R_X + R_B R_X}{R_E (R_B + R_X) + R_B R_X/\beta}$$

$$S_E = \frac{R_X}{R_E (R_B + R_X) + R_B R_X/\beta}$$

$$S_\beta = \frac{(E_{BB}R_X + I_{CBO} R_B R_X)(R_E R_B + R_E R_X + R_B R_X)}{[\beta R_E (R_B + R_X) + R_B R_X]^2}$$

Fig. 2-19. Bias circuit using resistors R_X and R_E to establish a good stability factor approximating 1.

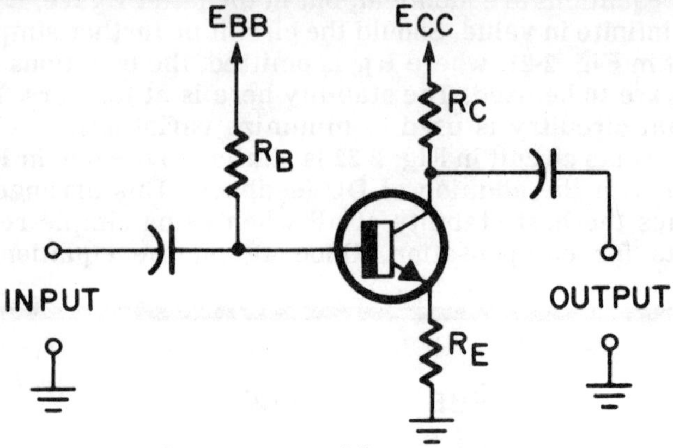

$$I_C = \frac{E_{BB} + (R_E + R_B)\, I_{CBO}}{R_E + R_B/\beta}$$

$$S = \frac{R_E + R_B}{R_E + R_B/\beta} \qquad S_E = \frac{1}{R_E + R_B/\beta}$$

$$S_\beta = \frac{E_{BB}\,(R_E + R_B) + I_{CBO} R_B\,(R_E + R_B)}{(\beta R_E + R_B)^2}$$

Fig. 2-20. The bias circuit in Fig. 2-19 with R_X omitted.

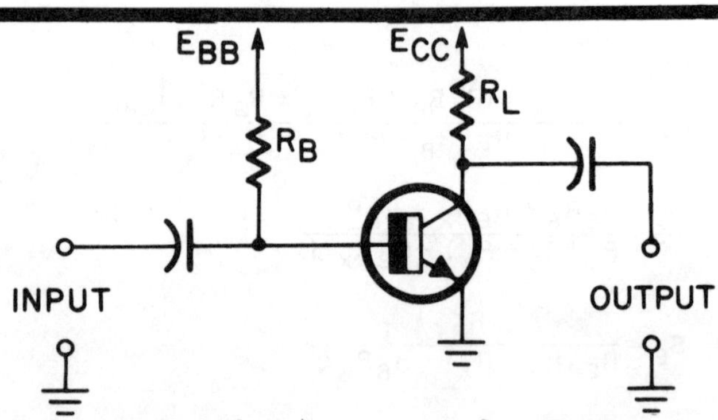

$$I_C = \frac{(E_{BB} + R_B I_{CBO})}{R_B/\beta} \qquad S = \frac{R_E + R_B}{R_B/\beta}$$

$$S_E = \frac{\beta}{R_B} \qquad S_\beta = \frac{E_{BB} + I_{CBO} R_B}{R_B}$$

Fig. 2-21. Further simplification of the circuit in Fig. 2-19. Both R_X and R_E are omitted. Stability is at its worst here.

stated in the figure. The circuit in Fig. 2-22 will still exhibit good stability characteristics if R_X is omitted, as in Fig. 2-23. The equations here are identical to those stated in Fig. 2-22, if R_X in that figure is considered infinite.

The collector current and stability equations for the emitter bias circuit in Fig. 2-24 are similar to the equations for the fixed bias circuit in Fig. 2-20. Just substitute E_{EE} for E_{BB}.

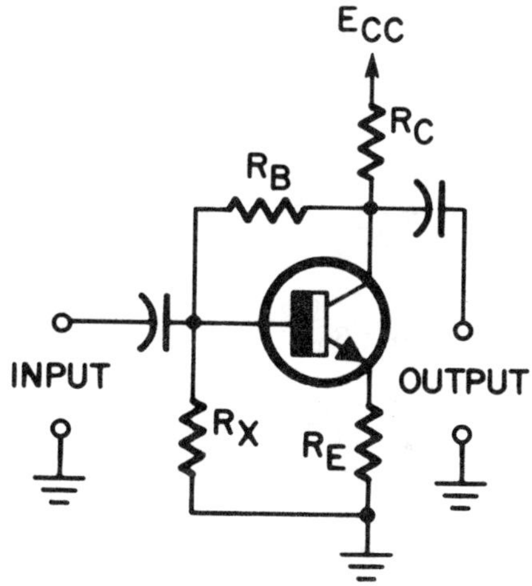

$$I_C = \frac{E_{CC}R_X + [(R_C + R_E + R_B)R_X + (R_C + R_B)R_E]\,I_{CBO}}{(R_C + R_B + R_X)R_E + R_X R_C + R_B R_X/\beta}$$

$$S = \frac{(R_C + R_E + R_B)R_X + (R_C + R_B)R_E}{(R_C + R_B + R_X)R_E + R_X R_C + R_B R_X/\beta}$$

$$S_E = \frac{R_X}{(R_C + R_B + R_X)R_E + R_X R_C + R_B R_X/\beta}$$

$$S_\beta = \frac{[(R_C + R_B + R_X)R_E + (R_C + R_B)R_X]\,[E_{CC}R_X + R_B R_X I_{CBO}]}{\left([(R_C + R_B + R_X)R_E + R_X R_C]\,\beta + R_B R_X\right)^2}$$

Fig. 2-22. Very stable bias circuit using DC feedback from the collector to the base.

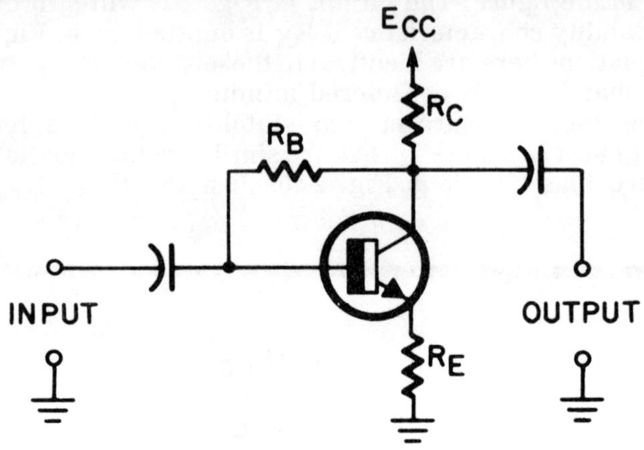

$$I_C = \frac{E_{CC} + (R_C + R_E + R_B)\, I_{CBO}}{R_C + R_E + R_B/\beta}$$

$$S = \frac{R_C + R_E + R_B}{R_C + R_E + R_B/\beta} \qquad S_E = \frac{1}{R_C + R_E + R_B/\beta}$$

$$S_\beta = \frac{(R_C + R_E + R_B)(E_{CC} + I_{CBO} R_B)}{[\beta(R_C + R_E) + R_B]^2}$$

Fig. 2-23. Circuit in Fig. 2-22 with R_X removed. Stability suffers to some degree.

One last bias circuit, discussed in the interest of stability, is shown in Fig. 2-25. This configuration looks very similar to the one drawn in Fig. 2-19, but with the addition of a forward-biased diode. The voltage across D1 is identical to that across the forward-biased diode junction composed of the transistor base and emitter. The voltage across diode D1 varies with temperature in accordance with the voltage change across the base-emitter junction. The stability equations in Fig. 2-19 can be used for this circuit, but the actual stability will be much better than indicated by the calculations.

Design Procedures

The stability equations are great for checking the circuits after they have been designed. But how should one proceed to

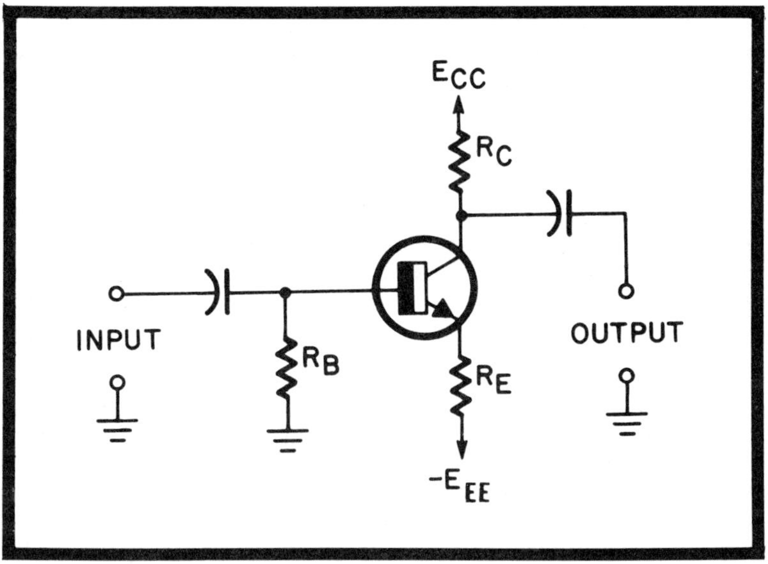

Fig. 2-24. Emitter bias circuit based on the same equations as in Fig. 2-20. Here, E_{EE} is substituted for E_{BB}.

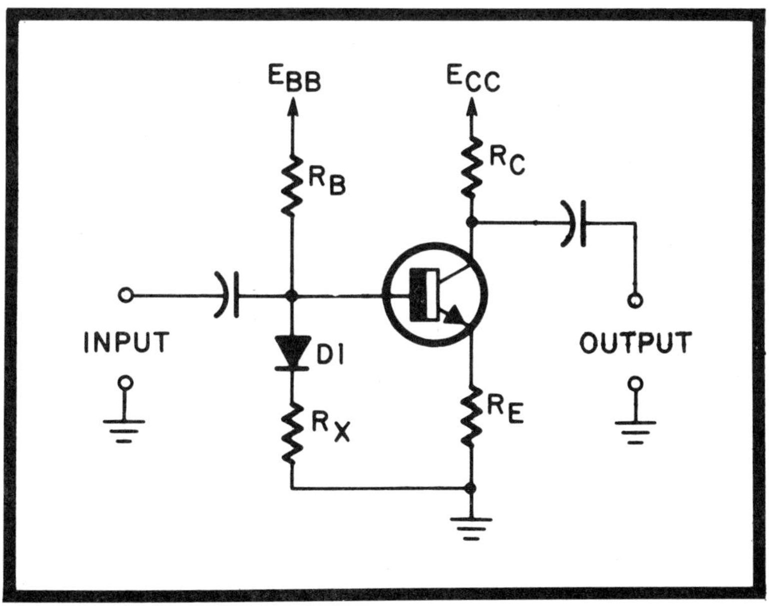

Fig. 2-25. Bias circuit in which diode D1 compensates for the forward-biased base-emitter junction voltage variation with temperature.

execute a good stable design? A few general rules can be deduced from the equations. In all bias circuits, R_X should be made as small as practical and R_E should be as large as possible.

If germanium transistors are used, where I_{CBO} is usually a very important factor affecting stability, S is reasonably small if R_C is large and R_B is minimized. V_{BE} and beta vary with temperature regardless of the transistor type. Furthermore, beta can have any of a large range of values within any specific transistor category. The stability factors concerning these items are best limited if R_B and R_C are large resistors.

Using these basic rules, a design example will illustrate procedures. Suppose you want to design an audio preamplifier using a 12-volt battery supply. The low-noise 2N3391A silicon transistor is to be used. The beta range is 250 to 500, with a mean average of 375, and the leakage current at 25 degrees C is 0.1 microampere (10^{-7} ampere or 0.0000001 ampere). The transistor is to be used in an ambient temperature ranging from 25 degrees C to 55 degrees C. The common-emitter circuit is to have a minimum AC voltage gain of 10 and a maximum output impedance of 6,000 ohms. The maximum input signal it should accommodate is 0.8 volt peak to peak and the output voltage must be capable of swinging a minimum of 0.8 x 10 or 8 volts.

Start with the design of the circuit by establishing the output load resistor value. The maximum output impedance is stated as 6,000 ohms. To assure this, put a load resistor of less than 6,000 ohms in the collector circuit. Using the next lower EIA standard value, let R_C equal 5600 ohms.

The minimum AC gain of the circuit has been specified at 10. AC voltage gain is approximately equal to the ratio of the resistor in the collector circuit to the resistor in the emitter circuit, or A_V equals R_C divided by R_E for high beta transistors. Consequently, for a minimum gain of 10. R_E must be less than 560 ohms. To compensate for losses, we will make R_E equal to 470 ohms. The approximate gain should then be about 5600 ohms divided by 470 ohms equals 12.

Now plot the load line. It is to represent the sum of R_C and R_E or 5600 ohms + 470 ohms or 6070 ohms. A plot of a 6000-ohm load line will suffice. See the plot drawn in Fig. 2-26A, based on a 12-volt supply. The actual transistor characteristic curves, unnecessary for this discussion, have been omitted from the drawing in the interest of clarity.

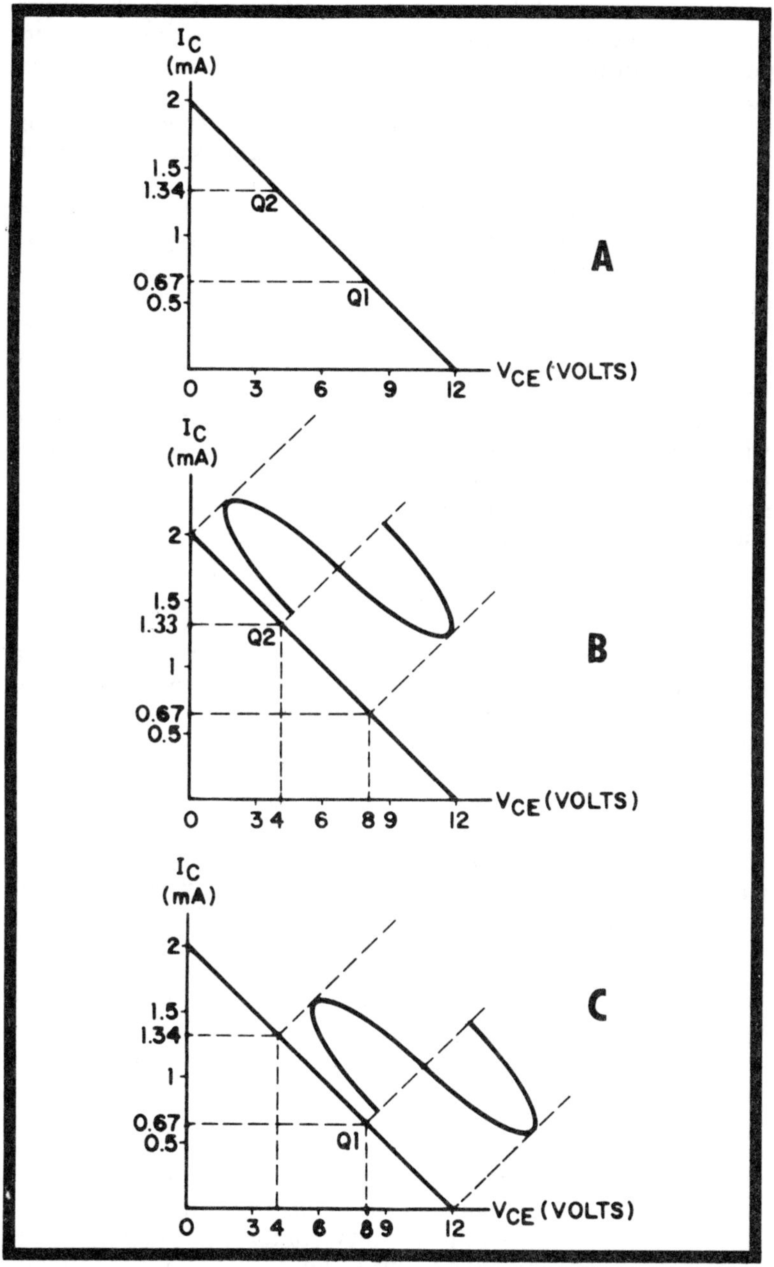

Fig. 2-26. Plot of the load line for the problem discussed in the text (A). Curve B is the quiescent point at one possible extreme position, and curve C is the second possible extreme position for quiescent point.

It has been stated that an 8-volt peak-to-peak output voltage swing is required to accommodate the 0.8-volt peak-to-peak maximum input signal. The peak-to-peak collector current swing is then 8 divided by 6000 ohms or 1.33 ma. The bias must be arranged so that the collector current can swing 1.33 ma. Should the bias be set at a minimum extreme position while accommodating the swing, the current will vary from 0 ma to 1.33 ma. At the opposite extreme position, the collector current will accommodate this swing if it will vary from 2 ma - 1.33 ma or 0.67 ma up to 2 ma. These two borderline conditions are shown in Figs. 2-26B and 2-26C, respectively. Obviously, the idling bias currents are at the middle of each curve, or (1.33-0) divided by 2 or 0.67 ma in one case and (2-0.67) divided by 2 + 0.67 or 1.34 ma for the second case.

Plot B may be considered the condition at one extreme when the transistor is operating in a 25-degree C ambient temperature and beta is at the minimum end of its specification range. In this case, I_{CBO} is at its minimum specified value. V_{BE}, the forward biased base-emitter voltage, is at the maximum of about 0.2 volt for germanium devices and 0.6 volt for silicon transistors.

Plot C, the other possible extreme, assumes the transistor is operating at the maximum temperature while beta is at a maximum of its range on the 2N3391A specification sheet. It also assumes that V_{BE} has dropped at about 2.5 mv (.0025 volt) for each degree the transistor ambient temperature has risen above 25 degrees C, while I_{CBO} has increased to its maximum value under the stated temperature limit.

From this analysis, we are then able to determine that the idling collector current change must be limited to 1.34 - 0.67 or 0.67 ma (the difference of the two idling currents), as the transistor temperature rises from 25 degrees C to 55 degrees C. We also determined that at 25 degrees C, the idling current must be adjusted to 0.67 ma (calculated above from Fig. 2-26B). The adjustment at 25 degrees C assumes the minimum beta and I_{CBO} and the maximum V_{BE}.

The remainder of the design procedure begins by surveying the various bias circuits that can be used. Initially, select the simplest logical one, design it using procedures outlined in the early parts of this chapter, and then substitute the component values in the appropriate stability equations. Notice the effect of changes of I_C on the initially designed circuit, considering each stability factor individually. Find the total change of I_C by summing the individual effects. Then, check if the resulting idling current (the initial quiescent current plus the sum of the ΔI_C values) is compatible with the plot in Fig. 2-26C. Thus, ΔI_C from all causes should not exceed 1.34 ma - 0.67 ma if the circuit is to operate properly under all desired conditions.

Should the change in I_C exceed the required value (0.67 ma in this case) compute the components needed for the next bias circuit in the complexity sequence. Once again calculate the stability factors and determine the ultimate effect on I_C. Add circuit complexities until you find an arrangement which will provide you with the desired results.

An alternate and somewhat more sophisticated design method uses extreme points in the engineering procedure. It starts by assuming the simple bias circuit in Fig. 2-27. Two equations are established for the base circuit. The first assumes one extreme of collector current, the minimum value, where I_C equals I_C(min) and the transistor operates under its minimum temperature condition. Here, I_{CBO} is at its minimum value, I_{CBO}(min), V_{BE} is at its maximum, V_{BE}(max), and beta is at its minimum, β(min). Assume I_C equals I_E. The base supply voltage is then equal to:

$$E'_{BB} = I_C\,(\text{min}) \left(R'_E + \frac{R'_B}{\beta\,(\text{min})} \right) - I_{CBO}\,(\text{min})\, R'_B + V_{BE}\,(\text{max}) \qquad (2\text{-}14)$$

A second equation assumes the maximum collector current, temperature and beta. The base-emitter voltage is at its minimum. The base supply voltage at this extreme condition becomes:

$$E'_{BB} = I_C\,(\text{max}) \left(R'_E + \frac{R'_B}{\beta\,(\text{max})} \right) - I_{CBO}\,(\text{max})\, R'_B + V_{BE}\,(\text{min}) \qquad (2\text{-}15)$$

The two equations are subtracted from each other. The various values of maximum and minimum collector current, beta, leakage current, and base-emitter voltage are then substituted into the resulting formula to establish a relationship between R'_E and R'_B. All these values can be determined from the data derived above. In some instances, such as for base-emitter voltage, actual data may not be supplied by the transmitter manufacturer. But actual data might not actually be required, because since Equation 2-14 was subtracted from Equation 2-15, the terms in the final equation will contain the difference of two values (such as V_{BE}(max) - V_{BE}(min)) rather than the absolute values. The difference of the two values of V_{BE}, for example, can easily be calculated from the temperature difference involved, since V_{BE}(max) - V_{BE}(min) is approximately equal to 2.5 millivolts multiplied by this temperature difference.

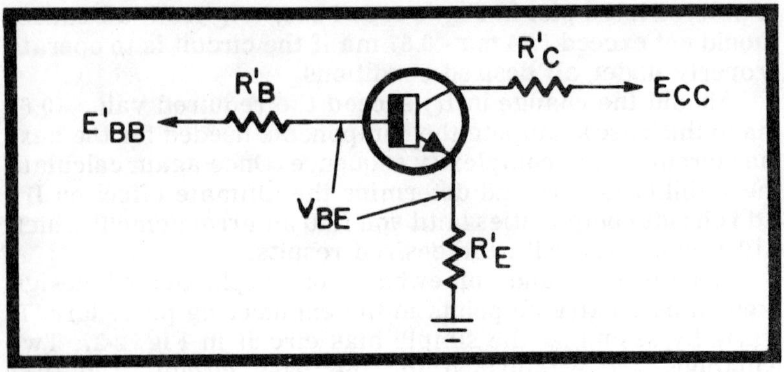

Fig. 2-27. Basic circuit used in the extreme point method of engineering stability into a design.

A specific value for either R'_B or R'_E may be dictated, limited by considerations such as gain, input impedance, distortion, etc. Determine the remaining resistor from the difference equation. Calculate E'_{BB} from Equation 2-14 or 2-15. Once these resistors and E'_{BB} have been determined, the circuit in Fig. 2-27 can be assumed to be but an equivalent circuit for any of the bias arrangements discussed in this chapter. Simple conversion can be made to any of the circuits by applying Thevenin's Theorem, but in a reverse manner.

Identical equations using this method of analysis can be written for the circuits in Figs. 2-19 and 2-22, if E_{BB} in Fig. 2-19 is connected to E_{CC}. They are:

$$R_X = \frac{E_{CC}\,E'_{BB}\,R'_B}{E_{CC} - E'_{BB}} \qquad (2\text{-}16)$$

$$R_B = E_{CC}\,R'_B/E'_{BB} \qquad (2\text{-}17)$$

$$R_E = R'_E \qquad (2\text{-}18)$$

However, for Fig. 2-22, $E_{BB}R_C$ divided by E_{CC} must be subtracted from R'_E to determine the value of R_E. Once the final circuit has been established, its performance should be double checked using the stability equations.

Biasing JFET Transistor Amplifiers

Chapter 3

The bias circuits designed for bipolar transistors serve two primary functions. The first is to set the quiescent base current with the ultimate goal of establishing a specific collector idling current. The second and equally essential intent is to stabilize the quiescent collector current against changes due to temperature variations, supply voltage fluctuations and parameter tolerances in specific transistor categories. FET bias circuits perform similar functions.

This discussion deals primarily with biasing the JFET, although many of the characteristics and procedures can be applied to the IGFET. The obvious partiality to the JFET is due to the extreme predominance of this type of FET in audio applications.

Only N-channel JFETs are discussed, but everything said with regard to this device can be applied to the P-channel JFET. However, the supply and element voltage polarities must be reversed. As for the bias mechanism, the gate-to-source idling voltage, V_{GS}, is set to establish a desirable quiescent drain current, I_D, in the output circuit. Stability of the required drain current must be maintained against variables similar, in some respects, to those plaguing the bipolar devices.

JFET CHARACTERISTICS

Fig. 3-1 shows a typical N-channel transistor schematic and a display of drain characteristic curves. The curves resemble those of the pentode vacuum tube. The display can be conveniently and functionally divided into two basic sections—the ohmic or **on** region and the pinch-off or **off** region. In audio amplifier applications, the ohmic region is just about ignored.

The curves are essentially horizontal for all gate-to-source voltages, V_{GS}, in the pinch-off region. It can be stated with reasonable accuracy that the drain current is virtually independent of the drain-to-source voltage, V_{DS}, although the

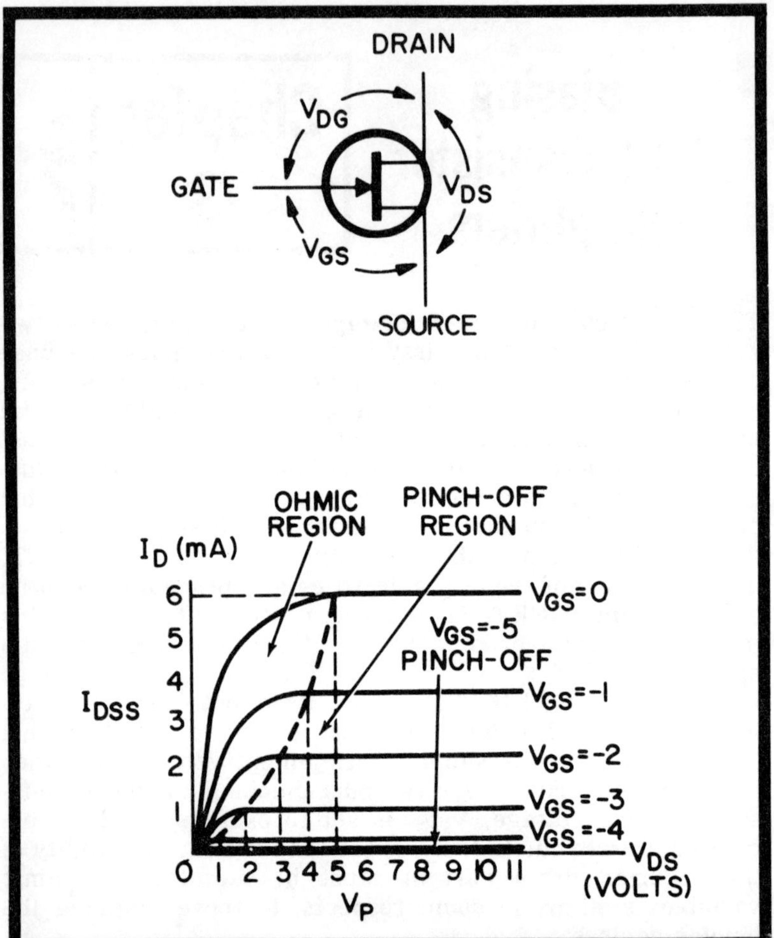

Fig. 3-1. The N-channel transistor symbol, associated voltages and a set of characteristic curves.

current does increase slightly with the voltage. The drain current, I_D, flowing when V_{GS} is equal to zero is assigned the special symbol I_{DSS}—the zero gate-to-source voltage drain current.

It is obvious from the curve that the gate of the N-channel JFET is made negative with respect to the source. The gate-source junction is reverse biased. As you may recall, a reverse-biased junction of any diode conducts practically no current. From Ohm's Law, the impedance of any device is the voltage across the device divided by the current flowing through it. Because the current flowing through a reverse-biased junction is just about zero, the impedance of the junction, by Ohm's Law, approaches infinity. The impedance of the reverse-biased gate-source junction similarly approaches infinity.

Despite not being shown on the graph, positive V_{GS} curves are applicable up to about +0.5 volt. The impedance remains high despite the forward-bias voltage applied to the junction. This follows from the diode characteristic where the forward-biased silicon junction does not conduct until that voltage exceeds about +0.5 volt.

As the gate-to-source voltage is made more negative, the collector current becomes very small, approaching infinitesimal proportions. The gate-to-source voltage generating this condition is referred to as the pinch-off voltage, V_P, of the transistor. This is similar to the concept of cut-off voltage in relation to the vacuum tube. As an acceptable standard, the gate-source voltage is considered at pinch-off when the drain current is one-thousandth to one hundredth of I_{DSS}, or less than 50 microamperes.

A broken-line curve is superimposed on the graph in Fig. 3-1. This trace is the sum of the drain-to-source voltage, V_{DS}, and the gate-to-source voltage, V_{GS}, with the polarity of both values ignored. The sum is equal to the pinch-off voltage. The mathematical statement for this curve is V_P equals V_{DS} + V_{GS}. It is obvious from the equation that when V_{GS} is equal to zero, V_P equals V_{DS}. This is a point on the V_{GS} equals 0 curve at the start of the horizontal pinch-off region. Pinch-off voltage can then be determined from this curve, as shown in Fig. 3-1, as well as from the value of V_{GS} which causes I_D to be very small.

Earlier, it was noted that the set of drain characteristics is separated into two regions. This broken-line plot of the equation is the dividing line. To be certain that you are designing audio amplifiers well within the linear pinch-off region of the plot, it is conventional that the minimum drain-to-source voltage be made 1½ times the specified pinch-off voltage.

The ohmic region to the left of this line is very valuable in some applications because each curve is a plot of a distinct drain resistance. The resistance is equal to the ratio of the change in drain-to-source voltage, ΔV_{DS}, to the change in drain current, ΔI_{DS}. Since this resistance is different for each V_{GS} display, the drain resistance of the JFET will vary with the gate-to-source voltage. This characteristic is very useful should you wish to design a circuit such as a volume compressor utilizing the JFET resistance change.

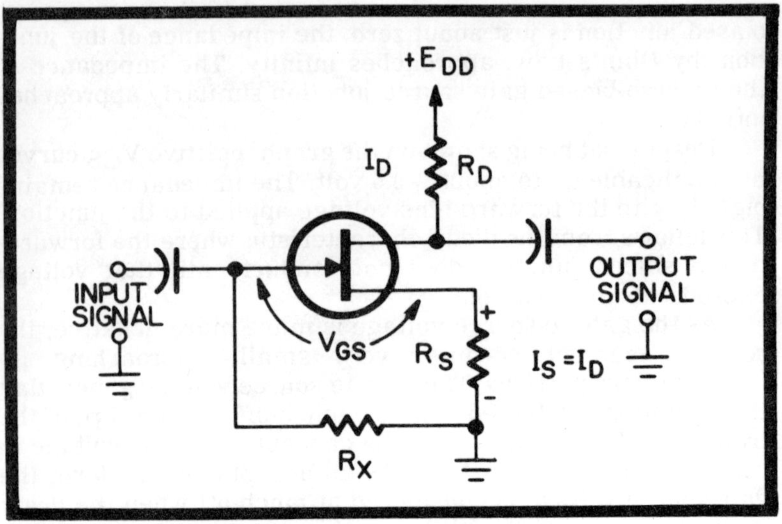

Fig. 3-2. Self-bias circuit for a JFET.

Getting back to the pinch-off region, it is obvious from the drawing that the quiescent drain current, I_D, is dependent upon V_{GS}. Although not equally as evident, drain current is related to V_P and I_{DSS}. The important relationship is:

$$I_D = I_{DSS}\left(1 - \frac{|V_{GS}|}{|V_P|}\right)^2 \qquad (3\text{-}1)$$

where V_{GS} is the absolute value of V_{GS} and V_P is the absolute value of V_P. As you know, the absolute value of numbers disregards the + or - signs in front of the numbers.

BIAS CIRCUITS

The quiescent drain or bias current can be calculated using Equation 3-1. I_{DSS} and V_P depend on the particular transistor used. As these factors remain relatively constant for any one transistor, the drain current depends only on specific values of V_{GS}. It is then necessary to find circuits to set the required V_{GS} to establish the desired quiescent drain current. The self-bias circuit in Fig. 3-2 performs this function simply.

The circuit is similar to the one most commonly used for vacuum tube bias. The drain current starts at the negative ground side of the supply, flows through R_S, through the transistor and R_D and on to the positive terminal of the supply $+E_{DD}$. Source and drain currents are essentially equal. A voltage with the polarity shown is developed across R_S and is equal to I_DR_S. No voltage appears across R_X (a statement to be modified below). The only voltage between the source and gate is that across R_S. Considering this, V_{GS} is equal to I_DR_S.

For stability considerations, it is desirable to make R_S large. A voltage supply bucking that developed across R_S can be placed in series with R_X, as shown in Fig. 3-3. The voltage between the gate and source is about equal to V_{GS} equals I_DR_S - E_{GG}. Since the gate-source voltage and the drain current must be identical to that determined for the previous circuit, the only variable factor in the equation is R_S. R_S must be larger in value here than it was for the circuit in Fig. 3-2 to compensate for the effect of the additional E_{GG} supply.

This circuit is not practical because a second power supply must be used for the E_{GG} voltage. An arrangement using but one supply is shown in Fig. 3-4. Voltage is developed across R_X due to the current, I_G, generated by $+E_{DD}$ which flows through R_G and R_X. It is of the proper polarity to buck the I_DR_S voltage. The circuit is frequently designed so that I_G is equal to one-tenth of the drain current. The gate-to-source voltage then becomes V_{GS} equals I_DR_S - I_GR_X equals I_DR_S - 0.1 I_DR_X equals $I_D(R_S$ - $0.1R_X)$. I_G may be much less than the one-tenth of the drain current, but it must be at least 10 times the leakage current to be discussed below.

Resistors R_G and R_X may, at times, be too low in value to satisfy the requirements of an input signal source feeding the

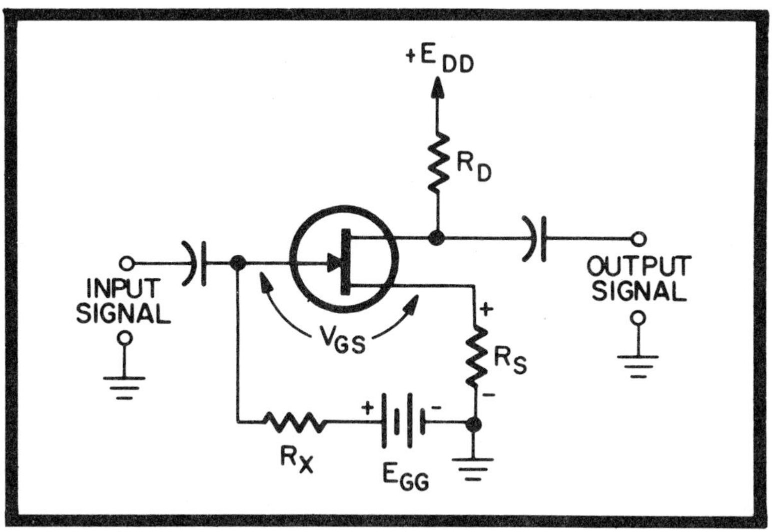

Fig. 3-3. Combined fixed- and self-bias circuit designed so that R_S can be made large.

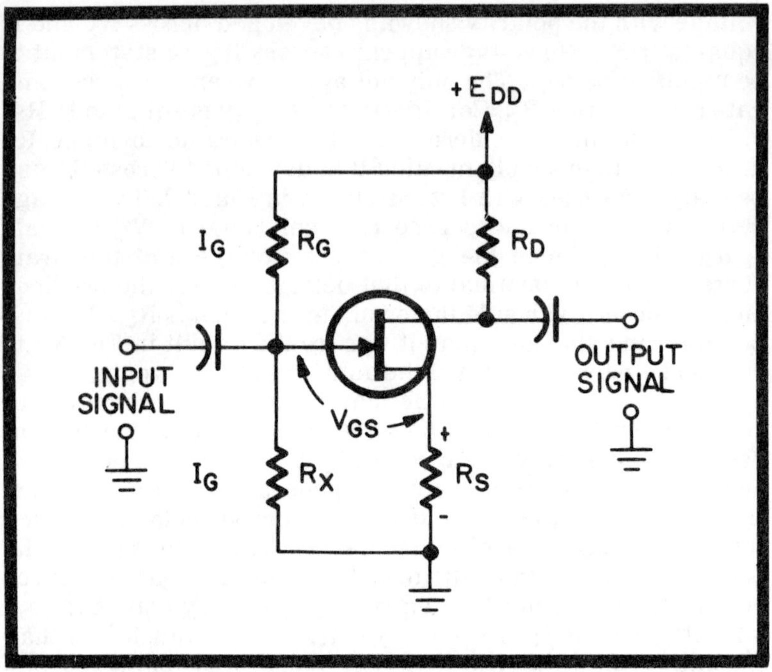

Fig. 3-4. Practical version of the circuit in Fig. 3-3 using only one power supply.

transistor. The input impedance is the parallel combination of R_G and R_X or $R_G R_X$ divided by $(R_G + R_X)$. Should the circuit be modified to that in Fig. 3-5, the input signal will see a load equal to R_A in series with the parallel combination. As no current flows through R_A, it is not a factor in determining the bias voltage.

In the interest of stability, one additional change is frequently added to the circuit in Fig. 3-4. Instead of connecting R_G to the power supply, it is connected to the drain as in Fig. 3-6. Any advantages are due to the DC voltage feedback from the drain to the gate through R_G. Of course, R_A can be added here as it was in Fig. 3-5 should additional resistance be required to increase the input impedance presented to a source.

LEAKAGE CURRENT

In the bipolar transistor discussion, we encountered a leakage current flowing in reverse-biased junctions. One leakage current is I_{CBO} due to the reverse current flowing between the collector and base. As for the JFET, the channel encompasses both the source and drain. The channel is reverse biased with respect to the gate. A leakage current flows between the channel and gate. This leakage current, I_{GSS}, is defined as the reverse current flowing from the gate when the source and drain are connected to each other. The current is proportional to the square of the voltage applied between the elements.

As is characteristic of all leakage currents, I_{GSS} increases with temperature. Depending upon the particular transistor, IGSS doubles with every 6-degree C to 15-degree C temperature elevation. In this book we will use 10 degrees C as a good average for all JFETs.

The leakage current for a JFET is usually very low. For example, at 25 degrees C, it can range from 0.1 nanoamperes (0.1×10^{-9} ampere) for a 2N5558 up to 1 nanoampere for a 2N4303. However, it is significant at high temperatures, especially if R_X in the gate circuit is large. This can be illustrated using the bias circuit in Fig. 3-2. The bias voltage, V_{GS}, was stated as equal to I_DR_S. A voltage is developed across R_X due to the fact that I_{GSS} is equal to $I_{GSS}R_X$. To calculate V_{GS} more precisely than before, $I_{GSS}R_X$ must be subtracted from the originally determined bias voltage.

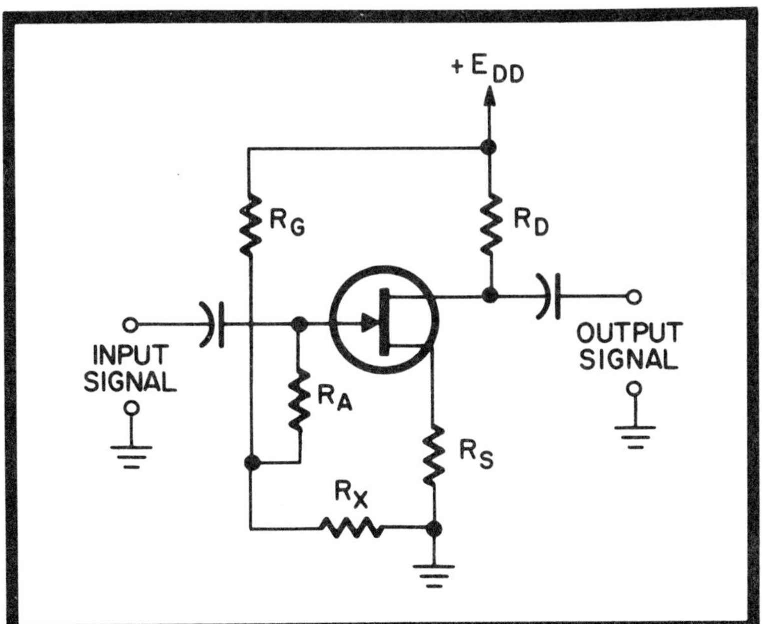

Fig. 3-5. The circuit in Fig. 3-4 is shown modified to present a higher impedance to the input signal.

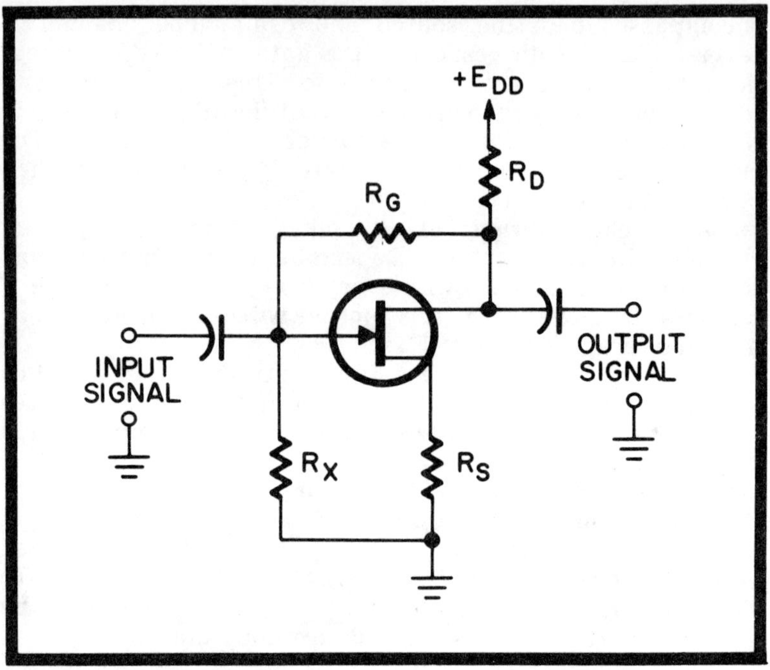

Fig. 3-6. The bias circuit is improved by adding DC feedback from the drain to the gate.

TRANSCONDUCTANCE

Among the important JFET parameters that change with temperature are I_{DSS}, I_{GSS} and g_m. The first two items were just discussed in detail while g_m, the AC transconductance, was defined in Chapter 1. Transconductance relates the output current to the input voltage by the equation:

$$g_m = \left(\frac{\Delta I_D \text{ ampere}}{\Delta V_{GS} \text{ volt}}\right) \text{(or mhos)} \qquad (3\text{-}2)$$

where ΔI_D is a change in drain current due to ΔV_{GS}, a change in gate-to-source voltage. This can be shown graphically with the help of Figs. 3-7 and 3-8. In Fig. 3-7, the plot of drain current against the gate-to-source voltage (the transfer characteristic curve) is derived from the drain characteristic curves.

At the left-hand side of Fig. 3-7, the drain characteristic curves in Fig. 3-1 have been redrawn. The horizontal drain current lines for each value of gate-to-source voltage have been extended to the vertical drain current axis in the graph at

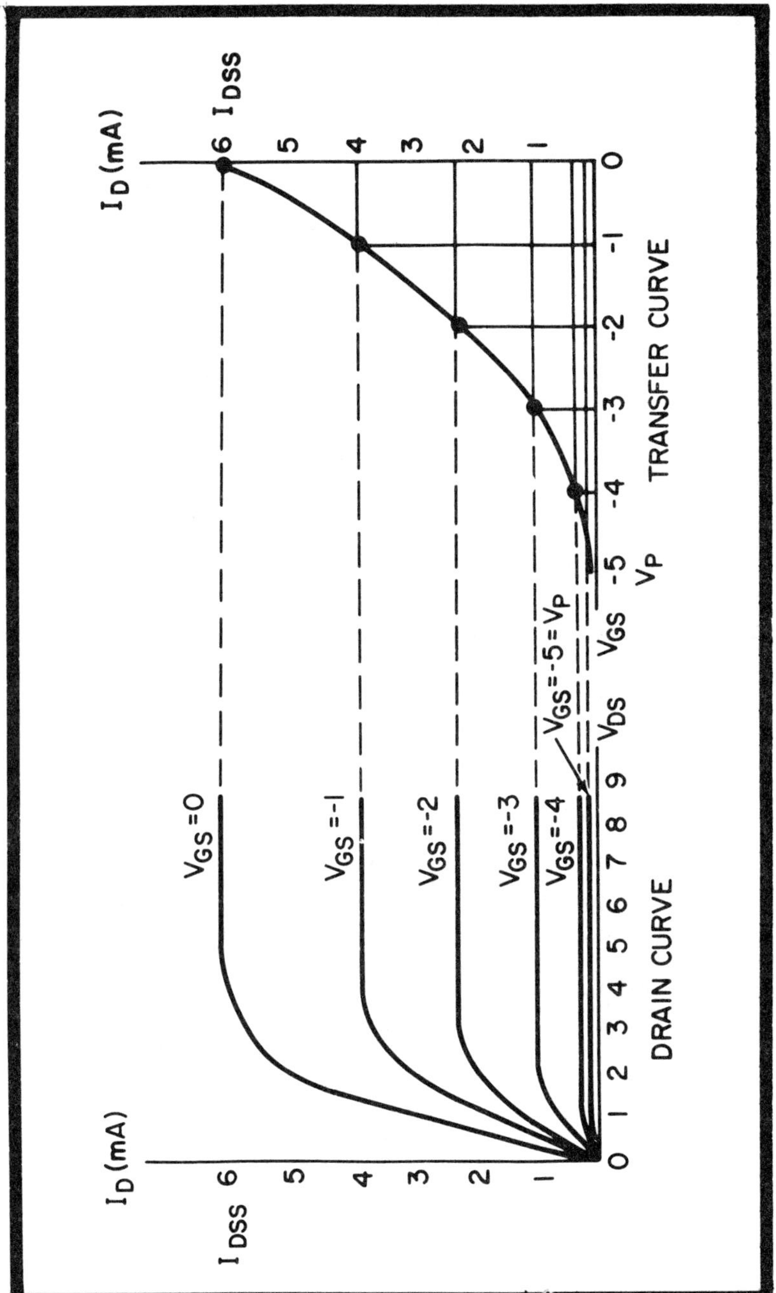

Fig. 3-7. The transfer characteristic curve is derived from the drain characteristic display in Fig. 3-1.

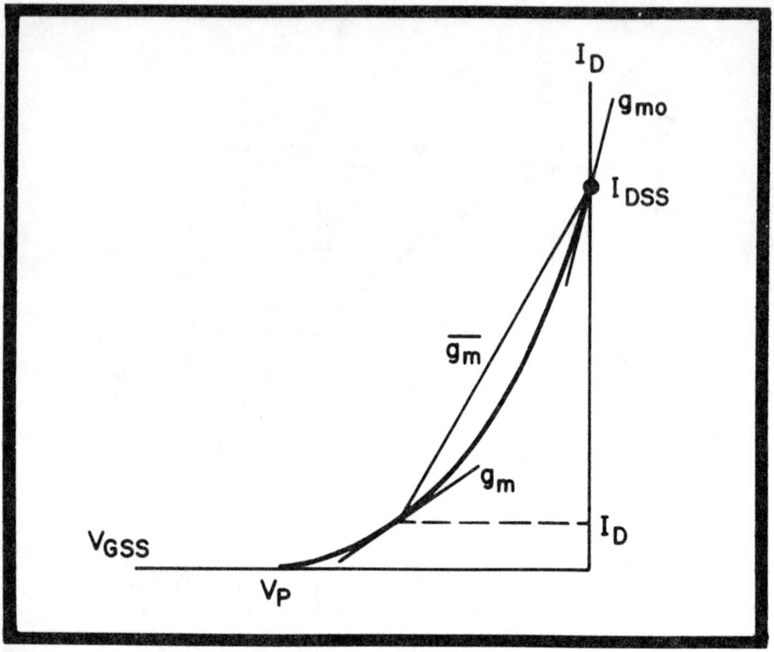

Fig. 3-8. Plot of transconductance on the transfer characteristic curve. Slope of the tangential lines are transconductances at various points on the curve. Slope of the line connecting I_{DSS} and I_D is the average transconductance.

the right of the figure. The horizontal axis of the second graph is calibrated in gate-to-source voltage. A vertical line is drawn from the gate-to-source voltage on the horizontal axis up to the horizontal current line which represents the drain current that flows for the specific V_{GS}. This procedure is repeated for each discrete value of V_{GS}. All necessary information is available in the original drain characteristic plot at the left. A dot is placed at the point where the related curves cross. All dots are connected to form the transfer characteristic curve.

Because the curve is a plot of the drain current against the gate-to-source voltage, the slope of the curve at any one point is the transconductance of the JFET for that particular current value. The variation of the slope from point to point on the curve indicates that the g_m varies with drain current.

Initially, two AC transconductance values are of interest to us. The first, referred to as g_{mo}, is the transconductance at I_{DSS}, where the gate-to-source voltage is zero (see Fig. 3-8). Assume the transistor is biased so that the quiescent drain current is I_D. Put a dot on the curve at this drain current. This is the second significant point at which the transconductance must be determined. Refer to this transconductance simply as g_m.

An average value of transconductance, $\bar{g}_m$, is (g_m + g_{mo}) divided by 2, the mean value of the two transconductances. An equation can be derived from Fig. 3-8 to determine $\bar{g}_m$ from other circuit and transistor characteristics. It requires that I_{DSS} be connected to I_D in Fig. 3-8 with a straight line. Applying equation 3-2:

$$\bar{g}_m = \frac{I_{DSS} - I_D}{V_{GS} - 0} \qquad (3\text{-}3)$$

or stated in a more useful form:

$$I_D = I_{DSS} - \bar{g}_m V_{GS} \qquad (3\text{-}3a)$$

We have by now defined all the transistor parameters necessary to design a stable circuit. It is necessary to stabilize the transistor against the wide tolerances that exist in the specifications of any particular type as well as against parameter variations with temperature and drain current. Let us now summarize the parameters that do concern us. They vary with temperature in the following manner.

1. For the JFET, I_{DSS} increases as the temperature decreases. As for the IGFET, it doubles for every 10 degree C rise in temperature.

2. g_m is an inverse function of temperature. It also changes with drain current.

3. V_P increases at about 2.2 millivolts for each degree Celsius that the temperature increases.

4. I_{GSS} doubles for about every 10 degree C rise of the JFET's temperature. As for the IGFET, I_{GSS} is minute and the change with temperature is negligible.

5. r_{DS} (on) is the resistance from the drain to source when the transistor is biased fully in the ohmic region. It increases with temperature.

Now let us turn to the specified transistor tolerances. Parameters such as I_{DSS}, V_P and g_{mo} are specified as having a range of values for any transistor type, rather than a specific value. For example, the I_{DSS} of a 2N4303 can be anywhere between 4 and 10 ma and still be specified as a standard 2N4303.

Applying these facts, we describe two procedures which can be used to design stable circuits. The first assumes a

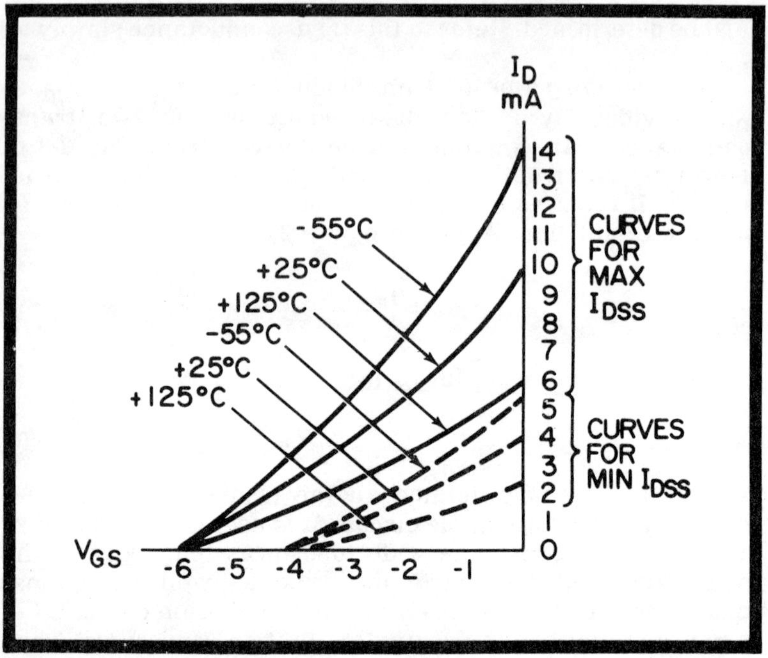

Fig. 3-9. The solid-line 25-degree C plot is the transfer characteristic curve for the maximum I_{DSS} value of a hypothetical JFET. The broken-line 25-degree C curve represents the transfer characteristic for the minimum value of I_{DSS}. The other curves are maximum and minimum values at extreme temperatures.

thoughtful manufacturer who provides the designer with all pertinent information and limits. In this case, the paper design will usually suffice. The second design procedure assumes that you have been provided with a minimum of detailed information. Many approximations must be made. The end result using any procedure should be checked in the laboratory. This is especially true of the latter, more inexact, approach.

STABILIZING A DESIGN USING COMPLETE DATA

The exact method assumes the manufacturer supplied the data for or the curves in Fig. 3-9. As explained in the figure, transistors are specified to be within any one category if their characteristics fall within a specific set of limiting values. The curves marked 25 degrees C represent the two extremes within which all transistors of a particular type must fall when the ambient temperature is 25 degrees C. Should the ambient temperature change, the curves are displaced accordingly.

The plots show the transfer characteristic curves within specific I_{DSS} limits at various temperatures. If these curves are not presented as such in the manufacturer's data, the plots at 25 degrees C may usually be derived from the drain characteristic curves using the procedures exercised in Fig. 3-7.

The design procedure is based on transistor operation limited to the boundaries of two of the curves. It can best be illustrated with a practical example. Assume that a circuit such as that in Fig. 3-3 is to be designed. The required output from the circuit is a minimum of 6 volts peak-to-peak across a 5600-ohm load resistor. The transistor must operate properly from +25 degrees C to +125 degrees C.

The temperature range is the first item to be considered. From the plots in Fig. 3-9, choose the 125 degree C broken-line curve for the lower extreme characteristics and the 25 degree C solid-line curve for the second limiting factor. All plots defining the characteristics of the particular transistor type between +25 degrees C and +125 degrees C are included between these curves. Hence, those curves as boundaries satisfy all temperature requirements. Redraw the limiting curves in Fig. 3-10. Now, place a 5600-ohm resistor in the drain circuit so that the output load will be the specified value. The 6-volt peak-to-peak requirement dictates a minimum drain current swing of 6 volts divided by 5600 ohms or 1.07 ma.

We proceed to plot a bias point on each of the two curves and set the extremes in idling current. It is undesirable to bias the transistor at or near pinch-off because this region is nonlinear and the transconductance is low. If we place the bias on the lower I_{DSS} curve at approximately 1.75 ma, we will operate at a reasonably linear portion of the curve. It will also allow the 1.07 ma peak-to-peak current to swing the maximum range without getting into pinch-off at one end and not reaching the zero V_{GS} limit at the other extreme. Call this point I_D(min) and plot it on the broken-line curve. Drop a vertical line from this point to the axis. The line crosses the axis at 0.75 volt gate-to-source voltage. Refer to this as V_{GS}(min) or 0.75 volt.

Since the required 1.07 ma excursion is a small portion of the possible transistor current variation, we have wide latitude in choosing the maximum value of drain current on the upper limiting curve. Select I_D as 4 ma and plot it on the solid-line curve. Extend the lines to the vertical and horizontal axis. The vertical line crosses the gate-to-source axis at 2.9 volts. The point on the upper curve is then defined by I_D(max)

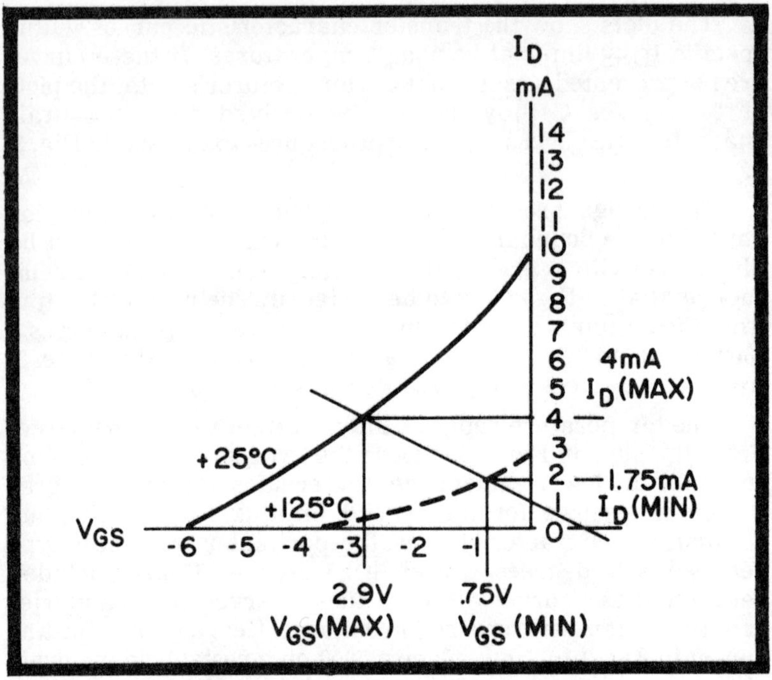

Fig. 3-10. Curve describing the limits for solving a problem using the exact design method.

equals 4 ma and V_{GS}(max) equals 2.9 volts. Connect the points plotted on the two limiting curves with a straight line.

In the next step, the source resistor, R_S, is to be calculated for Fig. 3-3. It is the slope of the line connecting the two points on the graph. R_S is equal to the difference in the gate-to-source voltages divided by the difference in the drain currents, or:

$$R_s = \frac{2.9 - 0.75}{4 \times 10^{-3} - 1.75 \times 10^{-3}} = \frac{2.15}{2.25 \times 10^{-3}} = 955 \text{ ohms} \tag{3-4}$$

Use the next higher EIA 10 percent resistor value of 1000 ohms.

The next task is to determine E_{GG} from the plot. Circuit theory dictates the relationship:

$$E_{GG} = I_D R_S - V_{GS} \tag{3-5}$$

Using the minimum values of V_{GS} and I_D, (V_{GS}(max) and I_D(max) can be used instead), we find:

$$E_{GG} = (1.75 \times 10^{-3})(1000) - 0.75 = 1 \text{volt} \tag{3-6}$$

R_X is chosen so that the voltage developed across it by I_{GSS} at the highest temperature is negligible compared to the E_{GG} of 1 volt. Assume that one-tenth of E_{GG} or 0.1 volt is a negligible value. If it were specified that I_{GSS} equals 1 nanoampere (10^{-9} ampere) at 25 degrees C, and that it doubles for every 10 degree C rise in temperature up to 125 degrees C, the maximum I_{GSS} is about 1 microampere or 10^{-6} ampere. Then the maximum R_X may be 0.1 volt divided by 10^{-6} ampere or 100,000 ohms.

The minimum drain supply voltage still remains to be determined. It is equal to the sum of three factors: 1) the maximum voltage across R_S, or $I_D(max)R_S$; 2) the minimum voltage across the transistor, $1.5V_P$, where V_P is the maximum pinch-off voltage for the transistor (V_P should be multiplied by about 1.5 to be certain that the transistor is operating in the linear portion of the pinch-off region); 3) the maximum voltage across R_D, or $I_D(max)R_D$. It should be noted that since R_D and R_S are 10 percent resistors, the values referred to in Steps 1 and 3 are 10 percent above the nominal values of both resistors. Thus, the R_S in question is 1000 ohms + 10 percent of 1000 ohms or 1100 ohms and R_D is 5600 ohms + 10 percent of 5600 ohms or 6160 ohms. In our example, the minimum supply voltage is:

$$\begin{aligned} E_{DD} &= (4 \times 10^{-3}\text{mA})\ \ (1100\ \text{ohms}) \\ &\quad + 1.5(6) + (4 \times 10^{-3}\text{mA}) \\ &\quad (6160\ \text{ohms}) \\ &= 4.4 + 9 + 24.64 \\ &= 38.04\ \text{volts} \end{aligned} \tag{3-7}$$

Use a 40-volt supply.

Should you wish to use the circuit in Fig. 3-4 instead of that in Fig. 3-3 and eliminate the second supply, the E_{GG} voltage must be developed across R_X. Assuming that I_G is about ten times the maximum I_{GSS} determined above, or 10 x 10^{-6} equals 10^{-5} amperes, and that E_{GG} is 1 volt as previously calculated, then R_X is 1 volt divided by 10^{-5} ampere or 100,000 ohms. Since E_{DD} is 40 volts, the remaining 39 volts must be developed across R_G. The same 10^{-5} ampere that flows through R_X flows through R_G. This makes R_G equal to 39 volts divided by 10^{-5} ampere or 3.9 megohms, a standard 10 percent EIA resistor value.

It should be noted here that the simple circuit in Fig. 3-2 can be used successfully to fulfill all stability requirements. Bias points on the two curves can be chosen to determine only a value for R_S and the E_{GG} supply is unnecessary. However, the complication was added to indicate a procedure rather than to solve a tight problem.

STABILIZING A DESIGN USING AN APPROXIMATE METHOD

Let us now turn to a second method of determining the components in the circuits. This time assume that the manufacturer did not furnish the curves in Fig. 3-9. To effect a solution, we start with the assumption that all temperature variations, with the possible exception of I_{GSS}, have little effect on determining the circuit components. Actually, changes with temperature are usually negligible when compared to the variations in I_{DSS} due to transistor tolerances.

Several pertinent relationships should now be established. From Fig. 3-8, we evolved Equation 3-3a to state a relationship between drain current, I_{DSS}, V_{GS}, and the average transconductance, $\bar{g}_m$. This equation, repeated here for convenience, is:

$$I_D = I_{DSS} - \bar{g}_m V_{GS} \tag{3-3a}$$

In many instances, we do not know the actual transconductance, $\bar{g}_m$, at a particular quiescent current. It can be determined mathematically, for it is related to g_{mo}, the transconductance at I_{DSS}, by the equation:

$$g_m = g_{mo}\left(\frac{I_D}{I_{DSS}}\right)^{1/2} \tag{3-8}$$

Equation 3-9 relates $\bar{g}_m$ to g_{mo} and g_m:

$$\bar{g}_m = \tfrac{1}{2}\,(g_{mo} + g_m) \tag{3-9}$$

for as an average value, $\bar{g}_m$ is midway between the values of g_m and g_{mo}. It should also be noted that g_{mo}, V_P, and I_{DSS} are related to each other by the equation:

$$g_{mo} = \frac{2\,I_{DSS}}{|V_P|} \tag{3-10}$$

where V_P is the absolute pinch-off voltage, disregarding polarity.

All factors in Equations 3-3a, 3-8, 3-9, and 3-10 are assumed to be average values. In order to apply these equations properly, the average values for the various parameters must be calculated from the limited data supplied by the manufacturer of the device. Initially, we must also assume that all operations take place at a 25-degree C ambient temperature.

We will apply the approximate method to the problem solved above using the complete data. In the problem, we assumed a transistor with a minimum $\bar{I}_{DSS}$ of 4 ma and a maximum of 10 ma. Both currents are specified at 25 degrees C. The average current is midway between the two limits, or ½(4 + 10) equals 7 ma or I_{DSS}. From the curves in Fig. 3-9, we notice that the pinch-off voltages range from 4 to 6. The average is $\bar{V}_P$ is 5 volts.

V_{GS} should never be allowed to swing to pinch-off to avoid the crowded region of the display and the consequent distortion. As a rule of thumb, V_{GS} should not be more than three-fourths of the minimum value of V_P. In this case, V_{GS} should be less than 3 volts. However, it may be allowed to swing somewhat beyond zero volts at the other extreme, for the transistor will not conduct until V_{GS} is more than +0.5 volt. An average gate-to-source voltage, $\bar{V}_{GS}$, would then be ½(3-0) or 1.5 volts. Substituting this into Equation 3-1, the average drain current is:

$$I_D = 7 \times 10^{-3} (1 - 1.5/5)^2 = 3.43 \times 10^{-3} \text{ ampere.} \quad (3\text{-}11)$$

If gain is important, $\bar{I}_D$ should be determined from Equation 3-3a using the $\bar{g}_m$ term, rather than determined from Equation 3-1 as above.

The average transconductance should be calculated from g_m and g_{mo}. Use Equation 3-10 to determine g_{mo}. It is $2I_{DSS}$ divided by $|V_P|$ equals $2(7 \times 10^{-3})$ divided by 5 equals 2.8×10^{-3} ampere per volt. Equation 3-8 dictates that the transconductance at the average drain current is g_m equals (2.8×10^{-3}) (3.43×10^{-3}) divided by $(7 \times 10^{-3})^{1/2}$ equals 1.96×10^{-3} ampere per volt. Hence, the average transconductance, g_m, is $(g_{mo} + g_m)$ divided by 2 equals $(2.8 \times 10^{-3} + 1.96 \times 10^{-3})$ divided by 2 equals 2.38×10^{-3} ampere per volt from Equation 3-9.

To summarize the important data, $\bar{I}_{DSS}$ equals 7 ma, $\bar{I}_D$ equals 3.43 ma, $\bar{V}_P$ equals 5 volts, g_m equals 2.38×10^{-3} mhos, $\bar{V}_{GS}$ equals 1.5 volts and I_{GSS} equals 10^{-6} amperes at 125 degrees C. (I_{GSS} at the elevated temperature was calculated in the previously discussed procedure.)

Using the above data, initially design the simple self-bias circuit in Fig. 3-2. R_D is 5600 ohms as specified for all circuits.

The entire gate-to-source voltage is developed across R_S and is equal to $\overline{I}_D R_S$, if we consider the drain and source currents to be equal. Since $\overline{V}_{GS}$ equals $\overline{I}_D R_S$, R_S equals $\overline{V}_{GS}$ divided by $\overline{I}_D$ equals 1.5 volts divided by 3.43×10^{-3} ampere or 438 ohms. Use a standard 470-ohm resistor. As before, R_X is 100,000 ohms, for the considerations here are the same as they were in the example with the more exact solution.

The circuit in Fig. 3-3 is of greater interest because the E_{GG} voltage which bucks $I_D R_S$ will permit R_S to be increased in size while retaining the desired V_{GS}. The equation for the gate voltage is:

$$\overline{V}_{GS} = \overline{I}_D R_S - E_{GG} - I_{GSS}(\max) R_X \quad (3\text{-}12)$$

Substituting the calculated values into the equation yields:

$$1.5 = (3.43 \times 10^{-3}) R_S - E_{GG} - (10^{-6})(10^5) \quad (3\text{-}13)$$

so that the gate supply voltage is related to the source resistor by:

$$E_{GG} = -1.4 + 3.43 \times 10^{-3} R_S \quad (3\text{-}14)$$

R_S must be bigger than determined for the circuit in Fig. 3-2. Doubling the 470 ohms to the EIA resistor value of 1000 ohms is a good estimate. Then E_{GG} will be —1.4 + 3.43 equals 2.03 volts from Equation 3-11. This is a logical solution not unlike that derived using the more exact methods. The results should be checked in the laboratory.

The required drain supply voltage should now be determined. The procedures used above also apply here. The voltage across the drain resistor is $\overline{I}_D R_D$ equals $3.43 \times 10^{-3} \times 5600$ ohms or 19.2 volts. The minimum voltage across the transistor is (1.5) $\overline{V}_P$, as derived above, or 1.5 (5 volts) equals 7.5 volts. The voltage across R_S must be at least that calculated for Fig. 3-2, or $\overline{I}_{DRS}$ equals $(3.43 \times 10^{-3})(438) + 1.5$ volts. The average supply voltage is the sum of these factors, or 19.2 + 7.5 + 1.5 equals 28.2 volts. Add about 30 percent to compensate for peak values in drain current and pinch-off voltage. The required supply will then be approximately 40 volts.

The practical circuit in Fig. 3-4 can be determined from the data which has already been accumulated. Although R_X remains at 100,000 ohms, R_G can be calculated as before. The voltage across R_G is E_{DD} less the voltage across R_X, or 40 volts - 2.03 volts ≈ 38 volts. (2.03 is E_{GG}, the voltage across

R_X.) As I_G, the current flowing through R_G is about the same as the current flowing through R_X or 2.03 divided by $10^5 \approx 2 \times 10^{-5}$ ampere. R_G is 38 volts divided by 2×10^{-5} ampere equals 1.9 megohms. Use a standard EIA 1.8-megohm resistor.

Should the more stable circuit in Fig. 3-6 be desired, the quiescent voltage across R_G is that at the drain, V_D, less, E_{GG}, the voltage across R_X. Because the quiescent drain current is 3.43×10^{-3} ampere and the resistor in the drain circuit is 5600 ohms, the voltage across R_D is $(3.43 \times 10^{-3})(5600) \approx 19$ volts. The voltage across R_G is $19 - 2.03 \approx 17$ volts. Assuming that I_G is equal to the 2×10^{-5} ampere determined above, R_G for this circuit is 17 divided by 2×10^{-5} equals 8.5×10^5 ohms. A standard 820,000-ohm resistor may be used here.

Chapter 4 Bipolar Amplifiers

The ultimate goal of an audio amplifier is to provide gain for the input signal. However important the bias and stability considerations are, a circuit must be designed to enable the amplifier stage to furnish voltage, current and-or power amplification. At the output, the signal delivered to a load should be an enlarged and undistorted version of the input.

Reference was made in Chapter 2 to the three most commonly used transistor circuits. These are, of course, the common-base, common-emitter and common-collector arrangements. For any circuit built around each of these arrangements, we must know the current, voltage and power gains, as well as the input and output impedances. These factors can be determined by plotting the appropriate circuit information on collector characteristic curves. Procedures of this type are quite accurate for any one transistor, but are at the same time cumbersome.

Small-signal amplifiers are more readily analyzed using equivalent circuits. These circuits involve the semiconductor device and the associated components. Two different equivalent circuits are widely used—the "hybrid equivalent" and the "equivalent T." Both circuits have been previously explained, although they may not have been recognized as such in the text. The hybrid equivalent is briefly discussed because some important parameters are based on this method of analysis. Greater emphasis is placed on the T-equivalent circuit methods, since low-frequency design requirements are easily satisfied using these procedures. Relationships will be established between the two sets of parameters.

HYBRID EQUIVALENT CIRCUIT

The hybrid model of the circuit assumes the transistor, in any configuration, has a pair of input terminals and a pair of output terminals. In Fig. 4-1, the input voltage and current fed to the transistor are e_1 and i_1 respectively. Similarly, the voltage and current at the output are e_2 and i_2. Four important ratios evolve from this block representation.

1. When a short is placed at the output, e_2 is zero. A ratio of short-circuit output current to the input current, i_2 over i_1, is assigned the symbol h_f or h_{21}. In the common-emitter circuit drawing in Fig. 4-3, this is referred to as the β_{AC} of the transistor. For the common-base circuit in Fig. 4-7, the ratio is $-\alpha_{AC}$. (For the remainder of the book, we will drop the AC subscripts after the alpha and beta symbols. Should DC conditions be considered, the DC subscript will be employed.) These current gain factors are discussed thoroughly in Chapter 1.

2. Maintaining a shorted output circuit, we can establish the relationship e_1 over i_1. The symbols for this ratio are h_i and h_{11}. These symbols represent the input impedance of the transistor—but only when the output circuit is shorted.

3. The output impedance can be determined from the condition when the input is open circuited and i_1 equals 0. It is the ratio e_2 over i_2. The reciprocal of this ratio, i_2 over e_2, the output admittance, is assigned the symbols h_o and h_{22}.

4. The final symbols, h_r and h_{12} also assume the input circuit is open and i_1 equals 0. They are equal to e_1 over e_2, where e_1 is the voltage fed back through the transistor from the output circuit to the input, and e_2 is the output voltage.

A hybrid equivalent circuit can be drawn for the transistor using the relationships just noted. Although all arrangements (common-base, common-collector and common-emitter) use the same equivalent circuit, the second subscript at the h-parameter symbol indicates which circuit is involved. In Fig. 4-2, the equivalent for the common-emitter circuit is used, so all h parameters end with an "e," such as h_{fe}. Similarly, a "b" added to the subscript indicates that a common-base parameter is involved, while the addition of a "c" means the parameter is for a common-collector circuit. The values of the parameters depend upon the particular circuit arrangement in which the transistor is being used.

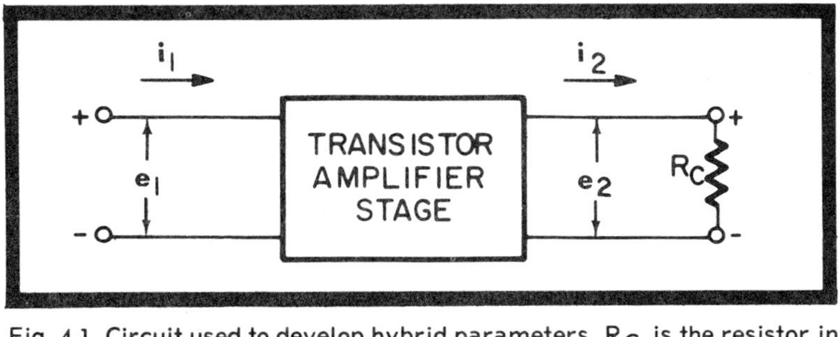

Fig. 4-1. Circuit used to develop hybrid parameters. R_C is the resistor in the transistor collector circuit.

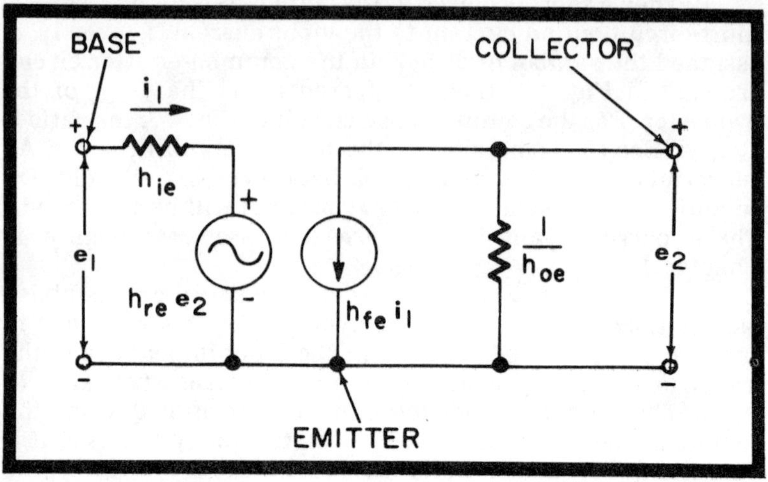

Fig. 4-2. Hybrid equivalent for a common-emitter transistor circuit. Only the equivalent circuit of the transistor proper is shown. Polarity markings show the relative phase at one instant.

In Fig. 4-2, only the transistor is represented in the equivalent circuit. The impedance looking into the base of the transistor is h_{ie} modified by the presence of $h_{re}e_2$. Similarly, the output impedance of the transistor itself is 1 over h_{oe} in parallel with the current source $h_{fe}i_1$. The voltage, current and power gains can also be determined from the two loops in Fig. 4-2.

The practical circuit in Fig. 4-3 has components which modify the characteristics of the transistor. For example, the output impedance presented by the transistor is 1 over h_{oe} in parallel with $h_{fe}i_1$. The output impedance presented by the circuit is the transistor output impedance in parallel with R_C. Similarly, the input impedance must be modified by the presence of R_G. The effect of the overall circuit can most easily be determined using the equivalent-T model.

EQUIVALENT-T CIRCUIT

In Fig. 4-4A, the equivalent-T model of the transistor is shown. It looks more like the actual transistor than does the hybrid equivalent network. The common-emitter equivalent is shown. In Fig. 4-4B, the input source and output load have been added to the transistor network. The input is applied between the base and emitter, while the output from the amplifier stage is developed between the collector and emitter.

An AC resistor representing the internal resistance of each transistor element is in series with each terminal inside the device. A current source, representing the gain of the device, appears across the collector resistor, r_d. b' is a point inside the transistor. All resistances from the transistor elements are connected to this point.

The emitter resistor, r_e, is 26 divided by I_E, where I_E is the quiescent emitter current expressed in milliamperes. r_b is the base resistance usually ranging from 200 to 800 ohms. The collector resistance is r_d for the common-emitter and common-collector circuits, usually assuming values between 10,000 and 100,000 ohms. The collector resistance for the common-base circuit is r_c. It is about equal to beta multiplied by r_d. The resistors are large in value when the input impedance to the transistor is small.

Various bits of data about the circuit in Fig. 4-3 can be determined by analyzing the T-equivalent circuit when it is connected to all resistors and sources in the associated circuit, as in Fig. 4-4B. The voltage source sees an impedance when looking into the transistor. It is equal to the base resistor, r_b, in addition to the sum of the emitter resistance, r_e, and the external emitter resistor, R_E, when the latter two are multiplied by beta. (All resistances, when reflected from the emitter into the base circuit, appear in the base circuit as the emitter resistors multipled by the beta of the transistor.)

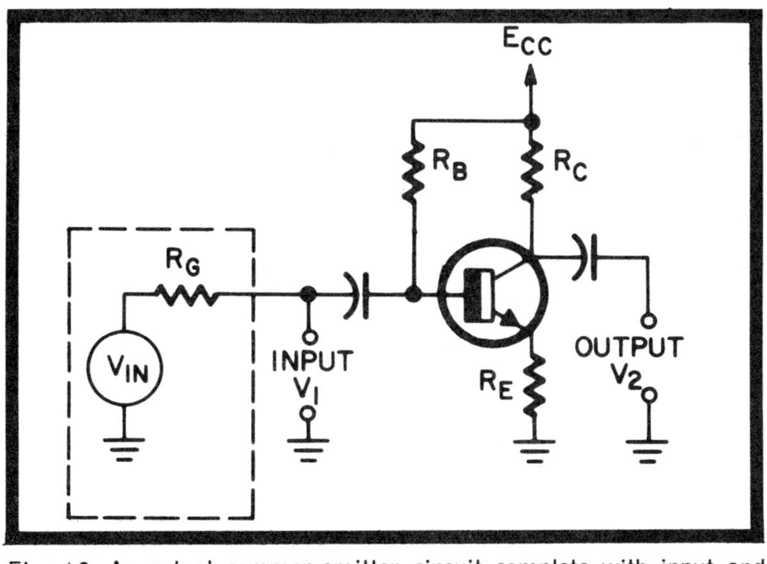

Fig. 4-3. An actual common-emitter circuit complete with input and output components and input signal generator.

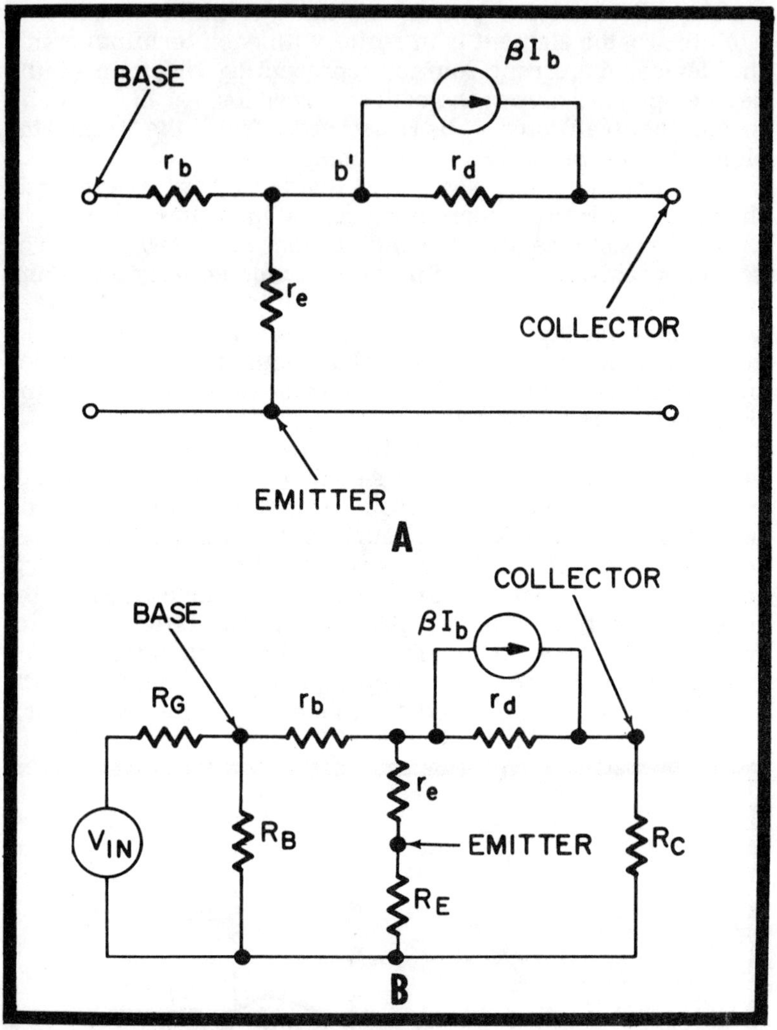

Fig. 4-4. T-equivalent circuit of only the transistor when used in a common-emitter arrangement. At B, the input and output components have been added to the transistor circuit for a complete equivalent of Fig. 4-3.

Consequently, the total transistor input resistance seen by the generator is:

$$R_{IN} = r_b + \beta\,(r_e + R_E) \tag{4-1}$$

R_{IN} is shunted by the bias resistor R_B in Fig. 4-3. The source feeding the transistor sees R_B in parallel with R_{IN}.

In the drawing, $R_C + r_d$ (the collector load resistor and the internal collector resistance, respectively) are shown as shunting $r_e + R_E$. However, they do not appear in Equation 4-1 since they are much larger than all emitter resistors. The effect of R_C and r_d on the parallel combination is indeed minor.

In a similar manner, with like considerations, all the other statistics for the common-emitter arrangement can be determined using the equivalent-T model. R_{OUT} represents the impedance the collector load resistor sees by looking back into the transistor, while A_V is the voltage gain of the circuit in Fig. 4-3, and A_i and G are the current and power gains, respectively. The approximate equations derived for the common emitter arrangements are:

$$R_{OUT} = \frac{r_d\,[(r_b + R_G) + \beta(r_e + R_E)]}{r_b + r_e + R_G + R_E} \qquad (4\text{-}2)$$

$$A_v = \frac{\beta R_L}{r_b + \beta\,(r_e + R_E)} \approx \frac{R_L}{R_E} \qquad (4\text{-}3)$$

$$A_i = \beta \qquad (4\text{-}4)$$

$$G = A_v A_i \approx \frac{\beta R_L}{R_E} \qquad (4\text{-}5)$$

In Equation 4-2, if the resistor in the bias circuit, R_B, is less than ten times the size of the source resistance, R_G, then the R_G stated in Equation 4-2 should be modified to $R_G R_B$ divided by $(R_G + R_B)$. This is the expression for R_G in parallel with R_B.

Now let us return to the hybrid parameters we described so carefully above. Some transistor manufacturers supply only these parameters in their data books. How can numbers be used for the various impedances in Equations 4-1 through 4-5 if only h parameters are provided? Simple! Let us set up equations relating the hybrid parameters to those in the equivalent-T circuit as follows:

$$r_b = \frac{h_{rb}}{h_{ob}} = h_{ie} - \frac{h_{re}\,h_{fe}}{h_{oe}} \qquad (4\text{-}6)$$

$$r_c = \frac{1 - h_{rb}}{h_{ob}} = \frac{h_{fe}}{h_{oe}} = h_{fe}\,r_d \qquad (4\text{-}7)$$

$$r_e = h_{ib} - \frac{h_{rb}\,(1 + h_{fb})}{h_{ob}} = \frac{h_{re}}{h_{oe}} \qquad (4\text{-}8)$$

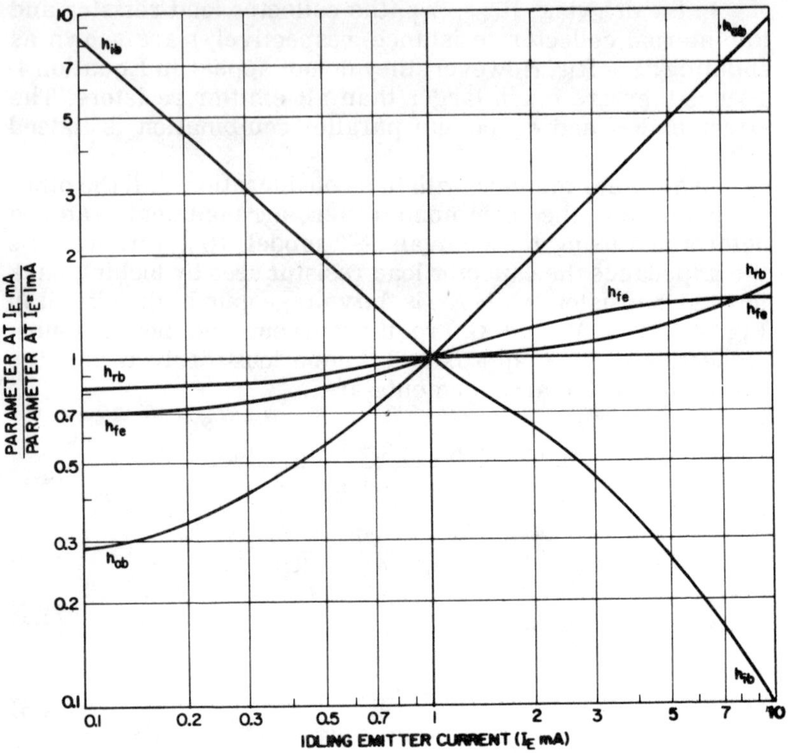

Fig. 4-5. Variation of h-parameters with the quiescent emitter current.

$$\alpha = -h_{fb} = \frac{h_{fe}}{1 + h_{fe}} \tag{4-9}$$

$$\beta = h_{fe} = \frac{-h_{fb}}{1 + h_{fb}} \tag{4-10}$$

Next, calculate the various equivalent-T elements from the relationship in Equations 4-6 through 4-10. Substitute the calculated values of r_b, r_c, r_d, r_e and β into the T-equations 4-1 through 4-5. Calculate the various statistics about a particular circuit using T-model parameters.

The various h parameters vary with emitter current and collector-to-emitter voltage. Curves such as those shown in Figs. 4-5 and 4-6 are normally supplied by manufacturers to describe how these parameters vary. As an example, assume the values for the h parameters are given: I_C equals 1 ma and V_{CE} equals 5 volts. If the value of h_{ob} at these quiescent conditions is 5 x 10^{-5} mhos, and it is required to know the h_{ob} at I_c equals 10 ma and V_{CE} equals 10 volts, the following mental gymnastics must be performed.

1. From Fig. 4-5, the ratios of the actual h_{ob} at 10 ma to h_{ob} at 1 ma is 9. Then due to the current deviation from the specified value, h_{ob} is 9 x (5 x 10^{-5}) equals 4.5 x 10^{-4} mhos.

2. From Fig. 4-6, the ratio of the actual h_{ob} at 10 volts to the h_{ob} at 5 volts is 0.7. Multiply this by the solution found in Step 1 and the required h_{ob} is 0.7 x (4.5 x 10-4) equals 3.15 x 10^{-4}.

3. Repeat this procedure for the other hybrid parameters before substituting these factors into Equations 4-6 through 4-10.

The vital statistics for the common-base and common-collector circuits in Figs. 4-7 and 4-8 respectively, are noted

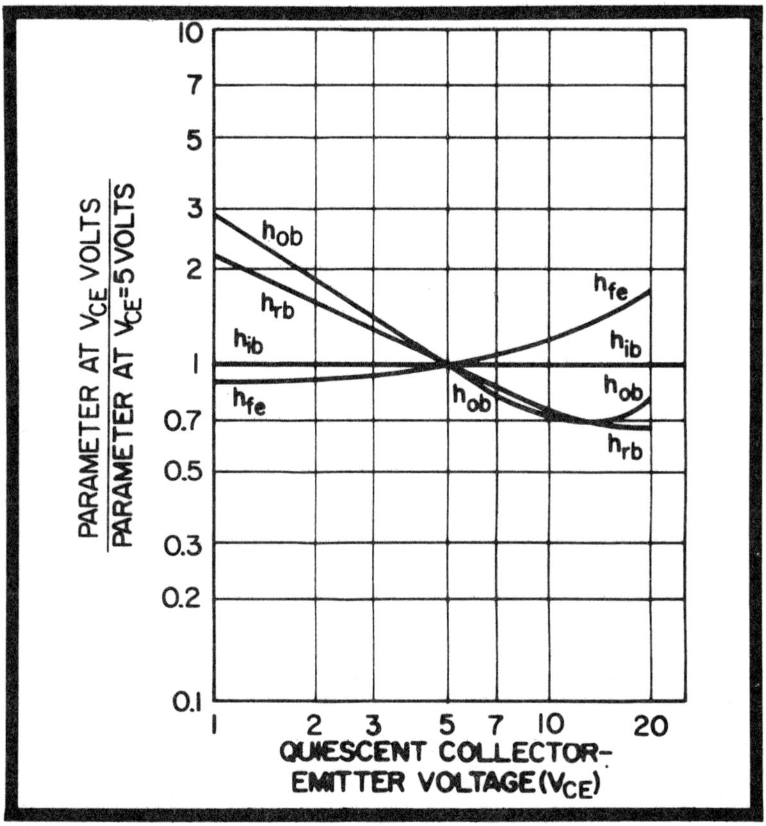

Fig. 4-6. Variation of h-parameters with the collector-emitter voltage.

below. The T-equivalent circuits are also shown in the figures. As for the common-emitter circuit discussed above, Equations 4-6 through 4-10 can also be used here to convert from the h to the T parameters if only the hybrid parameters are stated in the specifications for the device. For the common-base circuit the equations are:

$$R_{IN} = r_e + \frac{r_b + R_B}{\beta} \quad (4\text{-}11)$$

$$R_{OUT} = \frac{r_e\,[r_b + R_B + \beta\,(r_e + R_G)]}{\beta\,(r_b + R_B + r_e + R_G)} \quad (4\text{-}12)$$

$$A_v = \frac{\beta R_L}{r_b + R_B + \beta\,(r_e + R_G)} \approx \frac{R_L}{R_G} \quad (4\text{-}13)$$

$$A_i = \alpha \quad (4\text{-}14)$$

$$G = \alpha\, A_v \quad (4\text{-}15)$$

Common-collector circuits are described by the following equations:

$$R_{IN} = r_b + \beta\,(r_e + R_E) \quad (4\text{-}16)$$

$$R_{OUT} = r_e + \frac{r_b + R_G}{\beta} \quad (4\text{-}17)$$

$$A_v = 1 \quad (4\text{-}18)$$

$$A_i = \beta \quad (4\text{-}19)$$

$$G = \beta \quad (4\text{-}20)$$

As noted earlier, if R_B is comparable in size to R_{IN} or R_G in Equations 4-16 and 4-17, both R_{IN} and R_G should be modified. The actual R_{IN} and R_G are to be made equal to the parallel combination of the particular resistance with R_B. A resistor shunting the input in Fig. 4-7 dictates similar modifications in Equations 4-11 and 4-12.

EQUATIONS FROM LOGIC

The equations stated above should not be memorized, but with a little applied logic and some "rules of thumb" you can derive many of the equations when analyzing a circuit or doing a design.

Start with R_{IN}. We know that the CC (common-collector) circuit presents the highest impedance to a source, while the CB (common-base) circuit is the lowest. The CE (common-emitter) circuit falls somewhere between the two. We also know that any resistor in the emitter, such as r_e and R_E, when referred to the base circuit, is to be multiplied by the beta of the transistor. It follows logically that any resistance in the base circuit such as r_b, R_G, R_B, etc., is to be divided by β when reflected into the emitter circuit.

For the common-emitter circuit, the input resistance in the base circuit is the sum of the base resistance of the transistor, r_b (about 200 to 800 ohms and usually negligible), added to the product of beta with the sum of $r_e + R_E$ in the emitter

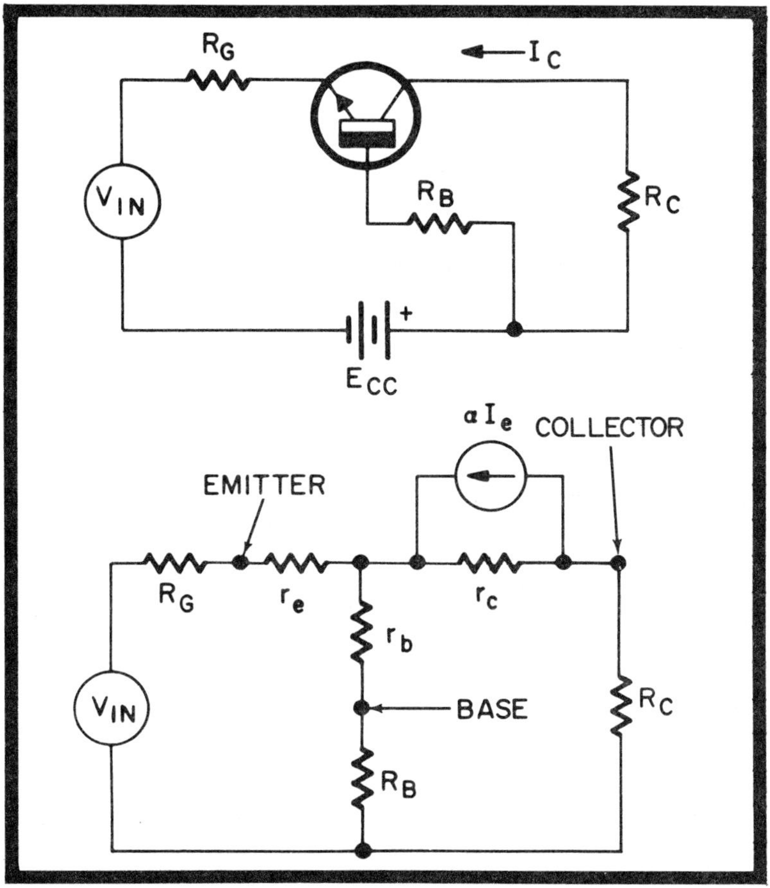

Fig. 4-7. Common-base circuit and its AC equivalent. Direct coupling is used to avoid complicating the analysis.

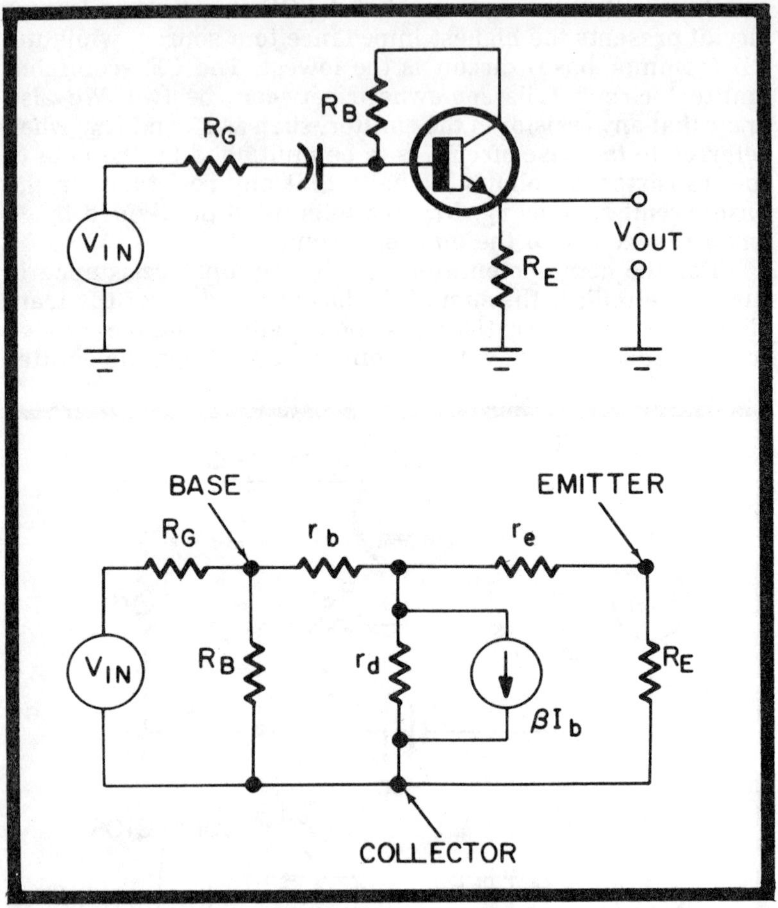

Fig. 4-8. Common-collector circuit and its AC equivalent.

circuit. The source sees these resistors shunted by the bias resistor, R_B. This is the same resistance the input source sees for the CC arrangement, only the value of R_E in this latter case is much larger than it is for the CE circuit. As for the CB circuit, $r_b + R_B$ are divided by beta when reflected back into the input emitter circuit, while also being added to the emitter resistance, r_e. The combination is in series with R_E, a resistor that may be placed in series with the emitter lead.

R_{OUT}, the output resistance of the transistor, is seen by the load resistor R_C when it looks back into the collector circuit in the CE and CB arrangements and by R_E when it looks back into the emitter circuit for the CC configuration. As for the latter, the load sees the resistors in the base circuit, r_b

+ RG, divided by beta (for they are reflected into the emitter circuit from the base circuit) added to the emitter resistance, r_e. Don't forget to put R_B in parallel with R_G if it is of comparable size. For the CE and CB circuits the R_{OUT} resistances are more difficult to determine, but they are somewhat less than r_d and r_c, respectively. Hence, the CB circuit has the highest output impedance, the CC arrangement has the lowest, with the CE circuit falling somewhere between the two.

The voltage gain is very close to 1 for the CC circuit (emitter follower), while it closely approximates the ratio of the resistor in the collector, R_C, to the resistance in the emitter, $r_e + R_E$ for the CE and CB arrangements. The current gain for the CB circuit is alpha, while for the CE and CC arrangements it is beta. The power gain is the product of the respective voltage and current gains for each circuit.

The calculated values for impedance and gain of the transistor itself are modified by external circuits. While we will analyze but one circuit in this chapter, in Chapter 9 we will see how the factors discussed here are affected by coupling the one transistor stage to another transistor as well as to other types of devices.

THE SINE WAVE

The basic function of the audio amplifier is to magnify the input so that the output voltage, current, or power is an enlarged version of the input signal. Most experimental work is done using the sine wave. Before proceeding with the solution of a practical problem, we will pause here to review the characteristics of the sine wave. The data accumulated in this section of the book will be useful in the design of all types of small-signal and power amplifiers as well as in the design of power supplies.

As you are aware, DC voltage and current are theoretically constant at all times. Disregarding aging, the voltage at the positive post of a battery is always a fixed amount above that at the negative terminal. The reverse is never true. This is not so with AC sinusoidal voltages and currents.

With respect to one voltage terminal of the AC generator, the voltage at the second terminal varies as shown in Fig. 4-9. The single cycle in Fig. 4-9 is repeated many times a second, depending upon the frequency involved. It is repeated 60 times a second for a 60-Hz power supply. As you see, each cycle is divided into 360 integral portions known as degrees. The

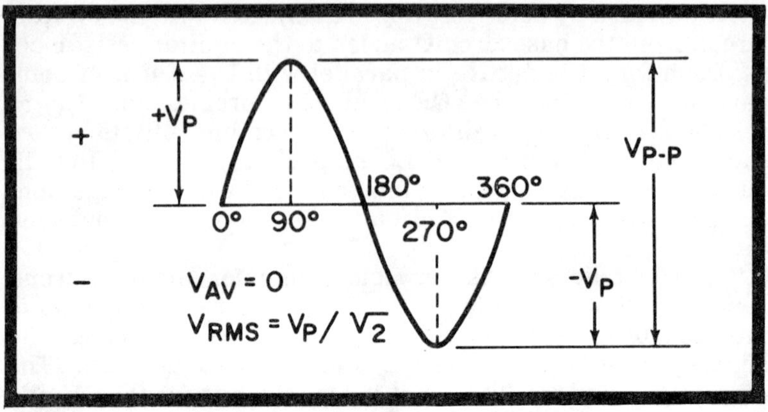

Fig. 4-9. Characteristics of the ordinary sine wave.

amplitude varies from zero at 0 degrees to a positive peak, V_P, at 90 degrees, back to zero and on to a negative peak at 270 degrees, then returning to zero.

This very basic description was introduced here to enable us to discuss, define and get a complete insight into the meaning of RMS, average, peak and peak-to-peak voltage.

The average DC voltage is the constant voltage at the terminals of a battery. The average AC voltage of a sine wave is zero over the complete cycle, because the two equal portions—the positive and the negative—are added vectorially and cancel.

The sine wave does deliver power despite the fact that the average voltage over the complete cycle is zero. The ability to deliver power is due to the availability of voltage (and current) in each individual half of the cycle. The power is delivered due to an effective or RMS voltage which is equal to V_P divided by $\sqrt{2}$ or .707 V_P.

Now let us assume that the negative half of the sine wave is flipped over so that it is always positive as shown in Fig. 4-10. Since the same varying voltage and current appear as before (only the polarity of one-half of the cycle is changed, not the amplitude) it can deliver as much power as the cycle in Fig. 4-9. The RMS voltage will remain unchanged at .707 V_P. But since it delivers current and voltage in one direction, there is also an average value for the voltage. This is $2V_P$ divided by π, where π is always equal to 3.14.

It is important to clearly understand the difference between average and RMS voltages. The average voltage has nothing to do with power. It just tells you what voltage you would have if the amplitudes of the sine wave were added at each point on the curve and then divided by the number of points considered. It is as if the sinusoidal voltage were flattened by squashing the upper part of the curve down to a flat level and the bottom part brought up to that level. The area under the flat curve is equal to that under the sinusoidal form.

On the other hand, the RMS voltage involves power and is derived from considering the power delivered to a load by the fluctuating sine wave. The DC voltage capable of delivering this power into the same load is considered as the basis of comparison. The AC and DC voltages are assigned identical amounts of volts and amperes when they deliver identical amounts of power. The AC voltage, known as RMS, is then compared to the peak sinusoidal voltage and a relationship between peak and RMS is established.

One last but important consideration: What happens if the negative pulse of the actual sine wave is entirely eliminated and only the pulse from 0 degrees through 180 degrees remains? The voltage is zero from 180 degrees through 360 degrees in each cycle. See Fig. 4-11. The average voltage is half that for the curve in Fig. 4-10, while the RMS voltage is that in Fig. 4-10 divided by $\sqrt{2}$.

A Typical Example

New magnetic phonograph cartridges frequently deliver less output voltage than do those of years ago. In our problem, we are assuming that you have a good amplifier requiring a 5 millivolt peak output from the cartridge. But if you have a new high quality record player and the excellent new cartridge you

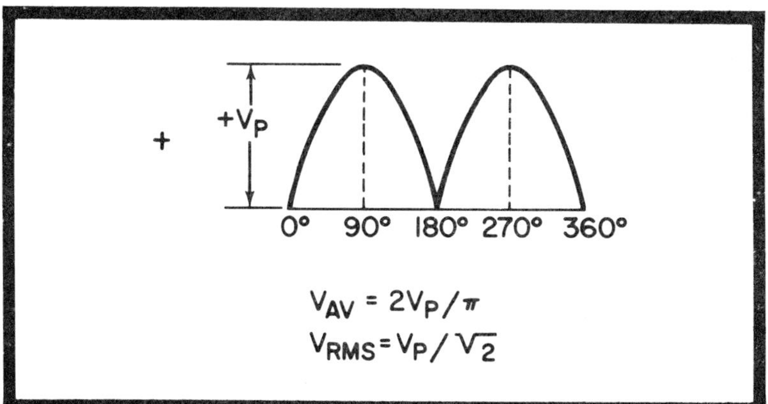

Fig. 4-10. Characteristics of a waveform consisting of repetitive positive sinusoidal pulses.

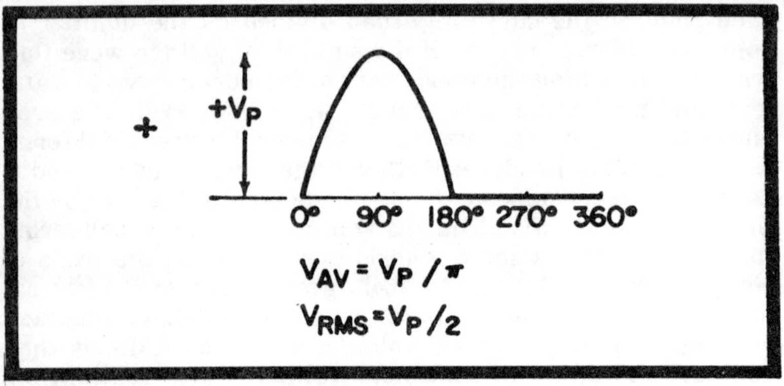

Fig. 4-11. Characteristic of a single sinusoidal pulse repeated only once each cycle.

chose delivers but 2 millivolts peak at an average velocity of 5 cm per sec, what should you do? Problems of this type can be solved by simply adding a stage of gain between the new cartridge and the amplifier.

The voltage gain of the amplifier stage to be designed must be at least equal to the ratio of the sensitivity of the old amplifier to the output from the cartridge, or 5mv divided by 2mv or 2.5. Adding some leeway, a gain of 3 will be fine. A low-noise transistor should be chosen for this application. The 2N3391A is a natural choice. The leakage current is just about negligible up to any practical temperature, the noise figure is low, and the beta ranges from 250 to 500. An average value of 400 can be assumed.

The average cartridge is designed to operate into 47,000 ohms. Hence, the stage we are designing must present a 47,000-ohm resistance to the cartridge. The output from this stage must also reflect the 47,000-ohm resistance in the old amplifier. To assure that the amplifier will not load our stage of gain, we will make the collector resistor in our design equal to one-tenth the resistance of the load it must see, or we set R_C equal to 4,700 ohms.

Now, choose a circuit. Since the voltage gain is to be more than 1, the CC circuit is not usable. The requirement of a fairly high input impedance makes the CB circuit impractical. We must then turn to the remaining CE arrangement in Fig. 4-3. You will find this circuit the most practical in many designs because of its gain and impedance characteristics.

The cartridge is the next consideration. The peak voltage output from an average size signal is 2 mv. For high amplitude signals, the peak voltage output from the cartridge may be as much as ten times this value or 20 mv. Across the 4,700-ohm load resistor, R_C, of the amplifier stage, the output voltage must be capable of swinging (plus or minus 20 mv) x 3 or plus or minus 60 mv. (As you recall, the gain of the stage must be at least 3.) The minimum collector current swing is then 60 mv divided by 4,700 ohms or 12 microamperes. Low noise considerations dictate that the collector bias current and voltage be very small. On the other hand, for a low distortion figure the quiescent current and voltage must be out of the leakage current and saturation regions of the collector characteristic curves. A good compromise is to use the idling conditions of V_{CE} equals 2 volts and I_C equals 0.5 ma. The quiescent voltage across R_C is then 0.5 ma x 4,700 ohms or 2.35 volts.

Since leakage current is no factor, the bias circuit can be very simple. The circuit in Fig. 4-12 can be used. The components and characteristics we have already determined are noted in the drawing.

An idling current of 0.5 ma means that the AC emitter resistance in the transistor is r_e equals 26 divided by I_E (see above) or 52 ohms. In order to have a voltage gain of 3, $R_E + r_e$ equals R_C divided by 3 (see Equation 4-3), so that $R_E + r_e$ equals 4700 divided by 3 or 1567 ohms. Subtracting r_e equals 52 ohms, and using the nearest EIA resistor value, we will make R_E equal 1500 ohms.

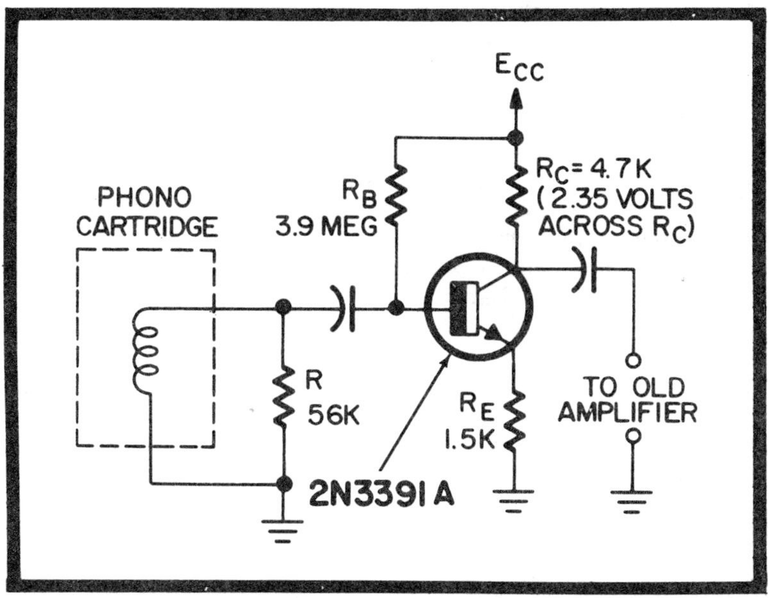

Fig. 4-12. Phono preamplifier designed for the text problem.

The idling voltage across R_E equals (1500 ohms) (0.5 ma) or 0.75 volts, for $I_C \approx I_E$. The supply voltage, E_{CC}, is equal to the sum of the quiescent voltages across R_E and E_C, added to V_{CE}, or 0.75 + 2.35 + 2 equals 5.10 volts. Use a small 6-volt battery for E_{CC} to power the circuit.

The bias resistor, R_B, can now be determined. The base current is the collector current, 0.5 ma, divided by the average beta of the 2N3391A, or 400. Doing the arithmetic, it is 1.25 microamperes (1.25 x 10^{-6} ampere). The current flowing through the base-emitter junction is E_{CC} divided by the sum of R_E reflected into the base circuit and R_B. Write the simple Ohm's Law relationship for I equals E over R:

$$1.25 \times 10^{-6} = \frac{6 - 0.6}{R_B + \beta R_E}$$

$$(\text{letting } V_{BE} = 0.6 \text{ volts}) \qquad (4\text{-}21)$$

$$1.25 \times 10^{-6}\,[R_B + 400\,(1500)] = 6 - 0.6 \qquad (4\text{-}22)$$

Solving for R_B, we find 3.72 x 10^6 ohms. Use a EIA standard 3.9-megohm resistor.

The minimum overall resistance of the base circuit should now be calculated. If this resistance is more than the required 47,000 ohms the cartridge must see, a resistor is to be placed across the cartridge in parallel with the base resistance. The combination of this resistor in parallel with the base circuit resistance must amount to about 47,000 ohms.

The impedance reflected into the base due to $r_e + R_E$ for the minimum beta of 250 for the 2N3391A transistor is 250 (1500 + 52) or 388,000 ohms. The cartridge sees this resistance in parallel with R_B, or approximately 350,000 ohms. If the cartridge is to see 47,000 ohms, a resistor must be placed across the cartridge to parallel the 350,000 ohms. Using the usual equations for parallel resistors (R_P is equal to the resistance of the two resistors, R and R_1, connected in parallel),

$$\frac{1}{R_P} = \frac{1}{R} + \frac{1}{R_1} \text{ or } R_P = \frac{R\,R_1}{R + R_1} \qquad (4\text{-}23)$$

we can calculate the resistor to be placed across the cartridge. Substituting component values, Equation 23 becomes:

$$47{,}000 = (350{,}000\ R)/350 + R \qquad (4\text{-}24\text{-})$$

R is equal to 54,000 ohms. Use a 56,000-ohm 10 percent EIA standard resistor.

Output impedance as well as current and power gains are not important in this problem. The gains can easily be calculated by substituting the numbers into Equations 4-4 and 4-5, as follows:

$$A_i = \beta = 400 \text{ (ranging from 250 to 500)}$$

$$G = A_i A_v = 400 \times 3 = 1200$$

R_{OUT} is more difficult to calculate because the transistor manufacturers do not supply all the data required for substitution into the equation. However, the old amplifier sees the collector load resistor, R_C, or the single stage shunting R_{OUT}. As far as the old amplifier is concerned, this stage has an output impedance of about 4,700 ohms because R_C is much smaller than R_{OUT}.

Chapter 5

JFET Amplifiers

Bipolar amplifiers can be conveniently divided into two groups—the small-signal voltage gain types and the large signal groups designed to deliver high power. As yet, there are no power FETs for use in entertainment equipment. We consider FET amplifiers as a cross between small-signal and power devices, but parallel the discussion on small-signal bipolar amplifiers.

Three basic circuit arrangements are frequently used: the common-source, common-drain (source follower) and common-gate. They are similar to the bipolar transistor counterparts. The most useful one, the common-source, is the center of most discussions here, but the characteristics of the remaining two circuits are not ignored.

CIRCUIT CHARACTERISTICS

As is the case with bipolar devices, different equivalent JFET circuits are possible. An approximate equivalent of the transistor itself is shown in Fig. 5-1. It will suffice for the common-drain mode of operation when audio frequency designs are considered.

At first glance, there is one obvious difference between this circuit and that drawn for bipolar devices in Chapter 4. Capacitors are integral components in the FET equivalent circuits. The reactive components have a negligible effect when compared to the resistive elements in the small-signal bipolar equivalents, and are thus omitted from the equivalent circuits. Because high impedance is an important attribute of the FET, the shunt and series capacitors gain new significance at audio frequencies.

Significant capacitance exists between the gate and source, C_{gs}, and between the gate and drain, C_{gd}. The input voltage, e_i, sees capacitance rather than resistance. Therefore, we talk about the input impedance, Z_{IN}, of a device rather than the input resistance. The above capacitances are instrumental in limiting the high-frequency response of an amplifier stage. It is fortunate that the capacitances are small, being but a few micro-microfarads (pf) in size.

Two capacitances other than C_{gs} and C_{gd} are frequently stated in the specifications. These are C_{iss}, the input capacity when the drain is bypassed to the source, and C_{rss}, the gate-to-drain capacity with the gate bypassed to the source. While C_{rss} is equal to C_{gd}, C_{iss} is the sum of C_{gs} and C_{gd}. The latter is obvious; by the definition of C_{iss}, C_{gs} is shunted by C_{gd} when the drain is connected to the source.

For JFETs, the output capacitance between the drain and source is negligible and is not shown in the equivalent circuit. In this chapter, we consider all capacitances in the equivalent circuit as negligible. This is not a wild approximation when designing circuits that apply to all but the top frequencies in the audio spectrum. Nevertheless, it is important to realize that the frequency limitations do exist; therefore, we must keep in mind the capacitance relationships just discussed.

The channel drain-to-source resistance, r_{ds}, and the current source, $g_m e_i$, are significant and important factors in determining the characteristics of the device in a circuit at any frequency. These factors, when added to components in the surrounding circuit, are used in various equations to describe the performance of a particular arrangement.

Input impedance, Z_{IN}, and output impedance, Z_{OUT}, are two important characteristics of the circuits. The only other characteristic of importance needed to describe an audio

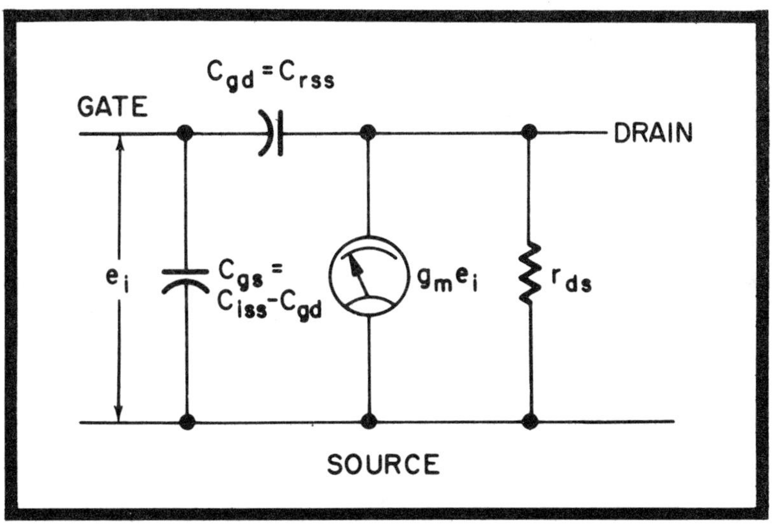

Fig. 5-1. Equivalent circuit of a JFET operating in the common-source mode.

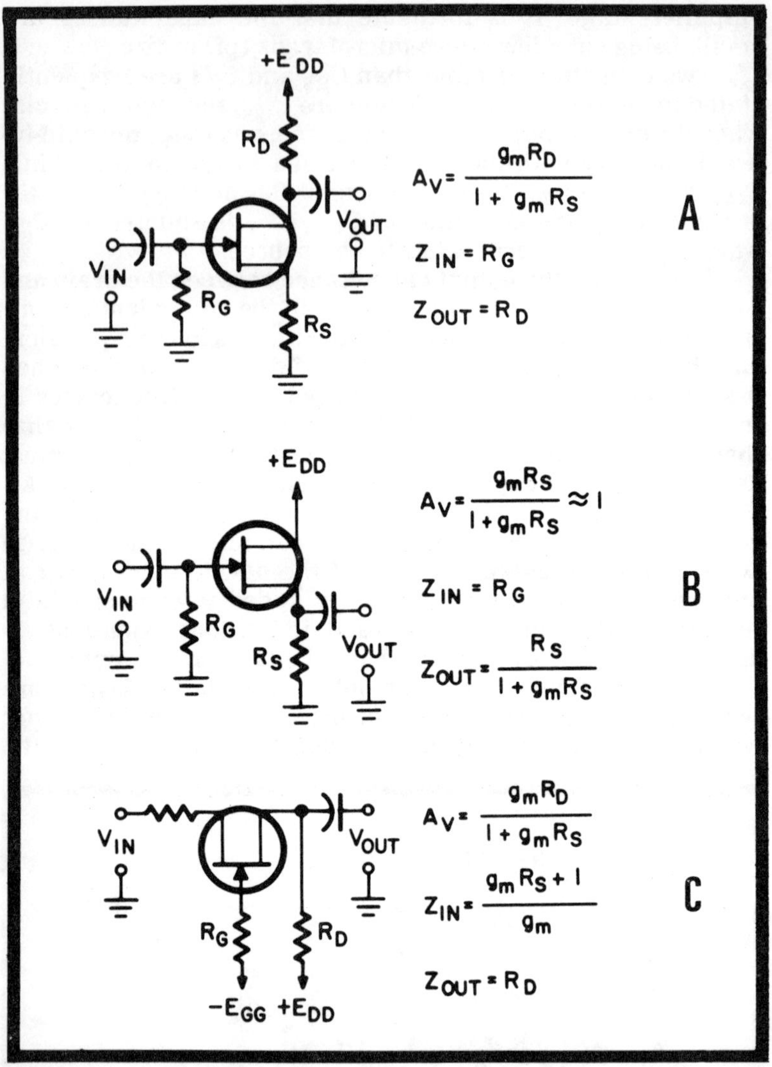

Fig. 5-2. FET circuit arrangement with formulas to calculate the characteristics of each circuit; common-source circuit, A; common-drain circuit (source follower), B; common-gate circuit, C.

JFET circuit is voltage gain, A_V. Unlike the bipolar device, current and power gains are insignificant factors. The effects of the capacitors in the equivalent circuits are discussed later.

The various circuit arrangements, and the significant approximate equations describing the circuits, are shown in Fig. 5-2. The input and output impedances are seen by looking into the transistor from the V_{IN} source and the V_{OUT} terminals, respectively. Various symbols in the equations require elaboration. Several modifications must also be included to account for circuit components not shown in the drawings.

Start with R_D. In the equations, it is the load resistor in the drain circuit. Should the transistor manufacturer specify a drain-to-source resistance, r_{ds}, that is comparable to R_D, then R_D in Fig. 5-2 is no longer the resistance to be used in the equation. Instead, substitute R_L for R_D, where R_L is a resistance equal to the parallel combination of R_D and r_{ds}.

Take this one step further. If the load at the output terminals is similar in size to R_D, it must also be considered as paralleling R_D and r_{ds} to form R_L. R_D is stated in the equation rather than R_L because in audio designs, r_{ds} and other paralleling loads are usually (not always) negligible when compared with R_D.

Now, let us turn our attention to R_S and R_G. Any resistor of comparable size shunting either of these will automatically make the equation results inaccurate. Shunting components must be added in parallel with these resistors before numbers are plugged into any of the equations. This calculation must be made regardless of the paralleling component, be it a shunting resistor, capacitor or inductor. In the latter two cases, the shunting impedance due to an inductance is 6.28fL and the impedance due to a capacitor is 1 over 6.28fC, where f is the frequency in Hertz, L is the inductance in henrys and C is the capacity in farads.

The final symbol of interest in the equations is g_m. g_m is the actual transconductance of the FET at a specific quiescent drain current, I_D. It is related to g_{mo}, the transconductance when the drain current is equal to I_{DSS}, by the equation:

$$g_m = g_{fs} = y_{fs}{}^* = g_{mo}\left(\frac{I_D}{I_{DSS}}\right)^{1/2} \qquad (5\text{-}1)$$

I_{DSS} is the drain current when the gate-to-source voltage, V_{GS}, is equal to zero. As a rule of thumb, g_{mo} is approximately equal to 1 over r_{dso}, where r_{dso} is the drain-to-source resistance of the V_{GS} equals 0 curve in the ohmic

---

* In Equation 5-1, y_{fs} is equated to g_m and g_{fs}. g_m and g_{fs} are identical. At low frequencies, they are equal to y_{fs}, the common-source transadmittance. Since y_{fs} includes the capacitor in the equivalent circuit, it is much larger than g_{fs} at high frequencies where g_{fs} is more significant.

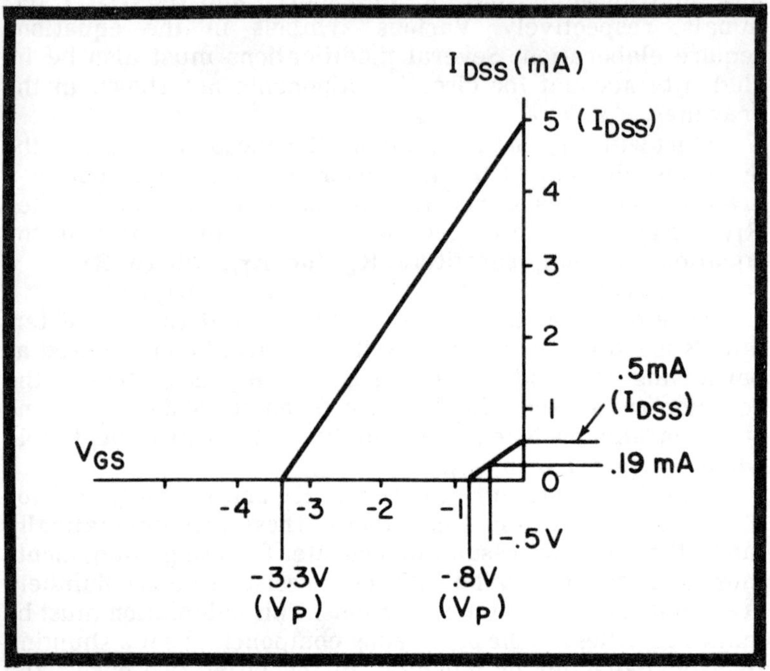

Fig. 5-3. Approximation of the transfer characteristic curves for the 2N4302.

region. Both g_{mo} and I_{DSS} can be found on transistor specification sheets.

A range rather than a single value of g_{mo} is usually stated in the specification sheets. Pinch-off voltage and I_{DSS} are treated in a like manner on data sheets. (See Fig. 3-1 for a plot describing V_P and I_{DSS}.) Should one or the other of these factors be missing in the data, it can be found from the equation:

$$(g_{mo})\,|V_p| = 2\,I_{DSS} \tag{5-2}$$

V_P is the pinch-off voltage with the polarity disregarded.

A Procedural Example

It would be simple here to pursue the standard procedures when presenting an example in the design of an amplifier stage. The design would follow a carefully chosen set of steps and all calculations would produce whole numbers. Unfortunately, little can be learned from this. No design is as straightforward as it appears on the printed page. In the example below, I present a method used by many design engineers. Solutions are tried and discarded for different reasons, even after many calculations. In the problem, it is first attempted to use a less expensive FET. Finally, a slightly more expensive device is found necessary to satisfy all requirements. In the final circuit, much of the original work is discarded in favor of a better, simpler and cheaper design.

This is a procedure concluding with a design. It is detailed here to present facts as well as to indicate to the reader the mental gymnastics he must go through to derive the compromise between the best and the most economical circuit. Parameter variations with temperature are disregarded since they are thoroughly discussed in Chapter 3.

A procedure detailed in Chapter 4 concerns the design of a stage of gain using a bipolar transistor. The basic design included a voltage signal source with a maximum 20 mv peak output for high amplitude signals feeding a voltage amplifier stage with a gain of 3. The voltage across the load resistor at the output of this stage was designed to be capable of a minimum of plus or minus 20 mv x 3 equals plus or minus 60 mv swing. The stage was required to present a 47,000-ohm load to the source and its output was to look into a 47,000-ohm resistor. Let us redesign this stage, but this time use an N-channel JFET.

In choosing a circuit, the common-drain arrangement must be excluded because the voltage gain here is less than three (actually it is somewhat less than one). The common-gate circuit cannot be used because the input impedance is about R_S, and R_S is normally much less than the required 47,000 ohms. By the process of elimination, only the common-source circuit remains. Start by designing for the circuit in Fig. 5-6A.

The drain resistor is chosen here on the same basis as the collector resistor was chosen for the bipolar device. The effect of any load on the circuit must be negligible. R_D, the drain resistor, should be less than one-tenth the size of the load it must feed, or one-tenth of 47,000 ohms. Let R_D equal 4700 ohms. The current swing across this resistor is at least plus or minus 60 mv divided by 4700 ohms or approximately plus or minus 13 microamperes. This is indeed a very small current.

Let us try an initial paper design by choosing an inexpensive 2N4302 to do the job. Due to transistor tolerances within this type, I_{DSS} can be any value between 0.5 ma and 5 ma, and the device can still pass as a standard 2N4302. From the curves supplied by the manufacturer, it is estimated that the pinch-off voltage is 0.8 volts when I_{DSS} is 0.5 ma and 3.3 volts when I_{DSS} is 5 ma. Using Equation 5-2, the respective g_{mo} values are 1.25×10^{-3} mhos and 3×10^{-3} mhos.

Now determine if this transistor can do the job. Draw the approximate transfer characteristic curves as shown in Fig. 5-3. This is executed by connecting the maximum I_{DSS} point to the respective maximum pinch-off voltage point on the V_{GS} axis with a straight line; then repeat this procedure for the minimum values of these characteristics.

The quiescent drain current and drain-to-source voltage are the next items to be considered. It is obvious from the drain characteristic curve in Fig. 3-1 that V_{DS} should not be allowed to drop below 1.5 times the pinch-off voltage if we are to operate within the linear pinch-off region of the curves. The idling gate-to-source voltage, V_{GS}, should be about two-thirds of the pinch-off voltage at a minimum, so that drain current changes with V_{GS} will be relatively linear.

The idling current is most critical on the lower curve in Fig. 5-4, because the current swing is limited to only 0.5 ma. The minimum idling V_{GS}, limited to two-thirds V_P can be set at 0.5 volt so the idling current becomes 0.19 ma. It can easily swing the required plus or minus 13 microamperes around this value without forcing the JFET into pinch-off.

The transconductance at a 0.19 ma drain current, from Equation 5-1, is:

$$\begin{aligned} g_m &= g_{mo}\,(I_D/I_{DSS})^{1/2} \\ &= 1.25 \times 10^{-3}\,(.19/.5)^{1/2} \\ &= .76 \times 10^{-3} \text{ mhos.} \end{aligned} \tag{5-3}$$

Using the common-source equation in Fig. 5-2, the maximum possible gain, assuming R_S equals 0 is $g_m R_D$ equals (0.76×10^{-3}) (4.7×10^3) equals 3.6. At first glance, the gain of 3.6 is satisfactory. Consider that the source resistor, R_S, must be shorted to obtain this gain. The source resistor is important. Distortion-reducing feedback voltage is normally developed across R_S. For example, if the gain were to be 6 rather than 3.6, half the gain may be discarded by developing a feedback voltage across the source resistor. This feedback will cause the distortion generated in this amplifier to be reduced by a factor of two. Although using a 2N4302 is nearly satisfactory, we should do better.

We can choose a somewhat higher gain and higher priced transistor such as the 2N4303. It has a minimum g_{mo} of 2000. Do the rough estimations to determine if this device will work.

The curves for the 2N4303 reveal the following characteristics. I_{DSS} can assume values anywhere between 4 and 10 ma. The respective pinch-off voltages are 2.9 and 5. Using Equation 5-2, it can be shown that g_{mo} is 2.75×10^{-3} mhos for a device in which the I_{DSS} is 4 ma and g_{mo} equals 4×10^{-3} for a 2N4303 with an I_{DSS} of 10 ma. Draw the approximate transfer characteristic curves for this transistor, as in Fig. 5-4.

The minimum quiescent V_{GS} is chosen on the lower curve at two-thirds of V_P or at about 2 volts. The idling current for the 2N4303 with this extreme characteristic is 1.25 ma. The g_m

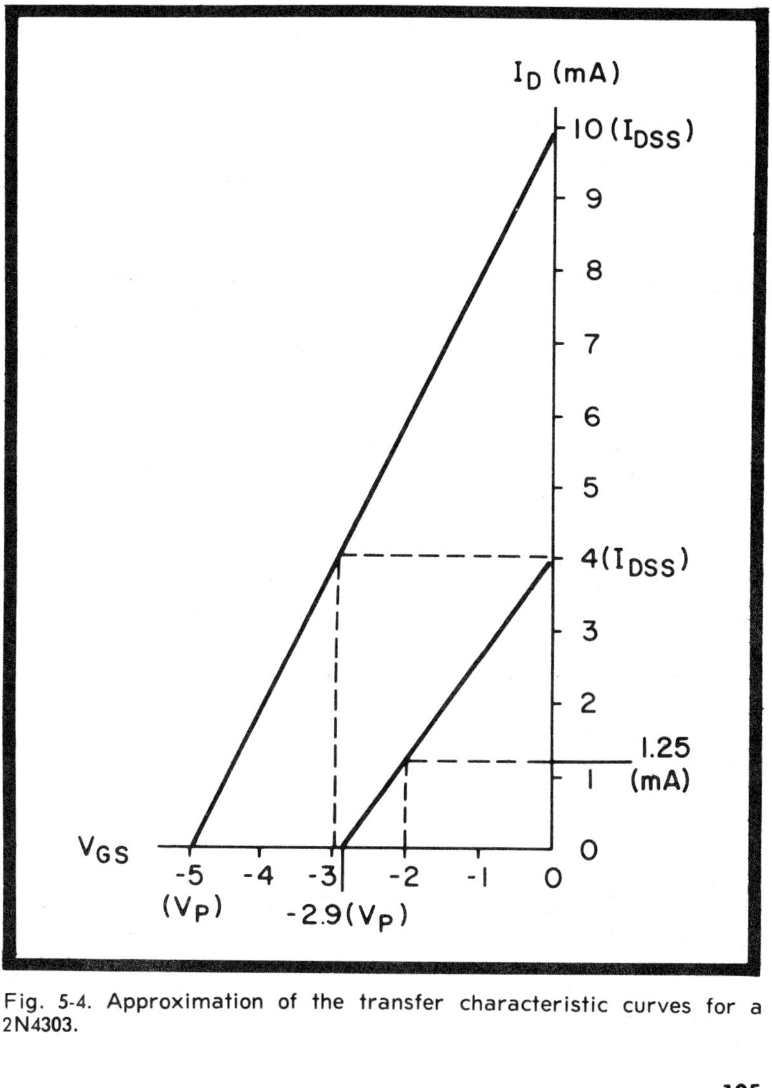

Fig. 5-4. Approximation of the transfer characteristic curves for a 2N4303.

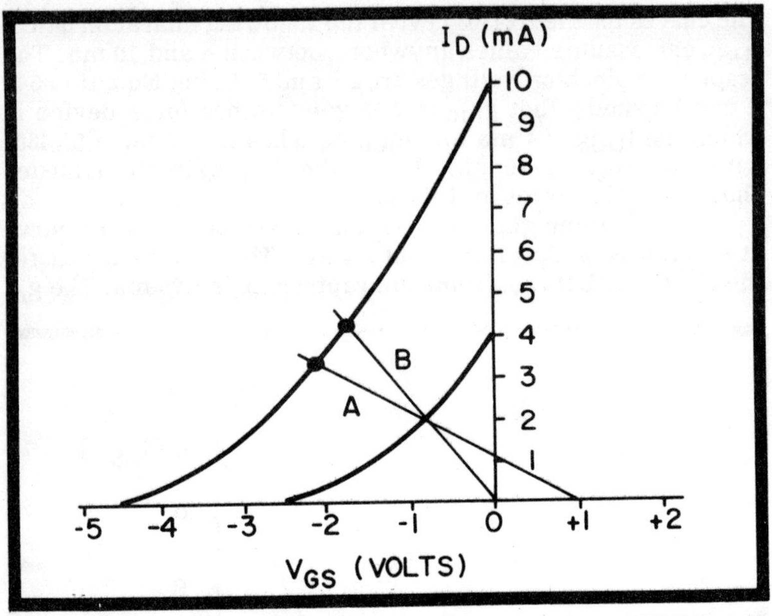

Fig. 5-5. Plot of reasonably exact transfer curves for a 2N4303 at 25 degrees C. The curves represent the two extremes of the characteristics for this type. Line A is a plot of 1 over R_S for the circuits in Fig. 5-6. Line B is a plot of 1 over R_S for the circuit in Fig. 5-7.

at this idling current is (2.75×10^{-3}) (1.25 divided by 2.9)$^{1/2}$ equals 1.8×10^{-3} mhos. The maximum possible gain, $g_m R_D$ is (1.8×10^{-3}) (4.7×10^3) equals 8.5. This gain is more satisfactory than was possible with the 2N4302.

As for a transistor with characteristics approximated by the upper transfer curve, the idling gate-to-source voltage at about two-thirds of V_P is 3 volts while the quiescent drain current is 4 ma. The g_m at this current is (4×10^{-3}) (4 divided by 10)$^{1/2}$ equals 2.5×10^{-3} mhos. The maximum possible gain for this particular device is, therefore, (2.5×10^{-3}) (4.7×10^3) or 11.8. The gain using a 2N4303 with an I_{DSS} of 10 ma is as satisfactory as the gain of the 2N4303 with the I_{DSS} at the lower extreme.

Convinced that we have a good chance to complete the design using a 2N4303, draw fairly exact transfer characteristic curves for this device. This can be accomplished using the equations shown in Chapter 3: I_D equals I_{DSS} (1 - $|V_{GS}|$ divided by $|V_P|)^2$, where I_D is the drain current that will flow when $|V_{GS}|$, the absolute value of the gate-to-source voltage, is applied between these leads. Substituting numbers into the equation yields the curves in Fig. 5-5. One curve is for a device where the I_{DSS} is at its maximum of 10 ma and the other for a 2N4303 with an I_{DSS} at 4 ma. We will assume operation is at 25 degrees C. See Table 5-1 for the various points to be plotted, as determined from the equation.

On the **lower** curve, the originally estimated quiescent conditions were V_{GS} equals 2 volts, I_D equals 0.38 ma. The actual g_m at this current is 2.75 x 10^{-3} (.38 divided by $4)^{1/2}$ equals 0.84 x 10^{-3} mhos and the maximum gain is (.84 x 10^{-3}) (4.7 x 10^3) equals 3.95. This gain is too close to the required 3. We should actually have a minimum gain of about 6 without feedback so that we can apply 6 db of feedback to reduce the gain and distortion in half. To establish this as the minimum gain on the lower curve, use the following procedure. Substituting numbers into the gain equation, A_V equals g_mR_D equals 6 equals g_m (4.7 x 10^{-3}), we find that g_m must be at least 1.28 x 10^{-3}. The minimum drain current for this g_m can be derived using Equation 5-1 where g_m equals g_{mo} (I_D divided by $I_{DSS})^{1/2}$ equals 1.28 x 10^{-3} equals 2.75 x 10^{-3} (I_D divided by $10)^{1/2}$. Solving for I_D, we find it must be at least 2.1 ma. Use I_D equals 2 ma and V_{GS} equals minus 0.82 volt as the bias point on the lower curve.

As for the **upper** curve, the originally estimated quiescent conditions were V_{GS} equals 3 volts, I_D equals 1.6 ma. 1.6 ma is

Table 5-1. See Figs. 5-4 and 5-5.

Table 1. Determination of transfer characteristic curves from equation $I_D = I_{DSS} (1 - |V_{GS}|/|V_P|)^2$ for 2N4303 transistor. The equation is used to derive the curves for the extreme conditions when I_{DSS} = 4 mA, $|V_P|$ = 2.9 volts and when I_{DSS} = 10 mA and $|V_P|$ = 5 volts.

V_{GS} volts	$I_{DSS} (1 - \|V_{GS}\|/\|V_P\|)^2$	I_D mA
For I_{DSS} = 4 mA, V_P = 2.9 volts		
0	$4 (1 - 0/2.9)^2$	4
1	$4 (1 - 1/2.9)^2$	1.71
2	$4 (1 - 2/2.9)^2$	0.38
2.9	$4 (1 - 2.9/2.9)^2$	0
For I_{DSS} = 10 mA, V_P = 5 volts		
0	$10 (1 - 0/5)^2$	10
1	$10 (1 - 1/5)^2$	6.4
2	$10 (1 - 2/5)^2$	3.6
3	$10 (1 - 3/5)^2$	1.6
4	$10 (1 - 4/5)^2$	0.4
5	$10 (1 - 5/5)^2$	0

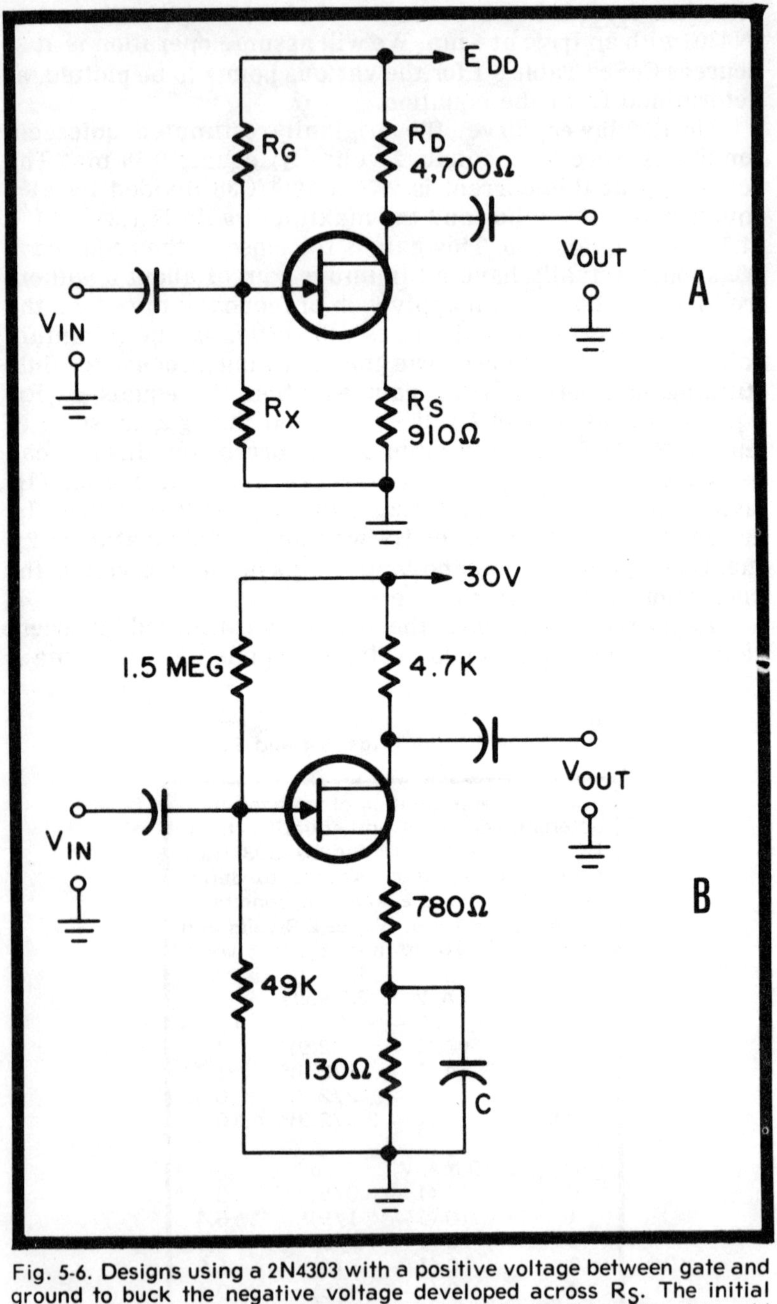

Fig. 5-6. Designs using a 2N4303 with a positive voltage between gate and ground to buck the negative voltage developed across R_S. The initial rough design is shown at A. At B the circuit is slightly altered to add feedback across the 780-ohm resistor.

less than the quiescent current just calculated for the lower curve. Proceed up the I_{DSS} equals 10 ma curve and select a point where the idling current will be higher than that chosen for the lower curve. Let us use, for example, I_D equals 3.4 ma and V_{GS} equals -2.1 volts. The g_m at this drain current is 4 x 10^{-3} (3.4 divided by 10)$^{1/2}$ equals 2.33 x 10^{-3} mhos so that the maximum gain for a transistor with characteristics along the maximum I_{DSS} curve is 2.33 x 10^{-3} (4.7 x 10^3) or 11.

Connect the point determined on the upper curve to the point on the lower curve with line A. The reciprocal of the slope of this line, (2.1 volts - .82 volt) divided by (3.4 ma - 2 ma) is 910 ohms. This is R_S in the transistor circuit. Extend line A to the horizontal axis at +1 volt. This positive voltage must be placed between the gate and ground. (See Chapter 3.) The circuit will initially take the shape shown in Fig. 5-6A. We must now proceed to determine the remaining components in the circuit.

The minimum drain supply voltage, E_{DD}, is the sum of the maximum voltage across R_L and R_S added to the minimum voltage that can be placed between the source and drain of the FET. To be certain that the FET is operating in the pinch-off region, the minimum voltage across the transistor should be $1.5V_P$(max) equals 1.5 x 6 equals 9 volts. (The maximum pinch-off voltage is specified by the manufacturer of the 2N4303 as 6 volts.) The maximum voltage across R_L and R_S is I_D(max) (R_L + R_S) equals 3.4 x 10^{-3} (4700 + 910) equals 19.1 volts. E_{DD} must be at least 19.1 + 9 equals 28.1 volts. Use a 30-volt supply.

Using the information that E_{DD} equals 30 volts and that 1 volt must be across R_X, we can now write two simultaneous equations to determine R_S and R_G. The circuit must present a 47,000-ohm resistance to the signal source. As far as the input is concerned, R_X is in parallel with R_G. Therefore, one equation is 47,000 equals $R_G R_X$ divided by (R_G + R_X). One volt is to be dropped across R_X. R_G and R_X form a voltage divider and 30 volts (E_{DD}) is applied across the two resistors. The second equation is 1 equals 30 (R_X divided by (R_G + R_X)). An approximate solution to the two equations yields R_X equals 49K and R_G equals 1.5 megohms.

The 910-ohm resistor in the source reduces the gain due to feedback. From the I_{DSS} equals 4 ma curve, where g_m is 1.28 x 10^{-3} mhos, the overall gain of the circuit using the equation in Fig. 5-2 is A_V equals (1.28 x 10^{-3}) (4.7 x 10^3) divided by 1 + (1.28 x 10^{-3}) (910) equals 2.77. This is somewhat less than the required 3. To get the gain of 3, the maximum R_S as determined from the same gain equation, 3 equals (1.28 x 10^{-3}) (4.7

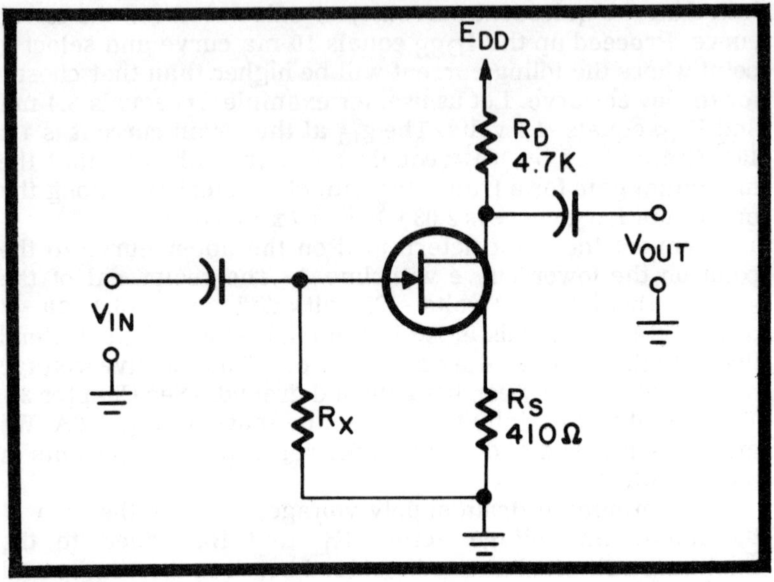

Fig. 5-7. This simplified circuit saves the cost of two resistors and a large electrolytic capacitor.

x 10^3) divided by ($1 + 1.28 \times 10^{-3}$ R_S), is 780 ohms. Since 780 ohms is less than the 910 ohms calculated from the curves, the 910-ohm resistor can be split into two parts, 780 ohms and 130 ohms. (See Fig. 5-6B.) The 130-ohm resistor is to be bypassed with a large capacitor so that it will not affect the gain but will still be in the circuit to help set the quiescent bias voltage. Feedback voltage affecting gain and distortion will be developed only across the 780 ohms. The capacitor across the 130-ohm resistor should be large enough to short the resistor at the lowest audio frequency to be amplified.

An alternate method can be used to eliminate resistor R_G. This is desirable, since any ripple in the power supply is fed through this resistor to the gate and amplified by the JFET. Extend line B in Fig. 5-5 from the point on the lower I_{DSS} curve to a zero V_{GS} point on the axis, so that no positive voltage will be required across R_X. The reciprocal of the slope of curve B is R_S equals (.82 - 0) volts divided by (2 - 0) ma equals 410 ohms. Since this is less than the maximum R_S equals 780 ohms calculated above, the gain will be more than satisfactory, although the feedback and distortion reduction will not be as pronounced. The simplified circuit in Fig. 5-7 can now be used, omitting C, the 1.5-megohm resistor, and one of the source resistors.

The input resistor, R_X, is the 47,000 ohms the source must see. The maximum idling current is 4.25 ma, determined by extending line B to the upper curve in Fig. 5-5. The minimum drain voltage supply is E_{DD} equals I_D(max) ($R_D + R_S$) + 1.5 V_P(max) equals (4.25×10^{-3}) (5110) + 1.5 (6) equals 30.8. A 30-volt supply will suffice; the JFET will just work a bit closer to the rounded pinch-off voltage portion of the characteristic curve in Fig. 3-1.

Which design is more desirable? Try both and decide for yourself. Some adjustment to both circuits may be necessary in the laboratory.

Chapter 6 Noise in Audio Amplifiers

Any undesirable voltage or current associated with a signal may be classified as noise. It is intolerable when it interferes with the enjoyment of the signal or completely obscures it.

Noise associated with amplifying devices can be due to environmental radiation from arcing electrical devices and phenomena, or induction from AC electric supplies. The methods of minimizing interference from sources of this type are discussed in Chapter 12.

The random noise generated by passive and active devices in a circuit is of more immediate concern. When designing a circuit, it is frequently necessary to keep noise at a minimum. This is especially true in the high-gain first stage of an amplifier, because noise generated here is boosted by all succeeding transistors.

Signal-to-noise ratio is a measurement of the relative amount of noise present in reproduced intelligence. Conventional methods of measuring this involve feeding a mid-frequency signal of average amplitude to the input of an amplifier, and noting the voltage, e_{so}, across the output load. Next, remove the signal, place a resistor at the input equal to the impedance of the transducer or signal source to be used, and once again measure the voltage. This second reading is the amount of noise voltage e_{no} riding along with the signal. The fraction of measured signal voltage to measured noise voltage is the signal-to-noise ratio for the amplifier.

Signal-to-noise fractions can also be power ratios. e_{so} and e_{no} are measured across the same output load resistor of the amplifier stage. Assume this load is R_L so that the signal power, P_{so} across the load is $e_{so}2$ divided by R_L and the noise power P_{no} is $e_{no}2$ divided by R_L. The signal-to-noise power ratio is P_{so} divided by P_{no} equals $(e_{so}$ divided by $e_{no})^2$.

The two signal-to-noise ratios, as defined, differ from each other. The decibel, db, is defined as 10 times the log of a ratio of two powers. Using a slight mathematical manipulation, db can also be defined as 20 times the log of the ratio of two voltages, when both voltages are measured across the same or equal resistors. The signal-to-noise ratio, when expressed in db, is the same regardless of the ratio involved, whether it be voltage or power. The equation relating db to these ratios is:

$$dB = 10 \log \frac{P_{so}}{P_{no}} = 20 \log \frac{e_{so}}{e_{no}} \qquad (6\text{-}1)$$

Table 6-1 is a chart that can be used with Equation 6-1 to relate any ratio in db to the power and voltage ratios. If the voltage or power ratio is not shown in the table, the db equivalent can easily be calculated from the data supplied. For example, a measured voltage ratio of 20:1 is not shown as such in Table 6-1. But 4:1 and 5:1 are stated, and the product of the two ratios is 20:1. The ratio in db for 4:1 is 12 db and for 5:1 is 14 db. These are **added** together—**not** multiplied as were the actual ratios. A voltage ratio of 20:1 is thus equivalent to 12 db + 14 db or 26 db. The rule is: Multiply two or more power or voltage ratios to equal a measured ratio. Then add all the db numbers representing these ratios.

A second example using identical procedures assumes a measured power ratio of 2.5:1. The 5:1 ratio divided by the 2:1 ratio will yield 2.5:1. The respective ratios in db are 7 and 3. Since one ratio is divided by the other, the numbers in db must

Table 6-1. DB voltage and power ratios.

POWER RATIO	VOLTAGE RATIO	DB
1:1	1:1	0
1.3:1	1.1:1	1
1.6:1	1.3:1	2
2:1	1.4:1	3
4:1	2:1	6
5:1	2.2:1	7
10:1	3.2:1	10
15.9:1	4:1	12
20:1	4.5:1	13
25:1	5:1	14
50.1:1	7:1	17
100:1	10:1	20
10000:1	100:1	40
10^6:1	10^3:1	60

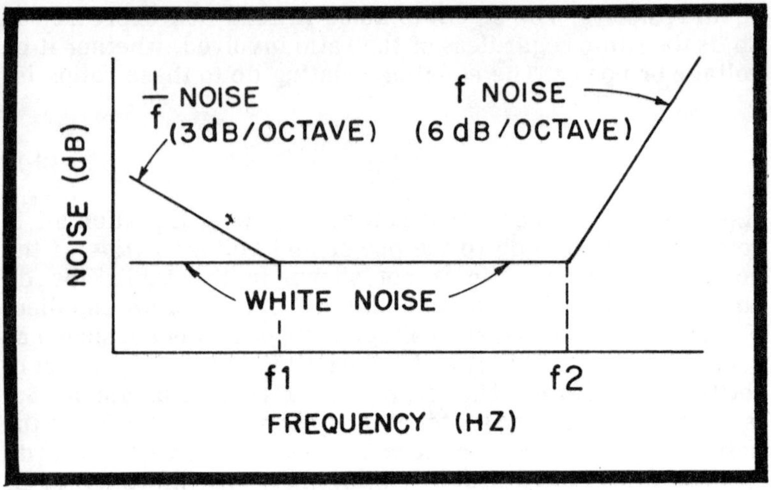

Fig. 6-1. Noise spectrum of a transistor.

be subtracted from each other. Hence a 2.5:1 power ratio is identical to 7 db - 3 db or 4 db.

ANALYZING THE NOISE SPECTRUM

The noise found in any transistor amplifier falls roughly into three groups. As shown in Fig. 6-1, there is one type of noise which covers the entire frequency spectrum. Identified as **white noise**, it delivers identical power at all frequencies. It can be generated by the random motion of particles in a conductor due to the energy imparted to these particles by heat. Because the noise is related to temperature, it is also referred to as **thermal noise**. Any noise due to a resistor in the base or gate circuit is amplified by the bipolar transistor or FET. Bias resistors at the input circuit, or even the base resistance of bipolar transistors, are important sources of this type of noise.

The noise power produced by any resistor is governed by the equation:

$$P_{ni} = 1.37 \times 10^{-23} (273° + °C)B \qquad (6\text{-}2)$$

where degrees C is the temperature in degrees Celsius and B is the effective noise bandwidth. B is not the bandwidth of an amplifier in the usual sense of the word. Normally, bandwidth is defined by $(f_H - f_L)$ where f_H and f_L are the high and low frequencies, respectively, at which the gain of the amplifier has dropped 3 db from its center value. Noise bandwidth as applied to the equation is equal to the sum of the products of the gain and the respective frequency at each point of the entire frequency range divided by the maximum gain of the amplifier. It is quite a difficult and length operation to derive B, even if integral calculus is used in the procedures. It is best to remember that in the usual case, where the rolloff of an amplifier is 6 db per octave, the noise bandwidth, B, is 1.57 multiplied by the bandwidth of the amplifier at the 3 db points.

When measurements are made, the meter readings will correlate with Equation 6-2 if B is assumed to be the bandwidth just defined. However, this may not relate to the audible noise. The characteristics of the ear must be considered. From data, it seems that the ear is most sensitive to approximately 2500 Hz and that the sensitivity decreases at frequencies on either side of this. Many tests have been performed and different standards established, each claiming to be the actual roll-off characteristic of the ear. From all of these, it would seem reasonable to let f_L equal 800 Hz and f_H equal 8000 Hz so that the audible B equals 1.57 (8000-800) or 11,300 Hz. However inaccurate this figure may be, we will assume it to be valid in this book. It is sure to agree with someone's auditory characteristics. Arguments can be presented against this assumption, but it must be remembered that no two people have identical hearing curves. All other factors in Equation 6-2 are constants.

Available noise power is the maximum power that can be transferred from the noisy resistor to the circuit in which it is used. This occurs when the resistor generating the noise power is equal to the resistance it sees in the circuit. In Fig. 6-2, an equivalent circuit shows the resistor as a source of noise. The noise-generating resistor is divided into two parts—a voltage noise source and a theoretical noiseless resistor—as an aid in the analysis. R_G can be the input circuit resistance of a transistor. For the maximum transfer of noise power from R_{NI} to R_G, R_{NI} equals R_G. R_{NI} and R_G form a voltage divider. Because R_{NI} equals R_G, the voltage across R_G and R_{NI} is e_{ni} divided by 2. The power across R_G is (e_{ni} divided by $2)^2$x1 divided by R, so that Equation 6-2 becomes

$$P_{ni} = e_{ni}^2/4R_G \text{ and}$$
$$e_{ni}^2 = 5.49 \times 10^{-23} \quad (273° + °C)(B)(R_G) \qquad (6\text{-}3)$$

The generated noise voltage is proportional to the square root of the size of the resistor creating it.

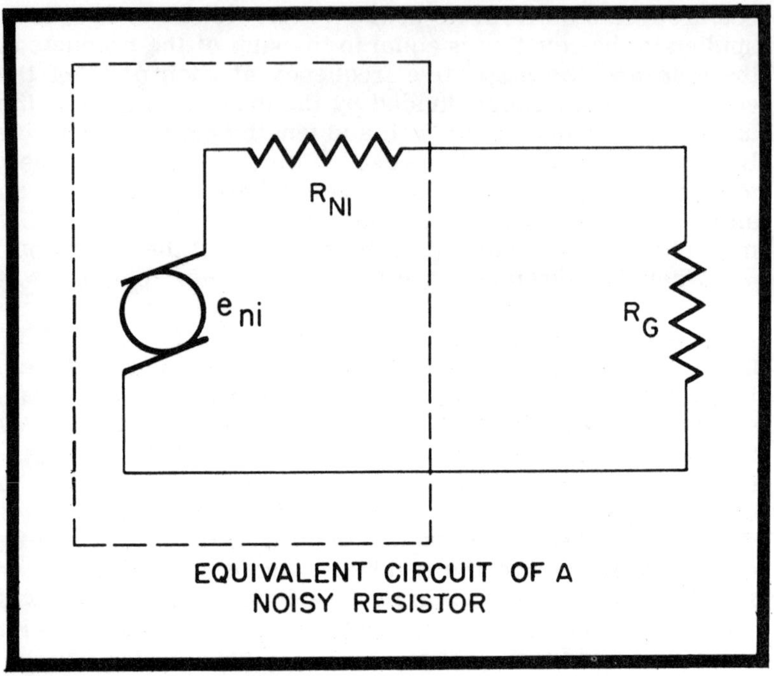

Fig. 6-2. Maximum noise power is transferred from the noise source when R_{NI} equals R_G.

Life would be very simple if Equation 6-3 could be used just as it is stated. However, R_G must be modified when it is shunted by some arrangement of capacitors and-or inductors. This is the case in Fig. 6-3 where the transistor impedance is so high that it presents a capacitance, C_{in}, to the input source rather than a resistance. The input can also be shunted by the capacitance of a voltage source such as a ceramic microphone or phonograph cartridge. The total capacity, C_t, across R_G is $C_s + C_{in}$. Should the reactance of C_t at the noise midfrequency be less than about 10 times the size of R_G in the figure, R_G for Equation 6-3 is not the same as R_G in Fig. 6-3. For use in the Equation, R_G must be only the resistive component of the input impedance of the circuit. This component is easily derived.

The reactance due to C_t is -j divided by ω C_t, where ω is 6.28f and f is the frequency (in Hertz) involved. The j notation indicates a 90-degree reactive phase shift where j equals $\sqrt{-1}$. The resistor in parallel with C_t is R_G. The formula used to determine the impedance of the combination is similar to the one used to determine the equivalent resistance of two resistors, R_1 and R_2, in parallel or R_1R_2 divided by $(R_1 + R_2)$. Here, however, the effect of the j operator must be included. The total impedance is:

$$Z = \frac{-R_G j/\omega C_t}{R_G - j/\omega C_t} \qquad (6\text{-}4)$$

Multiply the numerator and denominator by R + j divided by ωC_t and change j^2 equals $\sqrt{-1} \times \sqrt{-1}$ to -1, and you have:

$$Z = \frac{\dfrac{-R_G^2 j}{\omega C_t} + \dfrac{R_G}{\omega^2 C_t^2},}{R_G^2 + 1/\omega^2 C_t^2} \qquad (6\text{-}5)$$

Since R_G is assumed to be much greater than 1 divided by ωC_t, the 1 divided by $\omega^2 C_t^2$ in the denominator of the equation may be ignored when compared in magnitude to R_G^2. The denominator now becomes only R_G^2. Since only the resistive component of the impedance is required, the j term in the numerator can be disregarded. Then the R_G for the equation becomes:

$$R_G \text{ (for Eq. 3)} = \frac{\dfrac{R_G}{\omega^2 C_c^2}}{\dfrac{R_G^2}{1}} = \frac{1}{R_G\, \omega^2 C_t^2} \qquad (6\text{-}6)$$

The frequency f (from the ω^2 equals $(2\pi f)^2$ is the geometric center frequency of the noise bandwidth. Thus, if the noise bandwidth B includes all frequencies between the

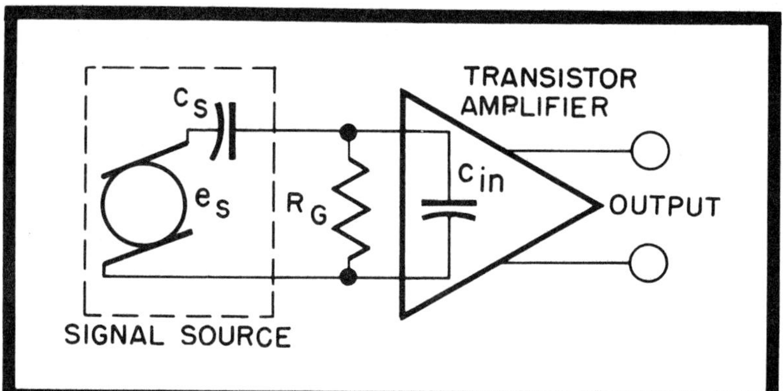

Fig. 6-3. Amplifier stage in which R_G is shunted by the input capacitance of the transistor, C_{in}, and the capacity of the signal source, C_S.

limit of a high one at f_H and a low one at f_L, the frequency to be used for determining R_G is f equals $\sqrt{f_L \times f_H}$. Using the derived formulas, adding a little calculus, and making several minor approximations, the information can be stated in a new formula similar to that in Equation 6-3:

$$e_{ni}^2 = 1.37 \times 10^{-24} (273° + °C) \times \left(\frac{B}{f_H f_L}\right)\left(\frac{1}{R_G C_t^2}\right) \tag{6-7}$$

where B, R_G and degrees C are identical to the values used in the previous equation; C_t, f_H and f_L have been discussed in the text.

Looking back over the discussion just completed, you recall that we started examining the reasons for the existence of the noise spectrum in Fig. 6-1. We started to discuss white noise and then described the effects of an important white noise generator, the resistor. We also discussed how much of this noise generated by the resistor can be transferred to a circuit in which it is located. Thermal noise is not the only cause of the white noise spectrum. The **shot noise** phenomenon in semiconductor junctions is also a source. It is noise generated by the random passage of charged particles through the junctions. Shot noise increases with the quantity of current flowing through the device.

The emitter current of bipolar devices divides between the base and collector, introducing **partition noise** which is similar in nature to white noise. Obviously, the noise is minimized when the beta of the transistor is large and the number of partitions is small.

In Fig. 6-1, the 6 db per octave increase in noise at the upper frequencies is due to the rolloff of beta at the high end of the band rather than to any phenomenon different from that discussed above. This noise increase is usually noted at some RF frequency and is unimportant when considering audio designs.

Semiconductor noise at the low frequencies drops at the rate of 3 db per octave. 1/f noise, as it is called, is due to surface phenomena and leakage current. It is minimized when the current flowing through the bipolar transistor and the voltage across the device are low. Minimizing temperature goes a long way in reducing noise from this factor. As for the JFET, this noise is least objectionable when the transconductance of the device is high.

The effect of 1/f noise is quite important in audio devices. While bipolar transistors are subject to this phenomenon up to 1,000 Hz, the effect becomes negligible at about 100 Hz when

JFETs are involved. Transistors have been designed where noise due to the 1/f effect may be disregarded at frequencies much below those noted.

NOISE FIGURE

The signal-to-noise ratio at the input of an amplifying device is higher than at the output because of the contribution of the stage of gain to the noise at the output. The **noise factor**, F, relates the two signal-to-noise ratios.

$$F = \frac{P_{si}/P_{ni}}{P_{so}/P_{no}} \quad (6\text{-}8)$$

where P_{si} and P_{so} are the input and output signal powers, respectively, and P_{ni} and P_{no} are the input and output noise powers, respectively. The optimum noise factor is equal to 1, for only then will the input and output signal-to-noise ratios be equal, suffering no noise contribution from the amplifying device.

Since the power gain of an amplifier is G equals P_{so} divided by P_{si}, the noise factor can be expressed as:

$$F = P_{no}/GP_{ni} \quad (6\text{-}9)$$

where GP_{ni} is the input noise power multiplied by the power gain of the device, or the output noise power due to noise present at the input.

Using Table 6-1, the number for noise factor F in terms of a decibel ratio, can easily be determined. When expressed in decibels, the ratio is referred to as **noise figure**, NF. For example, the db equivalent of a power ratio of 2:1 is 3 db. Hence, a noise factor of 2 is identical to a noise figure of 3 db.

The noise figure is specified in one of two ways. The more conventional data sheet indicates the **spot noise figure**. This is a maximum noise figure number at one frequency, usually 1000 Hz, for a bandwidth of 1 Hz. In an alternate method, the noise figure is specified over the entire band. This latter number usually supplies more exact information covering both the white noise and the 1/f regions of the curve.

EQUIVALENT NOISE GENERATORS

The noise associated with the transistor circuit can be analyzed as being caused by three noise generators, one real and two hypothetical, at the input to a theoretically noiseless

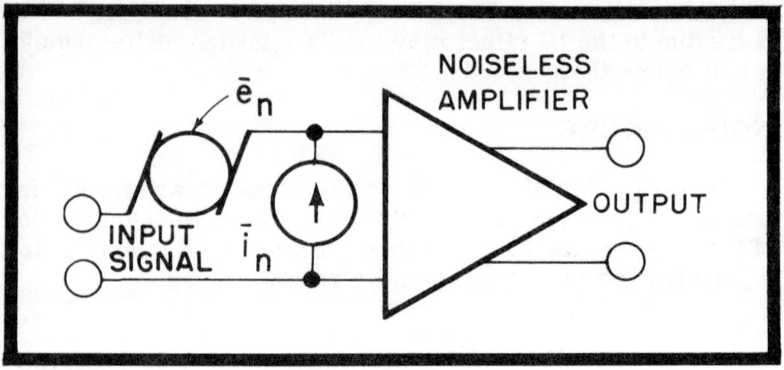

Fig. 6-4. Equivalent circuit to study noise generated in a transistor consists of a voltage noise generator, $\bar{e}_n$, and a current noise generator, $\bar{i}_n$.

amplifier. The real noise is generated by components connected to the input circuit and transferred to the amplifier stage. The noisy resistor generator discussed above is an example of this. The two hypothetical noise generators—a voltage generator and a current generator—are within the transistor. These two generators are mathematically referred to the input as sources of interference for a noiselsss amplifier.

Here, we are to mathematically determine the effect of the transistor-generated noise on the circuit. Hence, we will use an equivalent circuit consisting of hypothetical noise voltage and noise current generators placed at the input of a hypothetical noiseless amplifier. This is illustrated in Fig. 6-4. The noise-generating sources, $\bar{e}_n$ and $\bar{i}_n$ are separated from the actual transistor and placed at the input of the noiseless amplifying device. This is the equivalent circuit of an actual condition. The generators will cause the noise at the output of the transistor to be the same as it would be from an actual device in a real circuit.

The output noise is due to $\bar{e}_n$ when the input to the amplifier is shorted and is due to $\bar{i}_n$ when the input is open circuited. $\bar{e}_n$ can be calculated by using data obtained from two measurements—the noise voltage at the output when the input to the amplifier is shorted and the voltage gain, A_V. $\bar{e}_n$ is equal to the noise voltage at the output divided by A_V.

To determine $\bar{i}_n$, compare the output current when the input is an open circuit with the current present at the output with a known current source at the input.

The total noise power at the input, P_{nit}, consists of the actual thermal noise power in Equation 6-2, or 1.37×10^{-23} (273 degrees + degrees C)B, in addition to the noise power due to $\bar{e}_n$ and $\bar{i}_n$ in the transistor, or ($\bar{e}_n^2$ divided by $4R_G$ + $\bar{i}_n^2 R_G$ divided by 4). Since both factors are input noise sources, they must be multiplied by G, the power gain of the transistor, if the result of the calculation is to be the noise at the output.

The actual noise power at the input only is assumed due to resistor R_G, or is 1.37 x 10^{-23} (273 degrees + degrees C)B. Substituting all the data in these two paragraphs into Equation 6-9, the noise factor in terms of $\bar{e}_n$ and $\bar{i}_n$ becomes:

$$F = \frac{P_{no}}{GP_{ni}} = \frac{GP_{nit}}{GP_{ni}} = \frac{G\left[1.37\times10^{-23}(273^\circ+{}^\circ C)B + \frac{\bar{e}_n^2}{4R_G} + \frac{\bar{i}_n^2 R_G}{4}\right]}{G[1.37\times10^{-23}(273^\circ+{}^\circ C)B]} \qquad (6\text{-}10)$$

$$F = 1 + \frac{\bar{e}_n^2/R_G + \bar{i}_n^2 R_G}{5.49\times10^{-23}(273^\circ+{}^\circ C)B} \qquad (6\text{-}11)$$

The noise figure is, of course, equal to the noise factor found in Equation 6-10 and expressed in db.

From Equation 6-11 it can be determined that the noise factor is at a minimum when R_G equals $\bar{e}_n$ divided by $\bar{i}_n$. Substituting this into Equation 6-11, we find that the best noise factor that can be obtained for a particular transistor is about 1 + $\bar{e}_n\bar{i}_n$ divided by 2.74 x 10^{-23} (273 degrees + degrees C)B.

Equation 6-10 can be rewritten in the form showing that the total input noise power affecting the output of the transistor is:

$$\begin{aligned} P_{nit} &= P_{ni}F \\ &= F(1.37\times10^{-23})(273^\circ+{}^\circ C)B \\ &= \frac{\bar{e}^2_{nit}}{4R_G} \end{aligned} \qquad (6\text{-}12)$$

so that the total (real and hypothetical) input noise voltage that affects the output is:

$$\bar{e}_{nit} = [4R_G F(1.37\times10^{-23})\times(273^\circ+{}^\circ C)B]^{1/2} \qquad (6\text{-}13)$$

The input signal voltage must be compared with $\bar{e}_{nit}$ to determine the signal-to-noise ratio of a transistor circuit.

NOISE AND THE BIPOLAR TRANSISTOR

The variations of noise figure with collector current and source resistor, R_G, can be described by the set of contour curves in Fig. 6-5. This is most useful in audio amplifier design. The maximum noise figure stated on data sheets

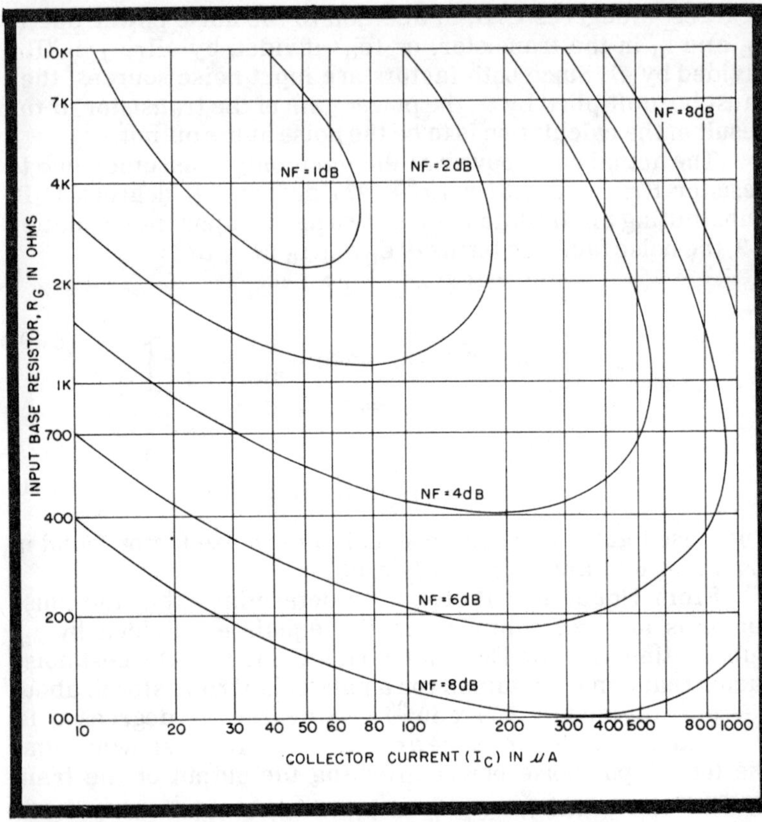

Fig. 6-5. A set of noise-figure contour curves. They are usually drawn at f equals 1 kHz, T_A equals 25 degrees C. (T_A is the ambient temperature.)

usually specifies the characteristic at a specific collector current, collector voltage, and frequency when a particular resistor is at the input. Using the curves, you can determine the noise figure for a coincident collector current and input resistor. Operation within the lowest contour assures you that the amplifier will deliver the minimum amount of noise to a load, within its capabilities.

In addition to the contour curves shown, other similar curves may be drawn. Some extremely wideband applications may require a set of contour curves showing how the noise figure is related to the input resistor at specific frequencies rather than at specific collector currents.

The R_G and I_c relationships can be split into two curves (Figs. 6-6A and B) showing the variation of NF with each of these factors. Another useful curve, in Fig. 6-6C, shows the

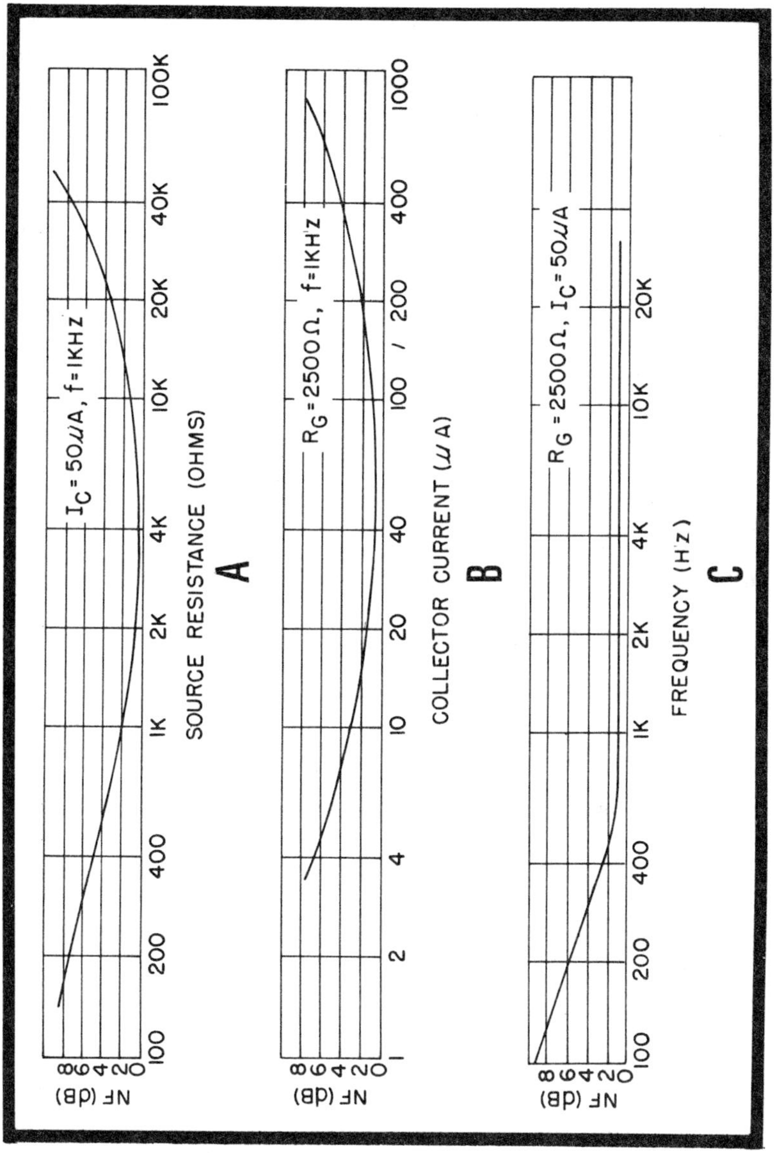

Fig. 6-6. These curves indicate how the noise figure varies with the source resistance, collector current and frequency for a hypothetical bipolar transistor. Curve A is a plot of the noise figure against the source resistance with the collector current and frequency held constant. Curve B is a plot of the noise figure against the collector current with the source resistance and frequency held constant. Curve C is a plot of the noise figure against frequency with the source resistance and collector current held constant.

variation of NF with frequency. All these curves are limited compared to the one in Fig. 6-5 because many factors must be specified before any one variable can be determined. Curves of this type supply information concerning one variable factor, while all others are usually held constant at their optimum values for minimum noise.

The design of an amplifier for low noise can be accomplished in several logical steps, beginning with the selection of the most suitable device. Should the source resistance be determined by the circuit requirements, characteristic curves of different transistors should be checked to determine which device will contribute least to the noise when a reasonable amount of collector current is flowing. On the other hand, if only a specific device is available, the circuit should be designed so that operation is within a low noise contour.

Once the device has been chosen, factors should be substituted into Equation 6-13 to calculate the noise voltage due to the transistor and the circuitry. The average input signal voltage to be fed to the device is compared with the generated noise, determined from Equation 6-13, to decide if the signal-to-noise ratio is satisfactory.

As an example, use a magnetic phonograph cartridge to feed a preamplifier stage. The resistance of the cartridge, R_G, may be taken at 3000 ohms. A collector current near the lowest noise contour curve in Fig. 6-5 coincident with R_G equals 3000 ohms is 50 microamperes. The noise figure is about 1 db. From Table 6-1, an NF of 1 db is equivalent to a noise factor, F, of 1.3.

Assume the amplifier is to reproduce 20 to 20,000 Hz within 1 db, so that the 3 db power points are at one octave beyond these frequencies, or at 10 and 40,000 Hz. The effective noise bandwidth is about 1.57 x 40,000 or 63,000 Hz. Substituting this into Equation 6-13, the noise voltage at 27 degrees C is:

$$\bar{e}_{nit} = (4 \times 3 \times 10^{3} \times 1.3 \times 1.37 \times 10^{-23} \times 300 \times 6.3 \times 10^{4})^{1/2} = 2.01 \times 10^{-6} \text{ volts.} \qquad (6\text{-}14)$$

The output voltage from a magnetic cartridge may be assumed to be 2 x 10^{-3} volts. The ratio of signal-to-noise is 2 x 10^{-3} divided by 2.01 x 10^{-6} which is approximately equal to 10^{3}. From Table 6-1, a voltage ratio of 10^{3}:1 is 60 db. This is a good specification for any amplifier.

The actual audible noise is less than the 2.01 x 10^{-6} volts multiplied by the voltage gain of the stage. The 2.01 x $10^{-6}A_V$ is the noise usually measured on test instruments at the output of the amplifier. This is for a noise bandwidth of 63,000 Hz. As discussed above, a practical audible noise bandwidth is 11,300 Hz. Substituting this into Equation 13 yields:

$$\bar{e}_{nit} = (4 \times 3 \times 10^3 \times 1.3 \times 1.37 \times 10^{-23} \times 300 \times 1.13 \times 10^4)^{1/2} = 0.85 \times 10^{-6} \qquad (6\text{-}15)$$

so that the signal-to-noise voltage ratio is 2×10^{-3} divided by .85 x 10^{-6} equals 2.35×10^3. Using Table 6-1, we can proceed as we did earlier in the chapter to determine this ratio in db. 2.35×10^3 can be separated into the product of 2.35 and and 10^3. The closest db figure for the 2.35:1 voltage ratio is 2.2:1 or 7 db. 10^3 is 60 db. Hence, the signal-to-noise ratio is 60 db + 7 db equals 67 db. The actual audible noise is 7 db better than the measured value.

As a general rule, bipolar transistor amplifiers should be designed with a minimum collector current and voltage as well as with the smallest practical source resistor in the base circuit.

NOISE AND THE JFET

JFET noise is not specified by using contour curves. Usually, individual curves are used to relate the noise figure to various factors affecting noise. Several typical curves spelling out these relationships are shown in Fig. 6-7.

Another set of curves relate $\bar{e}_n / \sqrt{\Delta f}$ and $\bar{i}_n / \sqrt{\Delta f}$, the noise voltage and current generators at the input to the hypothetical noiseless amplifier, to frequency. The curves are shown in Fig. 6-8. The $\sqrt{\Delta f}$ refers to the noise bandwidth of the circuit. The curves in Fig. 6-8 refer to the variations of $\bar{e}_n$ and $\bar{i}_n$ with frequency when the bandwidth is 1 Hz. Should the noise bandwidth be 2 Hz, the e_n divided by $\sqrt{\Delta f}$ or $\bar{i}_n$ divided by $\sqrt{\Delta f}$ read from the curve is to be multiplied by $\sqrt{2}$. In a similar manner, for a 100-Hz noise bandwidth, both readings are to be multiplied by $\sqrt{100}$ or 10.

Some specifications state $\bar{e}_n \sqrt{\Delta f}$ (sometimes written $\bar{e}_n \sqrt{\sim}$ or $\bar{e}_n \sqrt{Hz}$) for bandwidths other than 1 Hz. If, for example, it were specified for a 5-Hz bandwidth, the stated number should be first divided by $\sqrt{5}$ before proceeding to find the noise figure for the actual bandwidth involved. Once the correction is made, the procedure outlined in the paragraph above can be used. This also applies to determining $\bar{i}_n$ if a bandwidth other than 1 Hz is specified in the data.

In Fig. 6-3 it is indicated that input capacitors are associated with the transistor. These are important factors

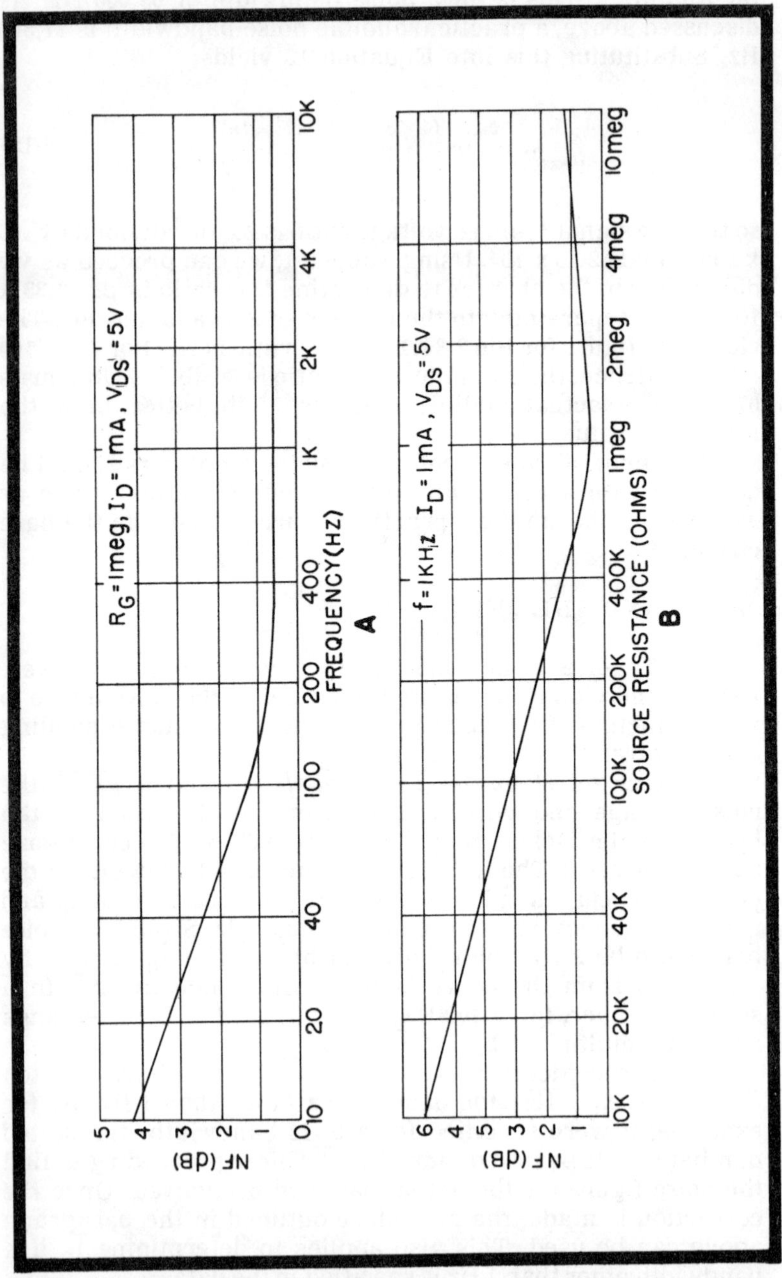

Fig. 6-7. Curves indicate how the noise figure varies with the size of the resistor at the input drain-to-source voltage, drain current and frequency for a hypothetical JFET. (See Fig. 6-8 for C and D.)

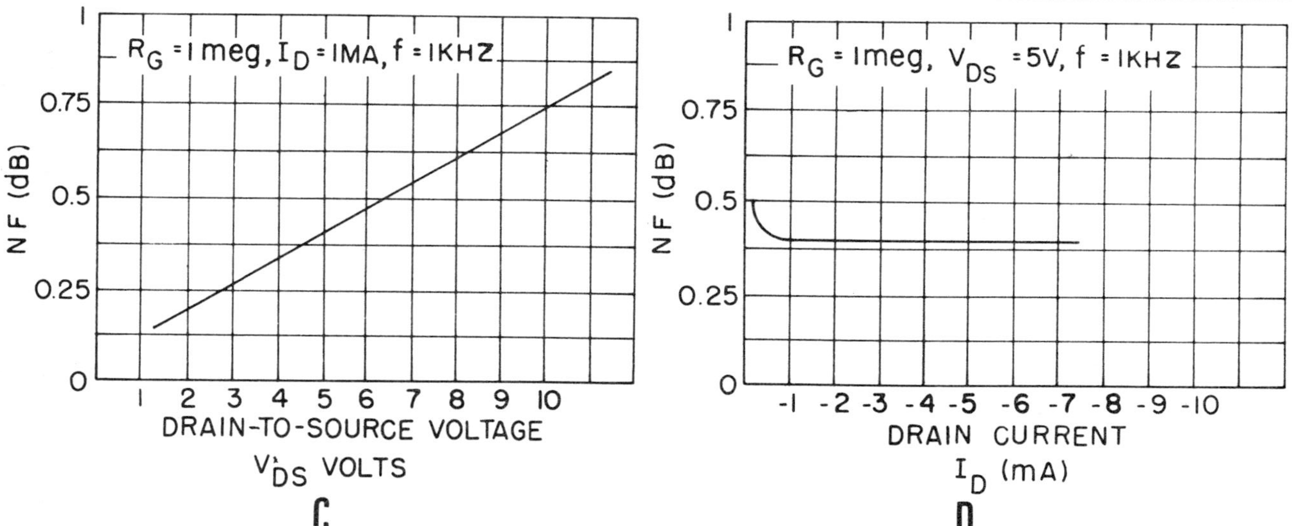

Fig. 6-8. Plot of noise figure against frequency with all other factors held constant at specified values, (A) in Fig. 6-7. Curve B is a plot of the noise figure against the size of the resistor at the input, with all other factors held constant at specified values. Curve C is a plot of the noise figure against the drain-to-source voltage with all other factors held constant at specified values. Curve D is a plot of the noise figure against the drain current with all other factors held constant at specified values.

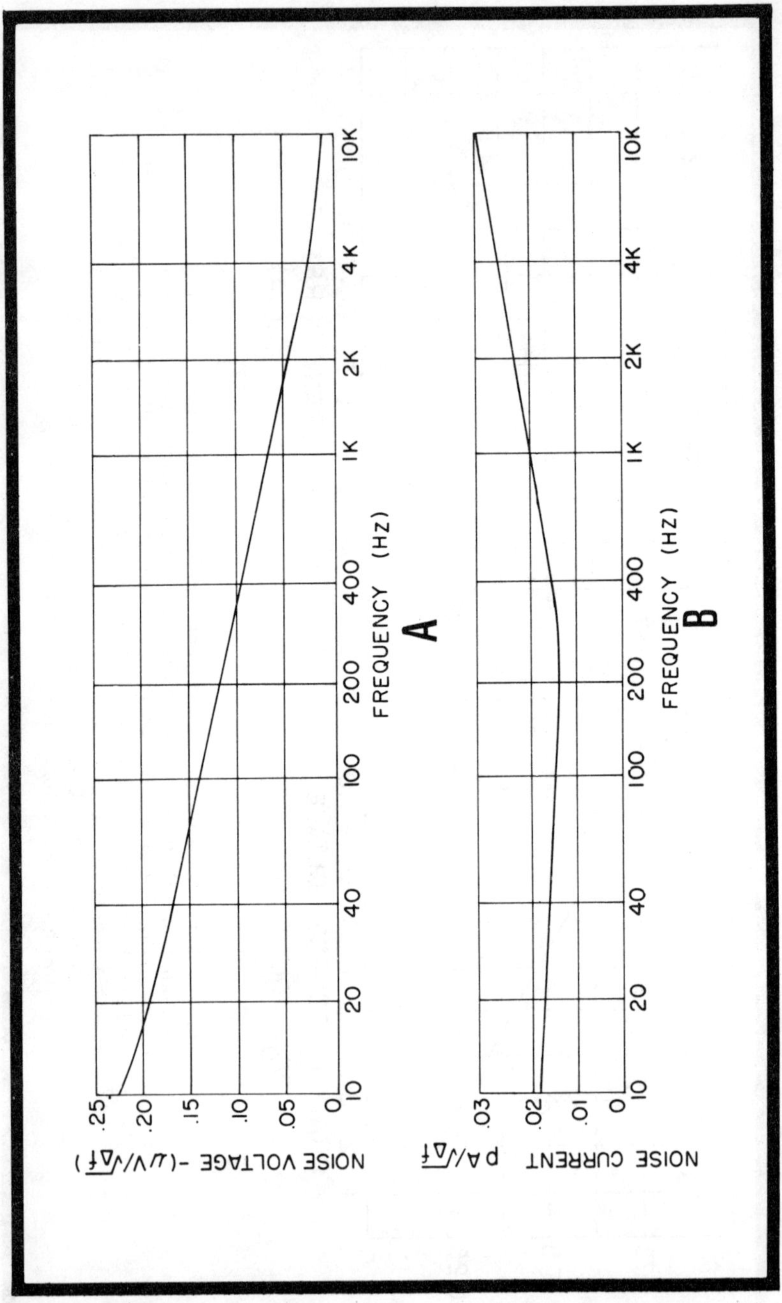

Fig. 6-9. Relationship of the hypothetical noise voltage to frequency (A), and the relationship of hypothetical noise current to frequency (B).

when calculating the noise associated with the JFET. It was indicated in Chapter 5 that there are two capacitances associated with transistors, namely C_{gs} and C_{gd}. C_{in} in Fig. 6-3 is equal to $C_{gs} + C_{gd}(A_v + 1)$ where, as you recall, C_{gs} is the gate-to-source capacity, C_{gd} is the gate-to-drain capacity, and A_v is the voltage gain of the transistor. The voltage gain in the common-source mode is, of course, approximately equal to the product of the drain resistor, R_D, and the transconductance, g_m. Along with C_{in}, the capacity, C_s, of a transducer feeding the amplifier is important in determining transistor noise characteristics.

The capacitive and resistive inputs to the circuit affect the magnitude of noise and signal voltages applied to the input of the hypothetical amplifier. $\bar{e}_n$, $\bar{i}_n$ and the input signal voltage, e_s, must be modified before they can be used to calculate the total noise and signal-to-noise ratio at the input of the JFET. (The input noise is, of course, directly proportional to all noise at the output.)

A circuit showing the various sources of noise referred to the input of the theoretical noiseless amplifier appears in Fig. 6-4. It is redrawn in Fig. 6-9 to include the input impedances along with the capacitance of the signal source. C_s and the

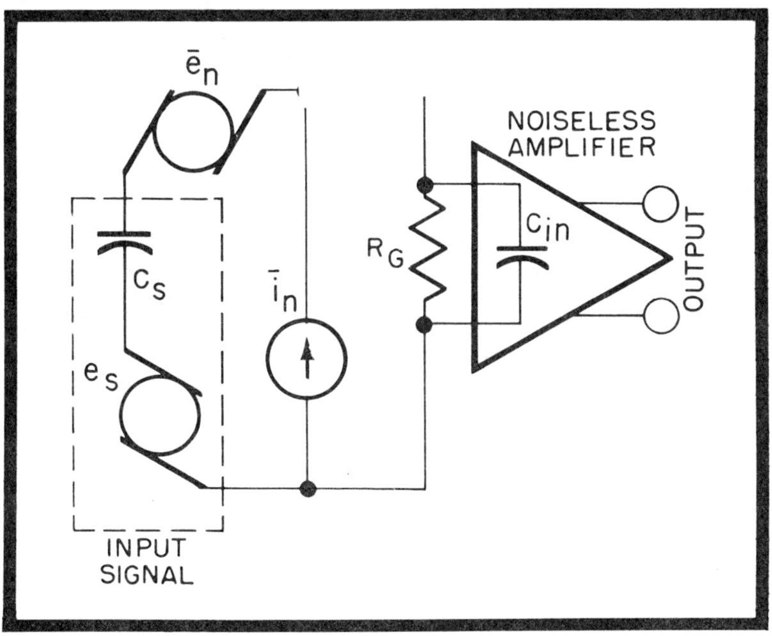

Fig. 6-10. Circuit drawn to include the effect of source and input capacitors in noise voltage and noise current calculations.

parallel combination of R_G and C_{in} form a voltage divider. If R_G is much larger than the reactance of $C_{in} + C_S$, the input signal voltage, e_{sa}, appearing at the amplifier input is:

$$e_{sa} = \frac{e_s C_s}{C_{in} + c_s} \quad (6\text{-}16)$$

while the noise voltage at the input is:

$$\bar{e}_{na} = \frac{\bar{e}_n C_s}{C_{in} + C_s} \quad (6\text{-}17)$$

The noise voltage at the input of the amplifier due to the current noise generator, $\bar{e}_{nia}$, is the product of $\bar{i}_n$ with all the impedances across the input, or:

$$\bar{e}_{nia} = \frac{\bar{i}_n}{6.28f(C_s + C_{in})^2} \quad (6\text{-}18)$$

where f is the frequency at the center of the noise spectrum and A_V is the voltage gain of the amplifier.

Should R_G not be much larger than the reactance of $C_{in} + C_S$, accuracy dictates that R_G be included in the derivation of e_{sa}, $\bar{e}_{na}$ and $\bar{e}_{nia}$. As a rule of thumb, disregard R_G if it is at least ten times larger than the parallel reactance of C_{in} and C_S. If the reactance of $C_{in} + C_S$ are at least ten times larger than R_G, use $\bar{e}_n$ and $\bar{i}_n$ without any of the modifications shown in the equations. When R_G and the reactance $C_{in} + C_S$ approach each other, the simplest procedure is to calculate noise twice, once assuming R_G is relatively large using the modifications in Equations 6-16, 6-17 and 6-18, and then assuming R_G is relatively small. In the second calculation, use $\bar{e}_n$ and $\bar{i}_n$ from the curves without any modification. The noise is somewhere between the solutions to the two calculations. Estimate the noise by noting whether the capacitive reactance or resistance dominates the input impedance, and by how much.

The total noise voltage due to all sources is the square root of the sum of the squares of the noise voltages due to each individual source, or:

$$\bar{e}_{nit} = (\bar{e}_{na}^2 + \bar{e}_{nia}^2 + e_{ni}^2)^{1/2} \quad (6\text{-}19)$$

The individual noise voltages are defined by Equations 6-17, 6-18 and 6-19, respectively. $\bar{e}_n$ rather than $\bar{e}_{na}$ should be used if

R_G is small compared to the reactance $C_{in} + C_s$. $\bar{e}_{nia}$, the voltage produced by the current generator, would in this case become $\bar{i}_n R_G$ rather than the value calculated from Equation 6-18. Likewise, e_s rather than e_{sa} will be used when determining the signal at the input of the amplifier.

It is now possible to determine the signal-to-noise ratio of an amplifying stage. As an example, assume that a 500-pf ceramic phonograph cartridge delivers 0.1 volt at 1 kHz to a 5.6-megohm resistor load at the input of a JFET amplifier. The voltage gain of the amplifier is 10 while C_{gd} equals 3 pf and C_{gs} equals 5 pf. C_{in} is equal to $C_{gs} + C_{gd}$ $(A_V + 1)$ equals $5 + 3(11)$ or 38 pf. The 3 db bandwidth from 10 to 40 kHz is the same as in the previous problem, so that the noise bandwidth is 1.57 x 40,000 Hz equals 63,000 Hz. The geometric center frequency for this bandwidth is $(10 \times 40{,}000)^{1/2}$ equals 632 Hz. Use the transistor described by the curves in Figs. 6-7 and 6-8.

The curves in Fig. 6-9A supply us with the information that $\bar{e}_n$ divided by $\sqrt{\Delta f}$ at 632 Hz is about 0.085 uV divided by $\sqrt{\Delta f}$ and $\bar{i}_n$ divided by $\sqrt{\Delta f}$ is about 0.0175 pA divided by $\sqrt{\Delta f}$. Δf is 63,000 Hz so that $\sqrt{\Delta f}$ is 250. The actual $\bar{e}_n$ for substitution into Equation 6-17 is 250 multiplied by 0.085 or 21.2 uV equals 21.2 x 10^{-6} volt. Similarly, $\bar{i}_n$ for use in Equation 6-18 is 250 x .0175 equals 4.4 pa equals 4.4 x 10^{-12} ampere. Considering that C_{in} was calculated at 38 x 10^{-12} farads and that C_s was stated at 500 x 10^{-12} farads, the total reactance at 632 Hz due to the capacitors is 1 divided by 6.28 x 632 x 538 x 10^{-12} equals 500,000 ohms. Because this is less than one-tenth R_G, the modifications dictated by Equations 6-16, 6-17 and 6-18 must be used.

$$\begin{aligned} e_{sa} &= 0.1(500)/(38+500) = 0.093 \text{ volt} \\ \bar{e}_{na} &= 21.2\times10^{-6}\,(500)/(38+500) = \\ &\quad 19.7\times10^{-6} \text{ volts} \\ \bar{e}_{nia} &= 4.4\times10^{-12}/6.28(632)\,(500+38) = \\ &\quad 2.07\times10^{-18} \text{ volts.} \end{aligned} \tag{6-20}$$

and at 27 degrees C e_{ni}^2 from Equation 6-7 is:

$$e_{ni}^2 = 1.37\times10^{-24}\,(300)\left(\frac{63{,}000}{10\times40{,}000}\right)\left(\frac{1}{(5.6\times10^6)\,(538\times10^{-12})^2}\right) = 4.0\times10^{-11} \tag{6-21}$$

Since $\bar{e}_{nia}$ is negligible compared to all other noise voltages,

$$\bar{e}_{nit} = \left[(19.7\times10^{-6})^2 + 4.0\times10^{-11}\right]^{1/2} = 20.8\times10^{-6} \text{ volts} \tag{6-22}$$

The signal-to-noise ratio is 0.093 divided by 20.8×10^{-6} equals 4.45×10^3. Using previously employed methods along with Table 6-1, the db figure for the voltage ratio is about 13 db + 60 db equals 73 db.

In general, FET noise is primarily due to $\overline{e}_n$ when impedances are high and $\overline{i}_n$ when impedances are low. The contribution from the input resistor is small, but is best when it is equal to $\overline{e}_n$ divided by $\overline{i}_n$. JFET noise will be at a minimum if the transistor is biased at about V_{GS} equals 0. g_m must be high if 1/f noise is to be relatively low.

Power Amplifiers

Chapter 7

FET and bipolar transistor data sheets present two groups of information. One group provides the designer with the electrical characteristics such as beta and leakage current for the bipolar device and transconductance and pinch-off voltage for the FET. These, as well as other characteristics, have been shown in detailing the designs of various circuits.

The second group presents information concerning the operating limits of the transistor. These ratings indicate, for example, the maximum base, emitter and collector currents that may be permitted to flow through the device. It discloses the maximum voltages that the designer may place across various junctions or combinations of junctions without causing the transistor to break down. The voltage allowed across any junction may vary with voltages applied at other junctions of the device, as well as with temperature. Limits also apply to the operating and storage temperatures, as well as to the amount of power a transistor may dissipate.

The maximum ratings were ignored in voltage and current amplifier discussions because these boundaries are seldom exceeded in small-signal preamplifier stages. However, it is essential to take these limits into consideration when choosing a transistor for any application. Economical power amplifier design cannot be completed without full cognizance of the permissible confines of transistor operation.

Many of the limits apply to both FET and bipolar devices. At this writing, the JFET is used almost exclusively as a small-signal device, although there is some work in progress on power field-effect transistors. This discussion of power devices and amplifiers is limited to arrangements using bipolar transistors only. Much of the discussion likewise applies to power JFETs should they become commercially available.

POWER RATINGS

Real power is developed across a device when current flows through it while a voltage exists across it. Heat is

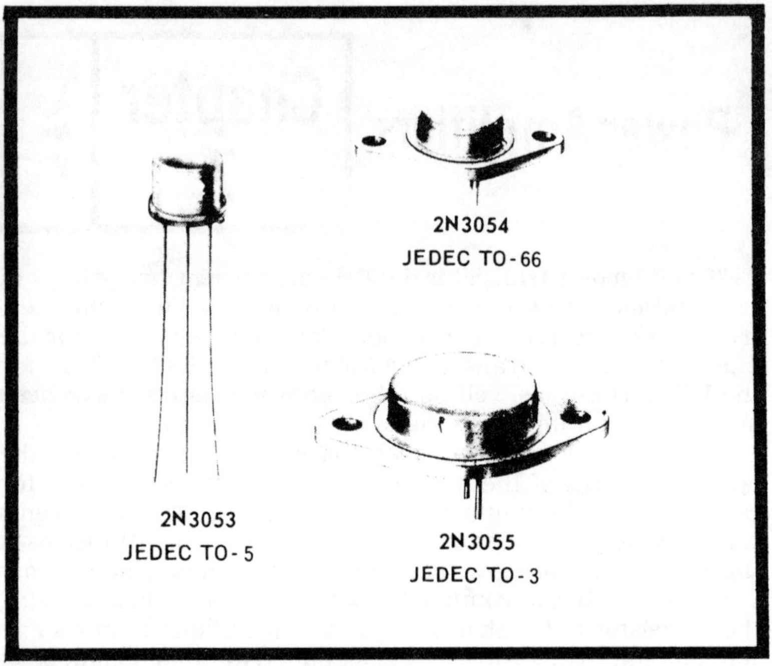

Fig. 7-1. Power transistors come in different case sizes and types. The three most popular case types are shown. (Courtesy RCA)

generated in the device due to the power dissipation. With a transistor, the amount of power it can dissipate depends upon the maximum temperature the junction may reach without damage. It also depends upon the ability of the finished package to pass generated heat from the junctions to the surrounding air.

Power transistors come in many different types of packages. The three most common ones are the TO-5, TO-66 and TO-3 shown in Fig. 7-1. The larger TO-3 devices can usually dissipate more power than the transistors in the small TO-5 package. Transistors in TO-3 cases, as well as those in the smaller size TO-66 cases, can be conveniently mounted on a large metal surface to facilitate the removal of heat from the junctions.

The amount of power a transistor can dissipate is related to the ambient temperature or to the temperature of the case. The curve in Fig. 7-2 shows this relationship for the popular 2N3055 transistor in a TO-3 case. Up to 25 degrees C, the dissipation is limited to 115 watts. The permissible dissipation drops when the case temperature exceeds 25 degrees C until zero watts may be dissipated at the upper thermal limit of the transistor case at 200 degrees C. Should the design, for example, require that 55 watts be dissipated by the transistor, the case temperature must not exceed 115 degrees C.

Power-limiting information should be plotted on the transistor collector characteristic curves. The characteristic curves are supplied by the manufacturer. Points on the power-limiting curve can be calculated using the Ohm's Law equation, I_C equals P_{CEM} divided by V_{CE}, where I_C is the collector current, P_{CEM} is the maximum power that can be dissipated by the transistor at the stated case temperature, and V_{CE} is the collector-to-emitter voltage.

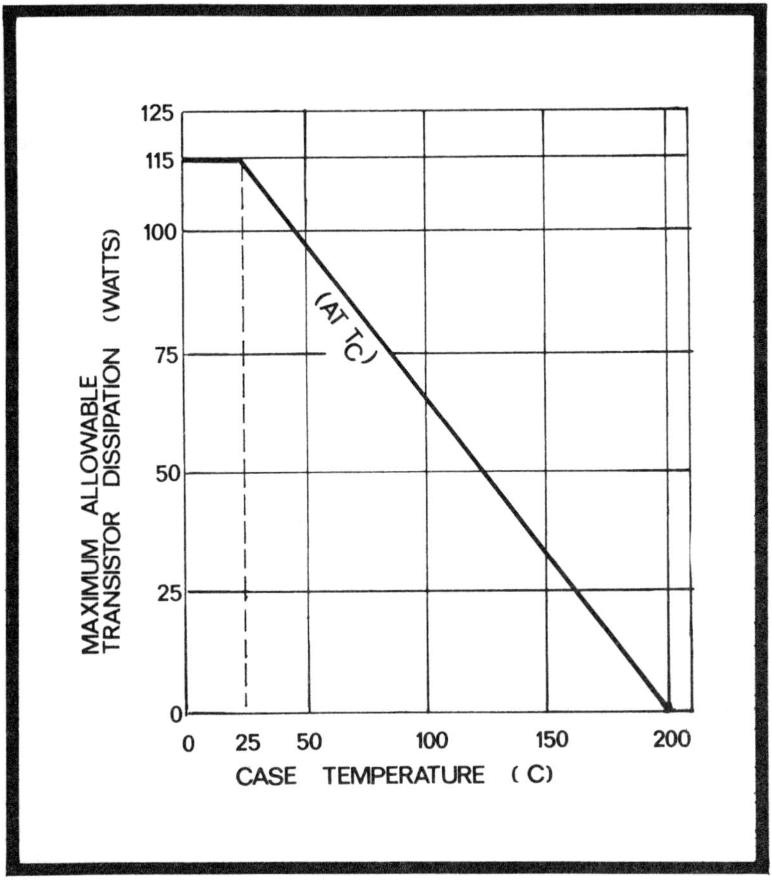

Fig. 7-2. Curve showing the maximum allowable power dissipation for the 2N3055 at various case temperatures. (Courtesy RCA)

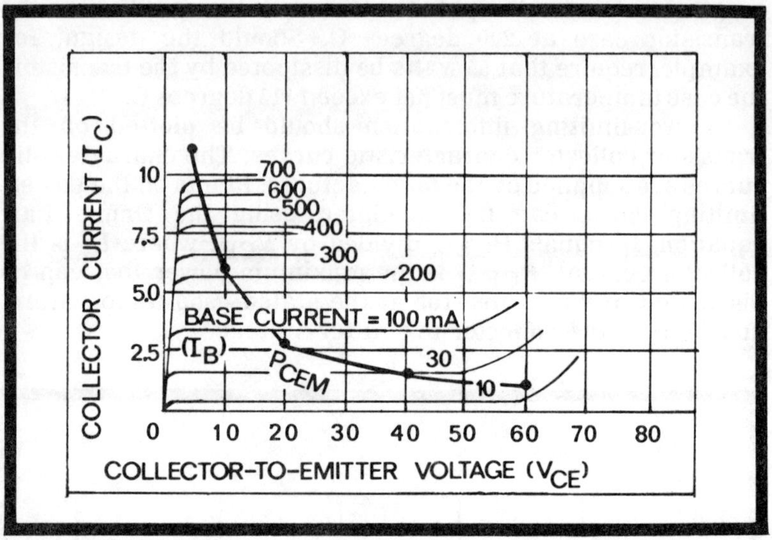

Fig. 7-3. Plot of collector characteristics along with a curve showing a power limitation. Operation is usually below the P_{CEM} curve. (Courtesy RCA)

For V_{CE} = 0 volts, I_C = 55 watts/0 volts = infinite amperes

V_{CE} = 5 volts, I_C = 55 watts/5 volts = 11 amperes

V_{CE} = 10 volts, I_C = 55 watts/10 volts = 5.5 amperes

V_{CE} = 20 volts, I_C = 55 watts/20 volts = 2.75 amperes

V_{CE} = 40 volts, I_C = 55 watts/40 volts = 1.38 amperes

V_{CE} = 60 volts, I_C = 55 watts/60 volts = 0.92 ampere

Plot these points on the collector characteristic curves and connect them with a smooth line curve as in Fig. 7-3. The circuit must be designed so that the transistor will operate within the limits set by the curve. Operation above the P_{CEM} curve may cause the device to overheat and be destroyed due to thermal factors. It should be noted that although the P_{CEM} curve is superimposed on the collector characteristics, it is independent of these curves. Wherever possible and in the interest of clarity, the actual collector curves are omitted from the drawings.

CLASS A POWER AMPLIFIER

A Class A amplifier is biased so that the collector current flows during the entire cycle. The output is limited only by the leakage and saturation characteristics of the transistor. Class A power amplifiers can be differentiated from Class A signal-voltage amplifiers by the magnitude of the output signal swing. This can be illustrated using the circuit in Fig. 7-4A and its load line.

You may recall that the DC equation of the output collector circuit of the transistor is:

$$E_{CC} = I_C R_C + V_{CE} \tag{7-1}$$

where V_{CE} is the voltage across the device. As was the case with the small-signal circuits, the equation defines the load line. One point on the load line is when V_{CE} equals 0 and I_C equals E_{CC} divided by R_C. A second point is when I_C equals 0 is V_{CE} equals E_{CC}. Plot the two points and connect them with a straight line. All coincident collector-to-emitter voltage and collector current conditions under which the transistor may operate must lie on this line.

Assume the transistor is idling at Point Q on the load line. The bias current is I_Q while the collector-to-emitter voltage is V_{CEQ}. In the absence of a signal, the transistor current and voltage are at I_Q and V_{CEQ} respectively. A small signal may cause the output current to swing from I_{CA} to I_{CB} or from A to B along the load line; with a large signal at the output, the collector current may swing the limits from 0 to I_{CD} or from C to D on the load line. In the former case, the amplifier may be designed using the small-signal equivalent circuits discussed in Chapter 4. Amplifier stages which must accommodate the larger swing should be designed using the power amplifier considerations detailed here. However, there is no definite point of demarcation where the signal amplifier stops and the power amplifier begins.

It is obvious from Fig. 7-4B that the collector current swing is limited to a minimum of 0 amperes and a maximum of E_{CC} divided by R_C amperes. Likewise, the collector voltage cannot drop below 0 volts or increase above E_{CC} volts. Other factors limiting the swing may be observed from the exaggerated collector characteristic curves in Fig. 7-5 and the load line on this drawing.

At the left, all curves seem to emanate from a reasonably straight line which starts at the vertex. This line is **not** coincident with the I_C axis. In this low collector-to-emitter

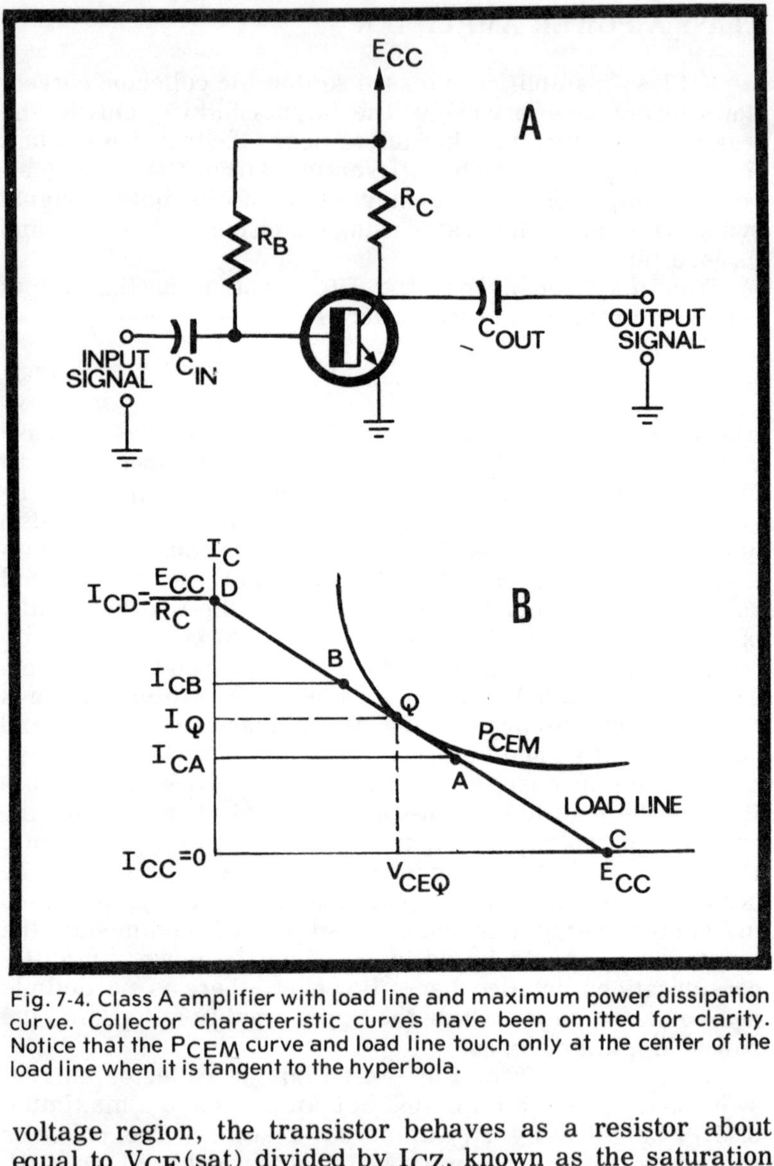

Fig. 7-4. Class A amplifier with load line and maximum power dissipation curve. Collector characteristic curves have been omitted for clarity. Notice that the P_{CEM} curve and load line touch only at the center of the load line when it is tangent to the hyperbola.

voltage region, the transistor behaves as a resistor about equal to V_{CE}(sat) divided by I_{CZ}, known as the saturation resistance. V_{CE}(sat), the saturation voltage, is specified at a specific collector current. In the drawing it is shown at the point of intersection of the characteristic curves and the load line. The voltage will differ with the saturation voltage at other collector currents. It may likewise not agree with the transistor data supplied by the manufacturer for he may specify $V_{CE}(sat)$ at a different collector current than the I_{CZ} in the drawing. However, the saturation voltage shown is the significant saturation voltage in this particular application. There is no amplification of any signal that may swing to the left of the saturation resistance curve.

At the right, the load line is limited by the I_B equals 0 ma curve. The output current may not go below this curve. Considering both limits, the actual maximum collector current swing is $I_{CZ} - I_{CEO}$ and the voltage swing is $V_{CEO} - V_{CE}(sat)$. In the initial design work, these limits may be ignored. However, they may be very significant in the final design stages.

In Fig. 7-4B, the load line just touches the P_{CEM} curve at point Q. Although Q has been used previously as the quiescent operating point, here it is assumed to be a point of contact between the P_{CEM} curve and the load line. The actual idling condition may be chosen at Q or at any other point along the load line. Point Q is the center of the load line. Significantly, if the load line touches (is tangent to) the P_{CEM} curve at any one point, the point of contact is always at the center of the load line.

Should a sinusoidal signal be applied to the input of the transistor, after amplification it will travel along the load line at the output, as shown in Fig. 7-6. Here it is assumed that Q is the quiescent bias point at the center of the load line. The transistor will dissipate the maximum power when it is idling at Q or during the instant in the cyclic variations when the current is I_Q and the collector-to-emitter voltage is at V_{CEQ}.

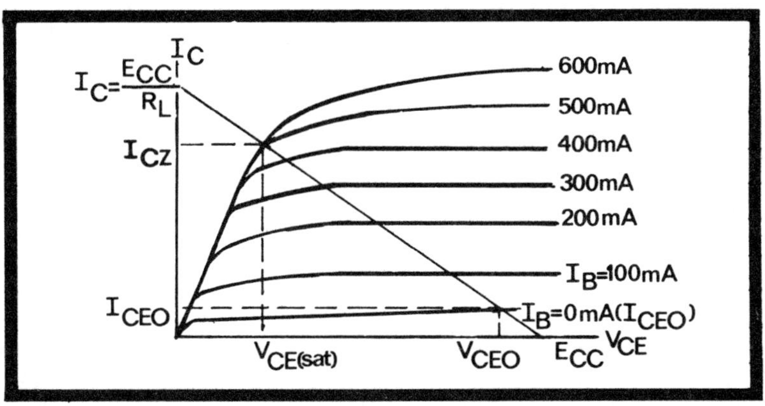

Fig. 7-5. Curve defining limits of collector voltage and current swings. The minimum useful voltage, V_{CE}, in audio applications is about $3V_{CE}(sat)$. It is usually sufficient to subtract a few volts to compensate for V_{CEO}.

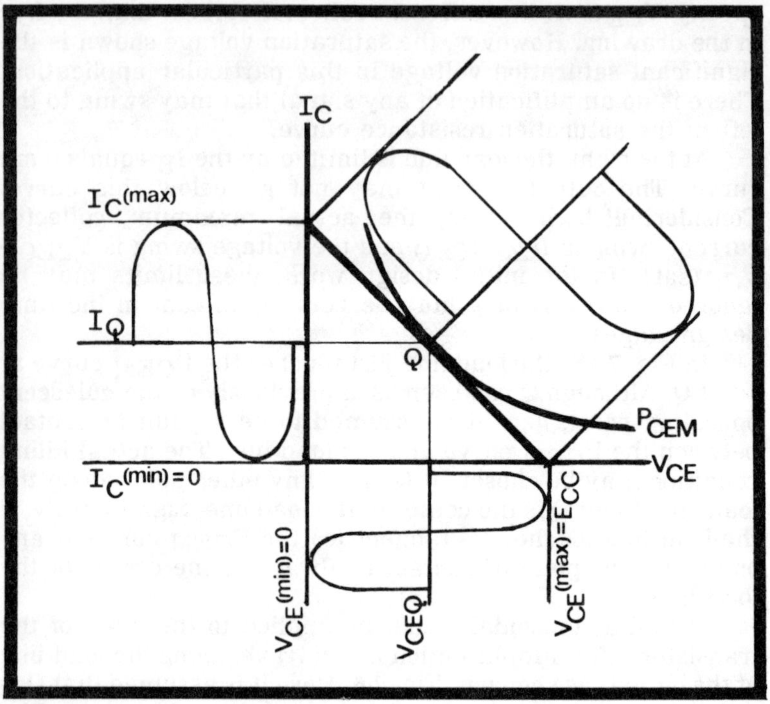

Fig. 7-6. The sine wave travels along the load line and the output current and voltage appear at the axis. The swing here is at a maximum, ignoring the limitations set in Fig. 7-5.

Since I_Q is equal to $I_C(max)$ divided by 2 and V_{CEQ} is E_{CC} divided by 2, the maximum power dissipated at any instant in the cycle is:

$$P_C\ (max) = \frac{I_C\ (max)\ E_{CC}}{4} \qquad (1)$$

$$= \frac{I_C\ (max)^2\ R_C}{4} = \frac{E_{CC}{}^2}{4R_C} \qquad (7\text{-}2)$$

where R_C, the load resistor in Fig. 7-4A, is equal to E_{CC} divided by $I_C(max)$.

If the load line is also equal to the maximum dissipation curve, $P_C(max)$ is also equal to the maximum permissible dissipation of the transistor. The product of $I_C(max)$ divided by 2 and E_{CC} divided by 2 on the load line is always the point in the cycle when the transistor dissipates the maximum power, independent of whether it is the actual quiescent point or whether or not the load line touches the P_{CEM} curve.

As discussed in Chapter 4, the RMS voltage of a sine wave is E_p divided by $\sqrt{2}$, where E_p is the peak voltage. It is also equal to $E_{p\text{-}p}$ divided by $2\sqrt{2}$ if we let $E_{p\text{-}p}$ equal the peak-to-peak voltage. It follows that the RMS current is $I_{p\text{-}p}$ divided by $2\sqrt{2}$. In Fig. 7-6, the peak-to-peak voltage across the load resistor is E_{CC} so that the RMS output voltage is E_{CC} divided by $2\sqrt{2}$. Similarly, the RMS current across R_C is $I_{C(max)}$ $2\sqrt{2}$. The signal power, P_{RC} developed across the load is the product of the RMS current and voltage or:

$$P_{R_C} = \frac{E_{CC}\ I_C\ (max)}{8} = \frac{E_{CC}^2}{8R_C}$$

$$= \frac{I_C\ (max)\ R_C}{8} \tag{7-3}$$

Comparing Equation 7-2 with Equation 7-3, it should be noted that the transistor can deliver only ½ the maximum power it can dissipate safely.

Assuming a perfect sinusoidal output, the quiescent point will not shift from Q. The average power dissipated by the transistor is then equal to the maximum instantaneous power as stated in Equation 7-2, or $E_{CC}2$ divided by $4R_C$. Since the average or quiescent voltage across the load resistor is E_{CC} divided by 2, the power dissipated across R_C is E_{CC}^2 divided by $4R_C$. Ignoring the power dissipated by the input circuit, the supply must furnish power equal to the sum of that dissipated by the transistor and that dissipated across the resistor load, or E_{CC}^2 divided by $2R_C$.

The efficiency of the circuit, defined as 100 multiplied by the ratio of the maximum AC power that can be delivered to the load to the power that must be provided by the supply, is 25 percent. To improve this, the wasted power across the load resistor must be minimized or entirely eliminated.

A circuit with an output transformer can be used to minimize the DC resistance in the collector circuit, thus diminishing the magnitude of DC power the supply must provide. In Fig. 7-7 an output transformer is shown between the transistor and the load. The effect of the load resistor, R_L, is reflected into the primary circuit. To determine the effect, multiply R_L by the square of the turns ratio of the transformer. If N_P is the number of turns in the primary winding and N_S is the number of turns in the secondary, the reflected AC load into the primary is R'_L equals R_L $(N_P$ divided by $N_S)^2$. R'_L is a component only of the AC load. There is also a DC load on the circuit.

Two load lines—DC and AC—must be drawn for the circuit. Assume zero resistance in the windings. Using the collector relationship E_{CC} equals $I_C R_C + V_{CE}$ stated above, one point on the DC load line is at I_C equals 0 and V_{CE} equals E_{CC} while the second point is at V_{CE} equals 0 and I_C equals

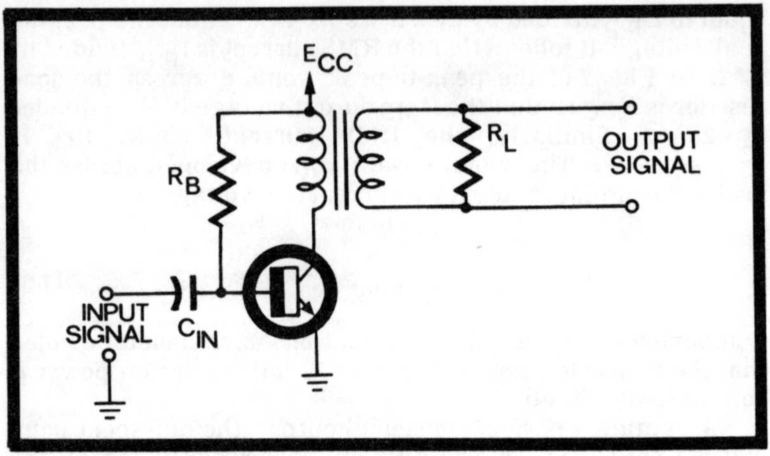

Fig. 7-7. Circuit using an output transformer.

infinity. This is a vertical line from E_{CC} in Fig. 7-8. Choose a quiescent point at Q where the DC load line crosses the P_{CEM} hyperbola. Draw an AC load line so that it passes through Q and is at a tangent to the P_{CEM} curve. The AC load line must then pass through two other points as well, namely through $2E_{CC}$ on the horizontal axis and through $2I_{CQ}$ on the collector current vertical axis.

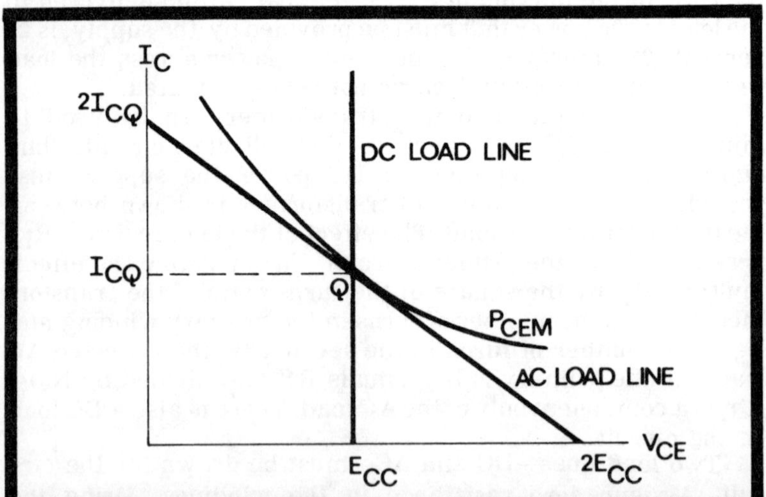

Fig. 7-8. AC and DC load lines for a Class A amplifier with transformer output. Zero resistance is assumed for the transformer and power supply. The swing here is at a maximum, ignoring the limitations set in Fig. 7-5.

Assume a sine wave across the load causes the collector voltage to swing from 0 to $2E_{CC}$ volts, while the collector current swings from $2I_{CQ}$ to 0 amperes. The power developed across the load is:

$$P_{R_L}' = \frac{4\ E_{CC}\ I_{CQ}}{8} = \frac{E_{CC}\ I_{CQ}}{2} \quad (7\text{-}4)$$

The maximum power that can be dissipated by the transistor is:

$$P_{CEM} = P_C(\text{max}) = E_{CC}I_{CQ} \quad (7\text{-}5)$$

As for the previous Class A circuit, the transistor can deliver but ½ the power it can dissipate safely, or P_{CEM} divided by 2.

There are three major advantages in using an output transformer. First, the circuit efficiency is almost doubled to 50 percent, since the DC supply must provide power only for the transistor. Little power is dissipated by the low DC resistance in the primary winding. Next, the transformer can be used to make a low-impedance load, R_L, such as a loudspeaker, appear as a higher impedance in the primary winding where it is most desirable. Finally, it isolates the speaker from the transistor collector current. It is undesirable to let DC flow through the speaker voice coil because it would limit the full and linear travel of the cone.

In most cases, the DC resistance in the collector circuit cannot be ignored. Should the resistance in the primary winding of the transistor be significant, the DC load line will no longer be perfectly vertical. It will be at an angle with the horizontal axis as in Fig. 7-9. The DC load line can be found using the methods applied to Fig. 7-4B.

The AC load line can be determined using the following logic. The quiescent point on the DC load line is Q. This is also one point on the AC load line. Two points are necessary to define any line. If we let R'_{LT} be the total resistance in the collector circuit (the reflected resistance through the transformer plus all resistance in the AC portion of the collector circuit), the second point, V_{CEA}, can be determined from the equation:

$$R_{LT}' = \frac{V_{CEA} - V_{CQ}}{I_{CQ}} \quad (7\text{-}6)$$

A PRACTICAL EXAMPLE

Our task is to design an audio amplifier capable of delivering 5 watts to a 3.2-ohm loudspeaker load. The tran-

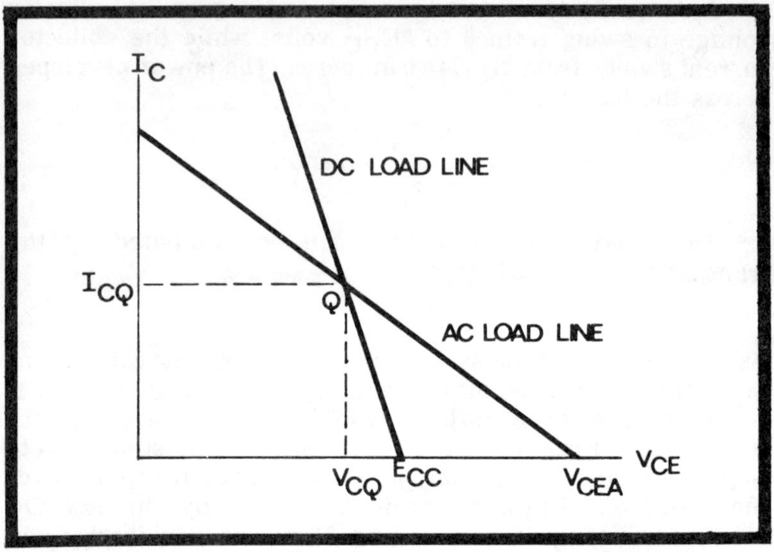

Fig. 7-9. Plot of load lines when the collector circuit has DC resistance.

sistor available is the 2N3055. Outline a step-by-step procedure using the 2N3055.

1. Draw a workable circuit. In order to keep the DC current out of the speaker winding, an output transformer should be used. For improved thermal stability, use a resistor in the emitter to provide DC feedback. If desired, bypass it with a capacitor to prevent AC degeneration and the reduction of power reaching the loudspeaker. A valid circuit arrangement is shown in Fig. 7-10.

2. Determine the actual power the amplifier must deliver. Although 5 watts is desired at the speaker, the 2N3055 must deliver more than this due to losses in the output transformer. Assuming 75 percent transformer efficiency, the power available from the 2N3055 across the primary of the transformer must be at least 5 divided by 0.75 or 6.67 watts.

3. Choose V_{CEA}, the voltage at which the AC load line crosses the horizontal axis. The following considerations apply. The maximum voltage that may safely be applied between the collector and emitter of a transistor varies with the size of resistor placed between the base and emitter. Breakdown voltages are specified on data sheets for different resistor values. There is no resistor used in the circuit in Fig. 7-10. Therefore, the BV_{CEO} rating applies. BV_{CEO} is the voltage that may be applied between the collector and emitter

of the transistor without destroying the device when there is no resistor between the base and emitter. BV_{CEO} is the breakdown voltage when the current flowing through the device is very small. As the collector current increases, the BV_{CEO} breakdown voltage is reduced to a more or less constant value, $V_{CEO}(sus)$, referred to as the sustaining voltage. For the 2N3055, $V_{CEO}(sus)$ is 60 volts. V_{CEA} should be equal to or less than 60 volts. Taking various factors into consideration, such as peak voltages due to the primary inductance of the transformer, the possibility of getting transistors at reduced prices when $BV_{CEO}(sus)$ is low, the economy of a low-voltage power supply, etc., choose V_{CEA} at 25 volts.

4. Calculate the power the transistor must dissipate. Because it must deliver 6.67 watts, the device must be capable of dissipating at least 2 x 6.67 watts or 13.3 watts. The 2N3055 is capable of dissipating 115 watts at room temperature so this 13.3 watts should be no problem. In fact, add 25 percent to the required dissipation capability to accommodate the limitation of current and voltage swings on the load line as well as signal power losses across the emitter resistor. The required dissipation capability should be 13.3 + (.25 x 13.3) or 16.6 watts.

5. Determine points on the P_{CEM} equals 16.6 watts parabola and draw a line from V_{CEA} equals 25 volts to the I_C axis touching the parabola at one point. Several points on the parabola are:

V_{CE} = 25 volts, I_C = 16.6 watts/25 volts = 0.66 amperes
V_{CE} = 20 volts, I_C = 16.6 watts/20 volts = 0.83 amperes
V_{CE} = 15 volts, I_C = 16.6 watts/15 volts = 1.11 amperes
V_{CE} = 10 volts, I_C = 16.6 watts/10 volts = 1.66 amperes
V_{CE} = 5 volts, I_C = 16.6 watts/5 volts = 3.32 amperes
V_{CE} = 2 volts, I_C = 16.6 watts/2 volts = 8.3 amperes

Draw the parabola using the data just calculated. See Fig. 7-11. Starting from the selected V_{CEA} equals 25 volts on the V_{CE} axis, draw an AC load line tangent to the parabola at V_{CEA} divided by 2 equals 12.5 volts and I_{CQ} equals 16.6 watts divided by 12.5 volts or 1.33 amperes. It will intersect the I_C current axis at $2I_{CQ}$ equals 2.66 amperes. The DC load line will also intersect the P_{CEM} curve at Q.

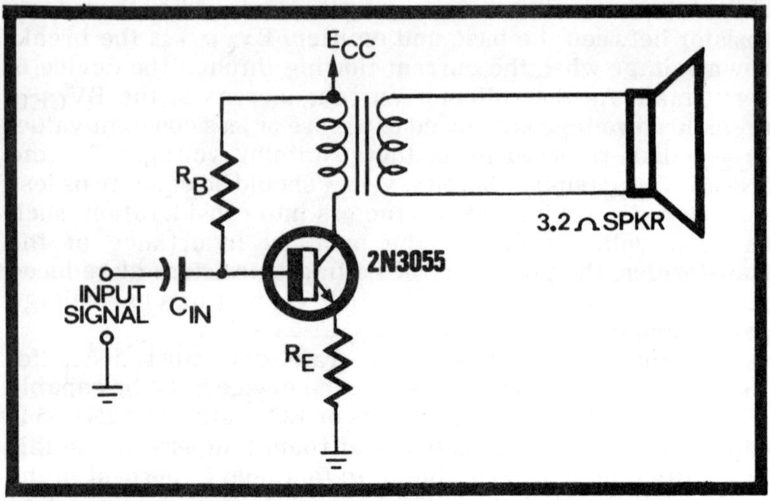

Fig. 7-10. Audio output circuit used in the text problem.

6. In Fig. 7-10, the voltage across R_E should be between 0.5 and 1 volt. Since the quiescent current has been chosen at 1.33 amperes, R_E may assume any value between 0.5 volt divided by 1.33 amperes or 0.375 ohm and 1 volt divided by 1.33 amperes or 0.75 ohm. Choose a relatively "standard" 0.56-

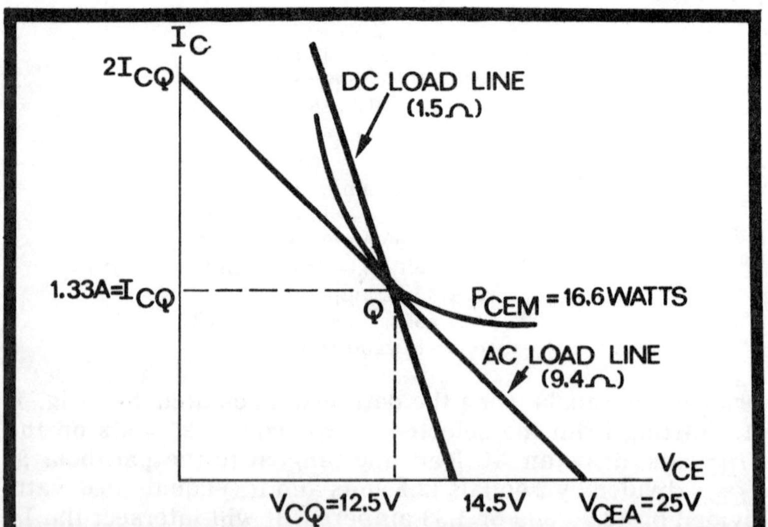

Fig. 7-11. Load lines and power limiting curves drawn for the text problem.

ohm 10 percent resistor. The power the resistor must be capable of dissipating is $I_{CQ}{}^2R_E$ equals $(1.33)^2(0.56)$ or 1 watt. Use a 2-watt component for safety reasons.

7. The AC resistance seen by the transistor is equal to the reciprocal of the slope of the AC load line, or V_{CEA} divided by $2I_{CQ}$ equals 25 divided by 2.66 or 9.4 ohms.

8. The DC load line is determined by the supply voltage, E_{CC}, the resistor in the emitter circuit, and the resistance of the primary winding of the output transformer. We know that R_E is 0.56 ohm. Choose a transformer so that the sum of R_E and the resistance in the transformer winding will be small, let us say, about 1.5 ohms. The slope of the reciprocal of the DC load line is equal to this resistance, so that 1.5 equals (E_{CC} - V_{CQ}) divided by I_{CQ} equals (E_{CC} - 12.5) divided by 1.33 (for V_{CQ}, when set at the center of the load line in this problem, is equal to V_{CEA} divided by 2). E_{CC} must then be 14.5 volts. Plot this point on the V_{CE} axis. Draw a line connecting the 14.5-volt point on the horizontal axis with Q on the AC load line. This is the DC load line.

9. The total AC resistance in the collector circuit is equal to the sum of the primary resistance of the transformer (1.5 ohms - 0.56 ohm or 0.94 ohm, the 0.56-ohm resistor in the emitter, and the resistance R_L', reflected from the secondary into the primary winding due to the 3.2-ohm speaker in the secondary. If the resistance of the secondary winding is not small compared to the 3.2-ohm load, R_L' is due to the reflection of the sum of the speaker resistance plus the resistance of the secondary winding. In Step 7, it was determined that the total AC resistance the transistor must see is 9.4 ohms. Thus, the reflected resistance into the primary of the transformer is 9.4 ohms - 0.56 ohm - 0.94 ohm or 7.9 ohms.

10. The impedance ratio of the transformer is 7.9 ohms:3.2 ohms. Since this is equal to the square of the turns ratio of 7.9 divided by 3.2 equals (N_P divided by N_S)^2, where N_P is the number of turns in the primary winding and N_S is the number of turns in the secondary, the turns ratio from the primary to the secondary must be about 1.6:1.

11. Check if sufficient power can be delivered to the load. The transistor characteristics set limits to the possible output current and voltage swings. At the one end, the saturation voltage of the transistor is stated at 1.1 volt at 4 amperes. Since our maximum current is 2.66 amperes, the saturation voltage is somewhat less. Let us assume that the minimum voltage across the transistor due to saturation is 0.75 volt. Let us also assume a 1-volt loss in swing at the other end of the load line due to leakage current limits. Thus, the signal voltage

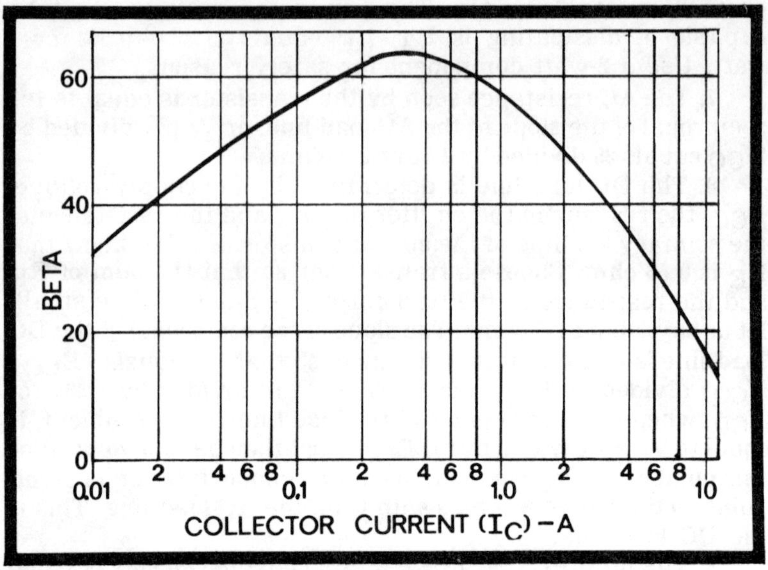

Fig. 7-12. Curve shows how DC beta varies with collector current when V_{CE} equals 4 volts. (Courtesy RCA)

swing is limited to 25 - 0.75 - 1 or 23.25 volts. The power the transistor can deliver is $(23.25 \text{ volts})^2$ divided by 8(9.4) equals 7.2 watts. The AC power developed across the emitter resistor is $(2I_{CQ})^2R_E$ divided by 8 equals $(2.66)^2(0.56)$ divided by 8 or 0.5 watts. The total power across the output transformer is 7.2 - 0.5 or 6.7 watts. If the transformer is 75 percent efficient, the power reaching the load is 0.75 (6.7) equals 5 watts.

12. DC bias conditions must now be established. The emitter resistance is 0.56 ohm. When reflected into the base circuit, it appears as a DC resistor equal to the beta of the transistor multiplied by R_E. The DC beta curve, reproduced in Fig. 7-12, shows the average beta is about 50 when I_{CQ} equals 1.33 amperes. Thus, the resistance at the input is 0.56 x 50 or 28 ohms.

13. The quiescent base current, I_B, must be I_{CQ} divided by β equals 1.33A divided by 50 equals 0.027 ampere.

14. Using the data from Steps 12 and 13, we can calculate R_B, for E_{CC} equals $I_B (R_B + \beta R_E)$ equals 12.5 equals 0.027 (R_B + 28 ohms). R_B is equal to 435 ohms. Use a standard 470-ohm resistor. The power dissipated by R_B is $I_B^2R_B$ equals $(0.027)^2(470)$ or 0.34 watts. A ½-watt resistor can be used here, although a 1-watt component is more desirable.

15. The AC resistance reflected into the base circuit from the emitter circuit is $(R_E + r_e)$, where r_e is the AC emitter

resistance of the transistor. In Step 12, the DC emitter resistance was ignored, since it is negligible. However, the AC emitter resistance was specified as equal to 26 divided by I_Q in Chapter 4, where I_Q is expressed in milliamperes. In this problem, r_e will be very small if calculated from the formula. In reality, it is seldom less than 1 ohm. We will assume it to be 1 ohm. The AC emitter resistance reflected into the base circuit is then 50(0.56 + 1) equals 78 ohms. (It is assumed the AC beta is equal to the DC beta.)

The input impedance is essentially this 78 ohms in parallel with the 470-ohm resistor calculated in Step 14, or 67 ohms. If this amplifier is designed to have a good frequency response down to 20 Hz, C_{IN} must be at least equal to 1 divided by 6.28(20)(67) or 120 mfd. This assumes that the impedance of the input signal source is zero. Calculations of this type will be elaborated upon in Chapter 9.

Distortion

In the problem, the transistor will deliver five watts only if the collector current and voltage will swing over its maximum capabilities. Because of nonlinearity of the collector characteristics, it is most desirable to limit the swing to the minimum. This is especially true near the saturation voltage lines as well as near the I_{CEO} curve where I_B equals 0. A good rule of thumb is to make the minimum collector voltage equal to three times the saturation voltage. Subtract another two volts due to the leakage current.

The above solution to the problem is marginal. In the laboratory, it will probably be found that more power output capability is necessary and improved circuit efficiency will be required. Some measures of improving efficiency include bypassing the 0.56-ohm emitter resistor with a large capacitor, using a more efficient output transformer, increasing E_{CC}, etc.

CLASS AB POWER AMPLIFIERS

The quantity of power that can be delivered by a Class A amplifier is extremely limited. A transistor can dissipate a specific amount of power. Should the transistor conduct current over the entire cycle as well as when it is idling, a considerable amount of power is wasted just keeping the device operating at its predetermined quiescent bias. This state of affairs exists in the Class A mode of operation.

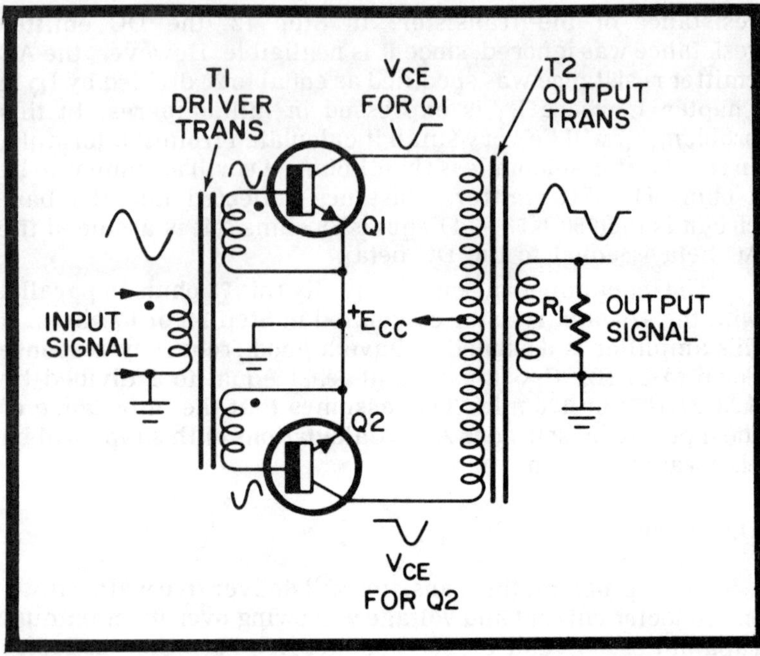

Fig. 7-13. Transformer coupled Class B pushpull circuit. Significant waveshapes are shown.

On the other hand, Class B power amplifiers are biased so that no collector current will flow when the transistor is idling. The transistor does not conduct until the applied signal is of a sufficient magnitude and polarity to put the device into its active region. Silicon transistors, for example, do not conduct until the signal applied to the base relative to the emitter is positive and greater in magnitude than the 0.6 or 0.7 volt necessary to turn on the base-emitter junction.

Even though the amount of power a transistor is capable of dissipating in Class B remains identical to its capabilities in Class A, in the former mode it will dissipate this power only when conducting useful audio currents. The portion of power wasted in maintaining the Class A bias is applied here to enable the Class B amplifiers to deliver more useful signal output.

Only one half of a pure AC sinusoidal cycle will turn on a transistor biased in the Class B mode. In order to reproduce the alternate half of the cycle, the circuit must use a second transistor which conducts during this latter portion of the cycle. Two transistors are required to reproduce a full cycle in Class B-biased amplifier circuits.

Class B Statistics

A pushpull Class B amplifier circuit is shown in Fig. 7-13 using two NPN transistors. Should a sine wave be fed to the input, the various waveshapes shown can be found at the specific points in the circuit when they are measured with respect to ground. The dots next to the ends of the input transformer windings indicate that the corresponding leads or terminals are identical in polarity with respect to those at the unmarked ends. The base of Q1 is in the same phase with respect to its emitter as is the "hot" lead of the input signal with respect to ground. To establish the proper status for pushpull operation, the polarity of the signal at the base of Q2 is 180 degrees out of phase with that at the base of Q1.

During the first half cycle, the base of Q1 is positive with respect to the emitter, so this transistor will conduct. Q1 will not conduct during the second half cycle because during this interval the base-emitter junction is reversed in polarity. The collector voltage is as shown, for the voltage across the transistor and the upper half of the transformer winding is reduced as the transistor goes into conduction.

The second half of the cycle turns Q2 on, since only then does the signal bias that base positive with respect to its emitter. The waveform of the output voltage at the bottom of the output transformer with respect to the center tap, as well as across the transistor, is as shown. The voltage across the entire primary is relatively sinusoidal, as is the output signal across the load R_L. Ideally, the output signal is a magnified version, power-wise, of the input signal.

As discussed for Class A amplifier, the two transistors see the actual load R_L reflected as a resistor, R_L', into the **entire** primary winding of output transformer T2. R_L' equals $R_L(N_P$ divided by $N_S)^2$ is an equation describing the relationship. In this formula, N_P is the number of turns in the **entire** primary winding and N_S is the number of turns in the secondary. Either one of the transistors sees a resistance, R_L'', across **one half** of the transformer winding. The number of turns each transistor sees is N_P divided by 2.

$$R_L'' = R_L\left(\frac{N_p}{2}/N_s\right)^2 = (R_L/4)\,(N_p/N_s)^2. \qquad (7\text{-}7)$$

Comparing the equation for R_L' with the equation for R_L'' we arrive at the important conclusion that R_L'' is equal to $\frac{1}{4}R_L'$ or that the AC load seen by either **one** transistor appearing across **half** the primary winding of the transformer is equal to

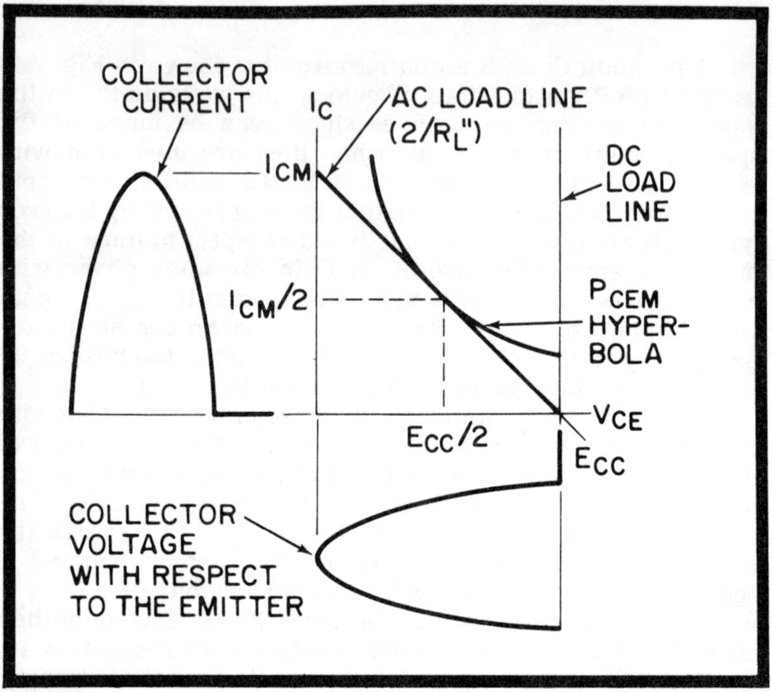

Fig. 7-14. Plot of load lines for Class B amplifier. A sinusoidal input and output signal is assumed.

one-fourth that seen by **both** transistors across the **entire** transformer.

Each transistor delivers power to the load. To determine the power delivered by the two transistors, we need only determine the power one transistor delivers and multiply this number by two. Using one transistor, draw the P_{CEM} maximum power dissipation hyperbola on the collector characteristic curves of the transistor, as shown in Fig. 7-14. As is our practice, we will omit the actual transistor curves to avoid cluttering the drawing. Assuming half the primary winding of T2 has zero resistance, the DC load line, defined by the equation V_{CE} equals $E_{CC} - I_C R_P$, is a vertical line up from the E_{CC} voltage point on the V_{CE} axis. R_P in the equation is the resistance of one-half of the primary winding of transformer T2.

The AC load resistance is R_L'' equals (R_L divided by 4) (N_P divided by N_S)^2. Assume that at its maximum the load line is tangent to the P_{CEM} curve. (We will deviate from the concept of not working above the P_{CEM} curve later on. Here, we assume the load line cannot cross this hyperbola.) If a sinusoidal signal is at the input, a half cycle of voltage and current will appear at the output due to one transistor. For the maximum output, the collector-to-emitter voltage will swing from E_{CC} to zero and the collector current will swing from its I_{CM} maximum to zero. Since the RMS values of a half sine wave are the peak voltage or current divided by two, the signal power delivered to the AC load, R_L'', is:

$$P_{R_L}'' = \left(\frac{I_{CM}}{2}\right)\left(\frac{E_{CC}}{2}\right) = \frac{E_{CC}^2}{4R_L''} = \frac{I_{CM}^2 R_L''}{4} \qquad (7\text{-}8)$$

This equation is valid regardless of where the AC load line is with respect to the P_{CEM} hyperbola. The maximum power is dissipated by the transistor at the instant the swing is at the center of travel or:

$$P_C\,(\text{max}) = \left(\frac{I_{CM}}{2}\right)\left(\frac{E_{CC}}{2}\right) \qquad (7\text{-}9)$$

This equation is also independent of the P_{CEM} curve.

It should be emphasized that $P_C(\text{max})$ is not the average power the transistor will dissipate. It is the power the transistor will dissipate only at the instant the swing is at I_{CM} divided by 2 and E_{CC} divided by 2. This is the maximum power it will dissipate at any point in the swing. If the load line is tangent to the maximum allowable power dissipation hyperbola, $P_C(\text{max})$ equals P_{CEM}. The power dissipated when averaged over the entire cycle is less than $P_C(\text{max})$ for it is $P_C(\text{max})$ for only two instants in the half cycle swing on the load line.

Comparing Equations 7-8 and 7-9, we notice that the transistor can deliver as much power as the maximum instantaneous power it will dissipate in the half cycle. This is double the power the same transistor was capable of delivering in Class A.

The average or DC current in a half sine wave of collector current is I_{CM} divided by π. This is discussed in Chapter 4. The power supply must provide this DC current. It must provide $E_{CC}(I_{CM}$ divided by π) watts to the transistor. Comparing this with the power delivered to the load, the percent efficiency of the circuit is $100P_{RL}''$ divided by P(supply) equals 100 π divided by 4 equals 78.5 percent—a decided improvement of the Class A case.

The transistor dissipates power during one-half the cycle only, for it does not conduct during the alternate half cycle.

Over a complete cycle, one transistor of the pushpull pair will dissipate:

$$P_{trans}(\text{for 1 cycle}) = 0.068 E_{CC}^2/R_L'' \text{ watts} \tag{7-10}$$

if the signal swings the output from zero to its E_{CC} maximum limit. The average power the device will dissipate over the cycle is higher if the swing is less than the maximum. Should the transistor deliver about 40 percent of the P_{RL}'' in Equation 7-8, it will dissipate more power than it will dissipate with any other degree of signal swing. The power it will dissipate is:

$$P_M(\text{av}) = V_{CC}^2/\pi^2 R_L'' \text{ watts} \tag{7-11}$$

if the power it delivers to the load is 40 percent of the maximum. Comparing this with Equation 7-7, the transistor can deliver about 2½ times the power it may dissipate.

Averaged over a complete cycle, the transistor may dissipate P_{CEM} watts. It will cross the P_{CEM} hyperbola and yet be within the power dissipation rating of the transistor when the dissipation is averaged over the 360 degrees. This differs from the Class A case, because here the idling and average power over the cycle were at P_{CEM} when the transistor was biased for minimum distortion. Since the power during idling (and hence the average during the cycle) was not permitted to exceed P_{CEM}, the load line was not permitted above the maximum power dissipation hyperbola.

Class B Design Procedure

Assume you are called upon to design a 60-watt pushpull amplifier which will drive an 8-ohm loudspeaker system. How would you proceed to specify the output transistors and transformer? Use the circuit in Fig. 7-13.

If the two transistors are to deliver 60 watts, each one must be capable of delivering half the power or 30 watts to the load. If the transformer is 25 percent efficient, the transistors must deliver 30 watts + 25 percent of 30 watts or 37.5 watts to half the output transformer primary. Add about 10 percent to compensate for losses due to saturation voltage and leakage current. Therefore, the circuit should be designed to be capable of providing about 42 watts.

A good power transistor for this application is the 2N3055. The maximum collector current that can safely flow through this transistor is 10 amperes. Adding some leeway, use 6 amperes as the maximum collector current. Since the RMS current of the half sine wave (see Fig. 7-15) is 6 divided by 2 or 3 amperes, the load the transistor must see at the primary of the transformer is R_L'' equals P_{RL}'' divided by I_{RMS}^2 equals 42 divided by 9 or 4.66 ohms. The supply voltage, from Equation 7-8, is E_{CC} equals $(4R_L''P_{RL}'')^{1/2}$ equals (4 x 4.66 x 42)$^{1/2}$ or 28 volts.

The maximum power is dissipated by the transistor when the power delivered to the load is about 40 percent of 42 watts, or 16.8 watts. When delivering this power, the transistor will dissipate V_{CC}^2 divided by π $2R_L''$ watts or 783 divided by (9.9)4.66 equals 17 watts. The 2N3055 can easily cope with this power dissipation requirement. It is interesting to stop for a moment and observe several things here. For the full swing, the transistor will dissipate $0.068V_{CC}^2$ divided by R_L'' equals (0.068) (783) divided by 4.66 or 11.4 watts. This is less than the power dissipated when 40 percent of the maximum power is delivered to the load.

The power a transistor will dissipate is equal to the power taken from the supply less the power delivered to the load. The power furnished by the supply can be calculated as follows. Based on Equation 7-6, and defining I_C(40 percent) and P_{RL}''(40 percent) as the collector current and output power, respectively, when the transistor delivers 40 percent below its

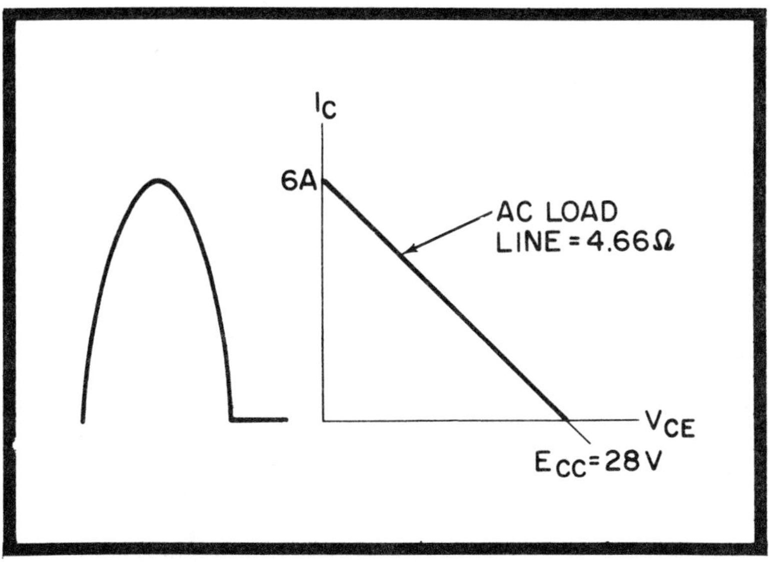

Fig. 7-15. Curves for Class B design procedure.

maximum capabilities, the square of the collector current swing is I_C(40 percent)^2 equals $4P_{RL}$" (40 percent) divided by R " for the case where the output is 40 percent of maximum. Hence, I_C (40 percent)^2 equals 4(16.8 watts) divided by 4.66 ohms equals 14.4 amperes^2. The current is then I_C(40 percent) equals $\sqrt{14.4}$ equals 3.79 amperes. The power from the supply is E_{CC} I_C(40 percent) divided by π) equals 28(3.79 divided by 3.14) or 33.9 watts. Subtract the 16.8 watts delivered to the output at 40 percent of the maximum output power from the 33.9 watts supplied by the power source, and the transistor must dissipate 17.1 watts. This is very close to the 17 watts solution determined from the V_{CC}^2 divided by π $2R_L$" equation above. This alternate method indicates a logical procedure used to determine the transistor power dissipation for any portion of the maximum power the device can deliver.

Since the impedance across one-half the primary of the output transformer is 4.66 ohms, the impedance from collector to collector (across the entire primary) is 4 x 4.66 ohms, or 18.64 ohms. The impedance ratio of the entire primary to the secondary is 18.64 ohms :8 ohms or 2.33:1. The turns ratio is the square root of the impedance ratio, or 1.51:1.

Class AB

A typical set of collector characteristic curves appear in Fig. 7-3 for the 2N3055. Notice that they are not evenly spaced. Should the collector current swing the full 10 amperes for the half cycle, the current gain differs at various points of the cycle. For example, at the point in the cycle where 2 amperes collector current is required, the base current at the input is about 40 ma. At the 4 ampere collector current point in the cycle, the base current is about 120 ma, while for 6 amperes it is about 250 ma. In the portion of the output curve where there is a 2 ampere collector current rise from 2 to 4 amperes, the equivalent base current increases about 120 - 40 ma or 80 ma, while for the same 2 ampere collector current increment from 4 to 6 amperes, the equivalent base current increase is 250 - 120 ma or 130 ma. Even though the increase in collector current remains at 2 amperes in both cases, the base current drive required is greater at higher collector currents. This information could, of course, have been derived from the beta curves in Fig. 7-12 which show that beta varies considerably with collector current. The AC and DC current gain is nonlinear, resulting in distortion.

Another curve, shown in Fig. 7-16, is a plot of the collector current against the base-emitter voltage. You may recognize this as a transconductance curve, where the definition of transconductance is not unlike that applied to the JFET. Notice that this curve, after the first 0.6 or 0.7 volt is a relatively straight line. The slope of this line is the AC transconductance. It is equal to ΔI_C divided by ΔV_{BE}, where the Δ indicates a change or difference. The AC transconductance is relatively constant with changes in collector current. Should the half cycle of 10 ampere collector current be at the output, the required input base-emitter voltage at the 2 ampere collector current point in the cycle is 0.9 volt, at the 4 ampere point it is 1.1 volts, and at 6 amperes it is 1.3 volts. While the collector current increase is in increments of 2 amperes, the base-to-emitter voltage increase in one case is 1.1 volts — 0.9 volt or 0.2 volt, and in the second case it is 1.3 volts — 1.1 volts or 0.2 volt. Hence, equal changes in base-emitter voltages produce equal changes in collector current. This is a linear situation favorable to the cause of low distortion.

Good power amplifiers are driven from low-impedance voltage sources where transconductance is the controlling

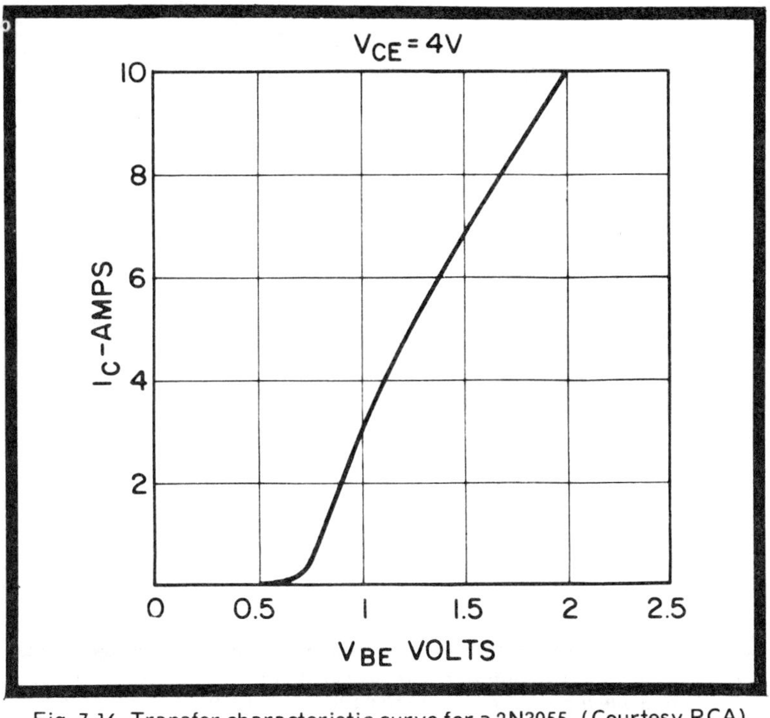

Fig. 7-16. Transfer characteristic curve for a 2N3055. (Courtesy RCA)

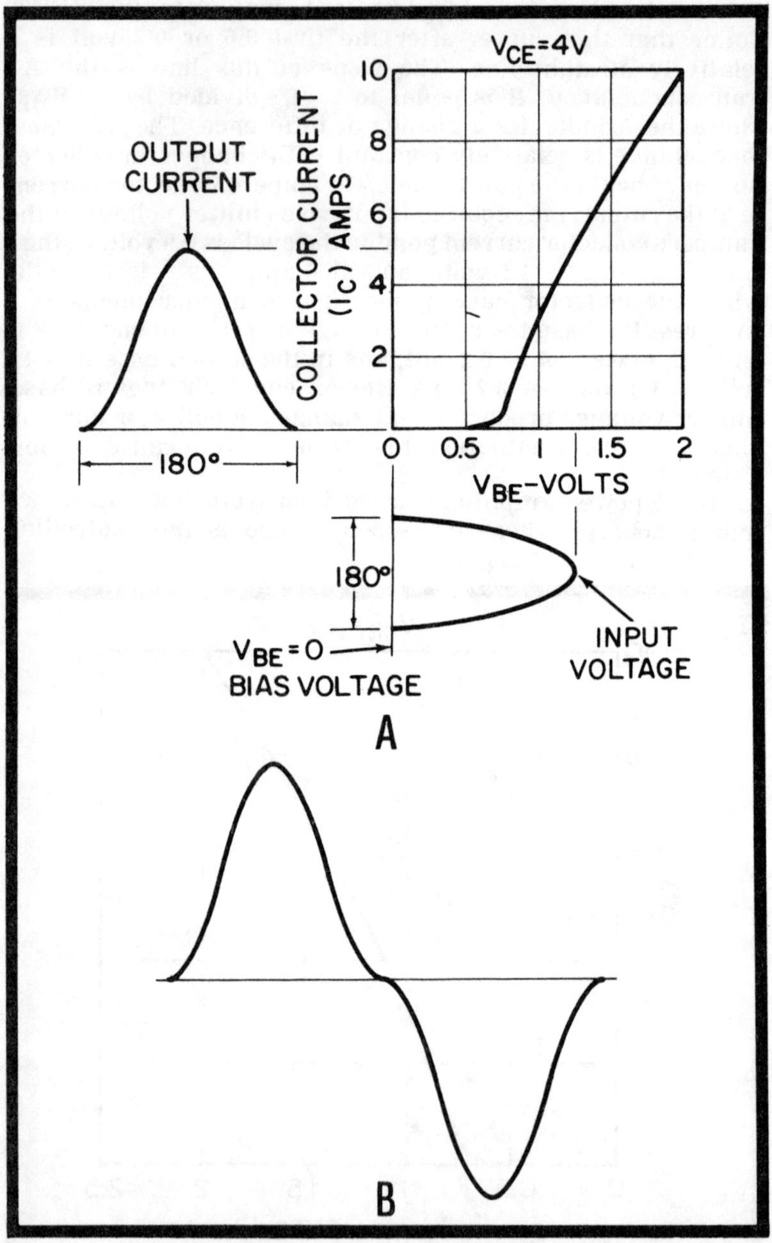

Fig. 7-17. Derivation of output curve. Curve A is the collector current with half a sine-wave input. Curve B is the collector current as it appears for a full cycle. Notice the crossover distortion. (Courtesy RCA)

factor, rather than high-impedance current sources where the varying beta determines the relatively distorted output. In Fig. 7-17A, we apply half a sine wave of voltage to the input between the base and emitter. The collector current at the output appears next to the I_C axis. Notice that there are portions in the cycle where there is little or no output current. This is known as the crossover region. For the 2-transistor push-pull circuit in Fig. 7-13, the output across the load, R_L, would appear as in Fig. 7-17B, rather than be perfectly sinusoidal. This signal has a considerable amount of odd harmonic distortion and intermodulation distortion. Even worse, in conjunction with transformer and speaker inductance, this abrupt crossover can cause sharp peaks in the circuit which may damage or destroy the transistor.

Crossover distortion can be minimized if the transistor is biased so that it is always conducting some minimal amount of

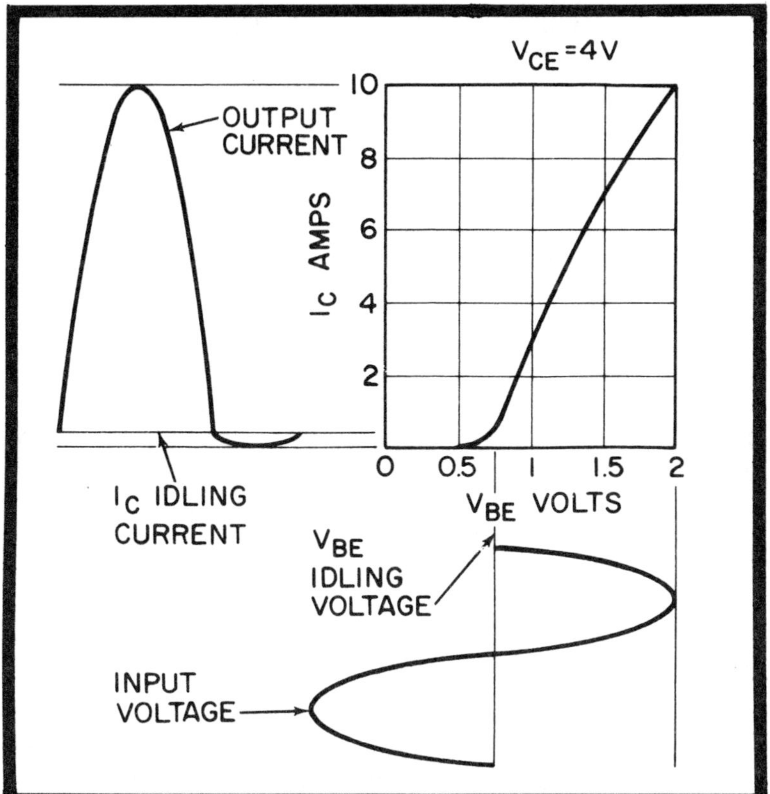

Fig. 7-18. Collector current curves when the transistor is biased for Class AB operation. (Courtesy RCA)

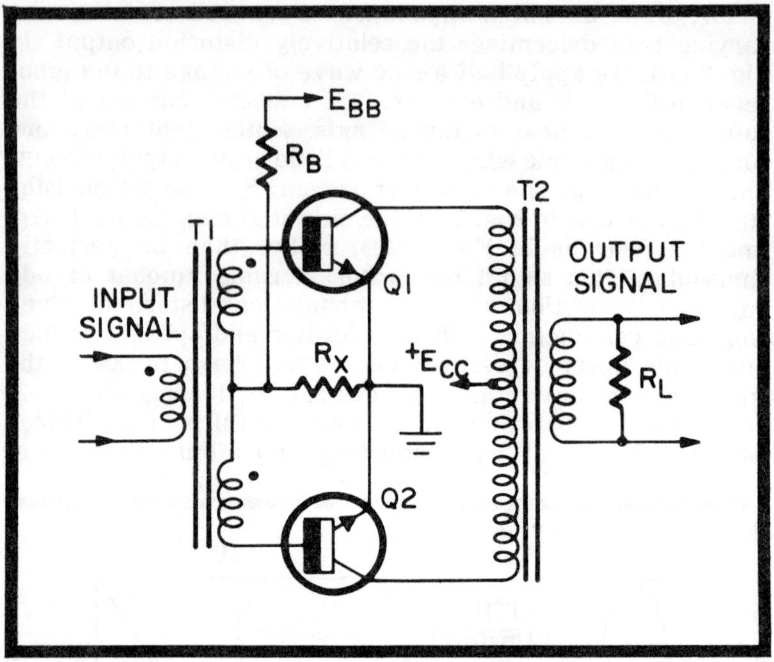

Fig. 7-19. The circuit in Fig. 7-13 biased for Class AB operation.

collector current. A normal procedure to determine the minimum base-emitter voltage for this type of operation is to extend the straight line of the curve to the V_{BE} axis. The curve in Fig. 7-16 crosses this axis at 0.75 volts, so the transistor should be biased at this voltage. It may be true that some power will be dissipated due to the idling collector current (about 40 ma at 0.75 volt), but the reduction in crossover distortion is well worth this minor expenditure of power. Some manufacturers allow the transistors to idle at much higher current to assure that the collector current will never be completely cut off. This type of biasing puts the transistor into what is referred to as Class AB operation.

A drawing of the output current when the transistor is biased at 0.75 volt is shown in Fig. 7-18. There is no portion of the half cycle during which there is an absence of collector current flow. The distortion is relatively low when compared to the Class B mode of operation. The output over the full cycle is a fairly good sine wave.

Should the circuit in Fig. 7-13 be biased for Class AB operation, it could take the form shown in Fig. 7-19. The bias voltage is developed across R_X, which is applied between the transistor bases and emitters. R_B and R_X form a voltage divider with E_{BB} as the source of the base bias supply voltage. The quiescent base current can be determined using Thevenin's equivalent circuit procedure described in Chapter 2. Do not forget to include the DC resistance of the driver transformer's secondary winding in series with the base circuit.

The design procedure in all other facets is identical to that described for Class B operation. A 10 percent factor was added there for contingencies such as saturation voltage and leakage current. Perhaps in Class AB, this factor should be increased to 15 or 20 percent.

Only one type of push-pull circuit using transformers is described here. It was used only for convenience. Modern designs obviate the need for this type of component at the input or at the output. The design considerations do not change in this type of circuit. Transformerless output circuits are described in detail in Chapter 11.

Chapter 8 Boundaries of Operation

Many transistor characteristics and the methods of utilizing these characteristics have been discussed in preceding chapters. Here, the discussion centers on information concerning the maximum ratings that define the various limits of the device.

BIPOLAR TRANSISTORS

In Chapter 1, we learned that if a diode is reverse biased (the anode negative with respect to the cathode), it will break down and start conducting large amounts of current if the applied voltage exceeds a specific level which varies from type to type. The same is true for transistor junctions.

The collector-base junction of a transistor is reverse biased in normal audio amplifier circuit applications. The junction will break down if the applied voltage is above an avalanche voltage referred to as $V_{(BR)CBO}$ equals BV_{CBO}. (Here and in the discussion below, the first symbol shown is the preferred JEDEC notation for a limiting value. Subsequent symbols are frequently used in the literature to indicate the same characteristic.) No voltage applied to any pair of terminals or leads or a transistor should exceed the $V_{(BR)CBO}$ rating.

A tighter limit imposed on the transistor is the breakdown voltage between the collector and emitter. The voltage is at its maximum when a reverse bias is applied between the base and emitter terminals. Here, the symbols for the breakdown voltage are $V_{(BR)CEV}$ equals BV_{CEV}. Starting with $V_{(BR)CEV}$, the highest collector-to-emitter breakdown voltage, other collector-to-emitter breakdown voltages in descending order of magnitude are:

$V_{(BR)CES}$ equals BV_{CES}, when the base-emitter junction is short circuited.

$V_{(BR)CER}$ equals BV_{CER}, when a resistor, R, is connected between the base and emitter.

$V_{(BR)CEO}$ equals BV_{CEO}, when the base-emitter junction is open circuited.

It is obvious that the smaller the resistor placed across the base-emitter junction, the higher the breakdown voltage will be. The voltage may even be specified for the presence of a particular circuit that may be placed between the base and emitter. In this case, the symbol is $V_{(BR)CEX}$ equals BV_{CEX}.

The various breakdown voltages occur at low collector currents. At high current, the breakdown voltage is reduced and is referred to as a sustaining voltage. It is simply noted by placing the symbol "(sus)" after the particular breakdown voltage notation. Sustaining voltage should be specified at the specific current used in a circuit. A hypothetical collector characteristic curve showing $V_{(BR)CEO}$ and $V_{(BR)CEO(sus)}$ is in Fig. 8-1.

The base-emitter junction forms a diode subject to breakdown when reverse biased. This voltage, $V_{(BR)EBO}$ equals BV_{EBO}, should not be exceeded at any time. Special precautions must be taken when operating in a pushpull Class AB or Class B mode, because a reverse signal is applied to this junction during a significant portion of the cycle.

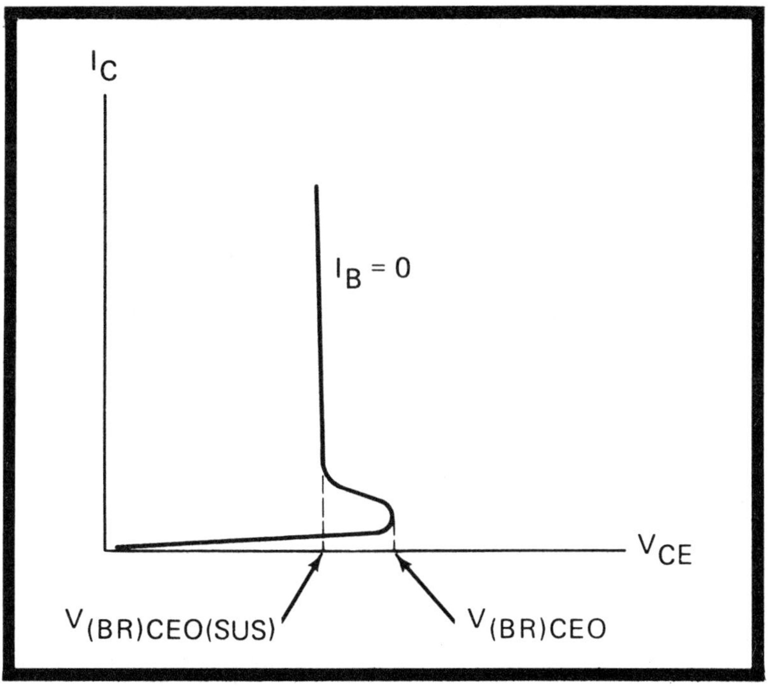

Fig. 8-1. Collector characteristic curve showing breakdown and sustaining voltage for a condition when the base is left open.

Other maximum ratings which must not be exceeded are the collector and base current limits, power dissipation at specific operating temperatures, and the specific maximum temperatures when storing, operating and soldering a device into the circuit.

JUNCTION FIELD-EFFECT TRANSISTOR

Similar to bipolar devices, the JFETs are encumbered by limiting factors. The obvious limiting voltages are:

$V_{(BR)GSS}$ equals BV_{GSS}, the gate-to-source breakdown voltage with the drain connected to the source.

$V_{(BR)DGS}$ equals BV_{DGS}, the drain-to-gate voltage with the source connected to the drain.

$V_{(BR)DSS}$ equals BV_{DSS}, the drain-to-source voltage with the gate connected to the source.

$V_{(BR)DSX}$ equals BV_{DSX}, the drain-to-source voltage with the gate at a specified voltage with respect to the source.

The maximum allowable drain power dissipation, drain current, and the various temperatures are normally stated on data sheets and should, of course, not be exceeded in any design.

SECOND BREAKDOWN

JFETs and small-signal bipolar transistors should not be forced to operate outside the limits set by the maximum power dissipation hyperbola discussed in Chapter 7. More stringent restrictions are placed on medium- and high-power bipolar devices by a phenomenon known as second breakdown.

A transistor is subject to second breakdown when there is a substantial amount of collector current while the collector-emitter voltage, although within all ratings, is relatively high. This phenomenon may occur even if the transistor is operating well within its maximum power dissipation rating. At one specific combination of I_C and V_{CE}, the collector-emitter voltage will suddenly drop and the collector current will rise rapidly. A curve illustrating this is shown in Fig. 8-2, where at $V_{(BR)(sus)}$ and I_{SEC}, the transistor goes into a second breakdown state. The voltage across the device in this state is V_{SEC}.

Second breakdown is usually explained by stating that large amounts of current flow through a tiny area of semiconductor material cause local overheating and breakdown. Once a device is in second breakdown, the transistor is usually destroyed. Several precautions should be taken to minimize the chances of second breakdown:

1. Keep the collector-emitter voltage to the minimum consistent with the design requirements.

2. Use single diffused rather than double or triple diffused devices, if this limitation is at all compatible with the circuit.

3. Check the frequency requirements of the circuit carefully. Use the transistor with the lowest high-frequency cutoff specification that will satisfy a design. The frequency characteristics of a transistor are discussed in the next two chapters.

4. Watch out for inductive loads! When a transistor is reverse biased during a portion of a cycle, it will theoretically not conduct current. However, when the voltage across the transistor is high, an inductive load will discharge current through the transistor. Current will flow through a small area of the semiconductor while V_{CE} is large—the perfect situation

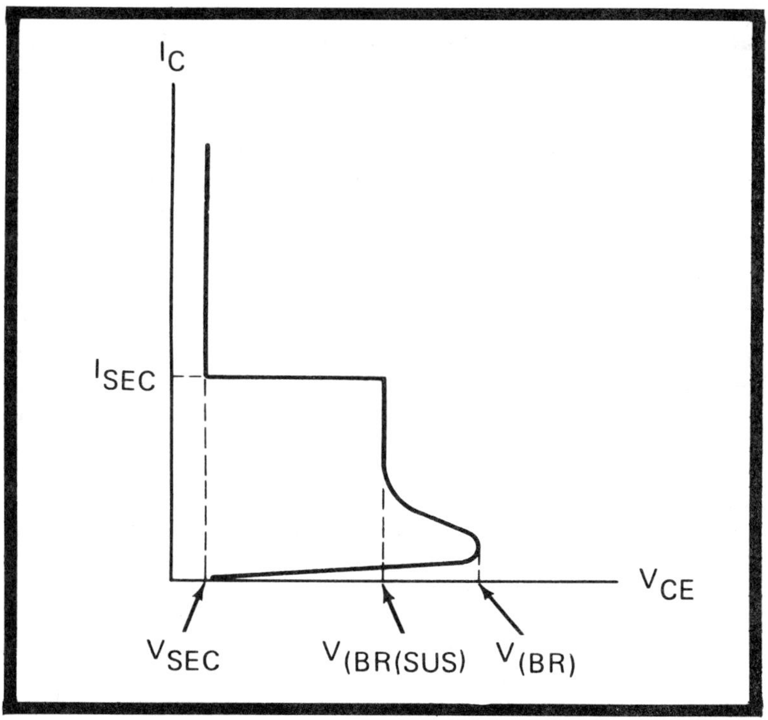

Fig. 8-2. Curve showing collector characteristics during a second breakdown.

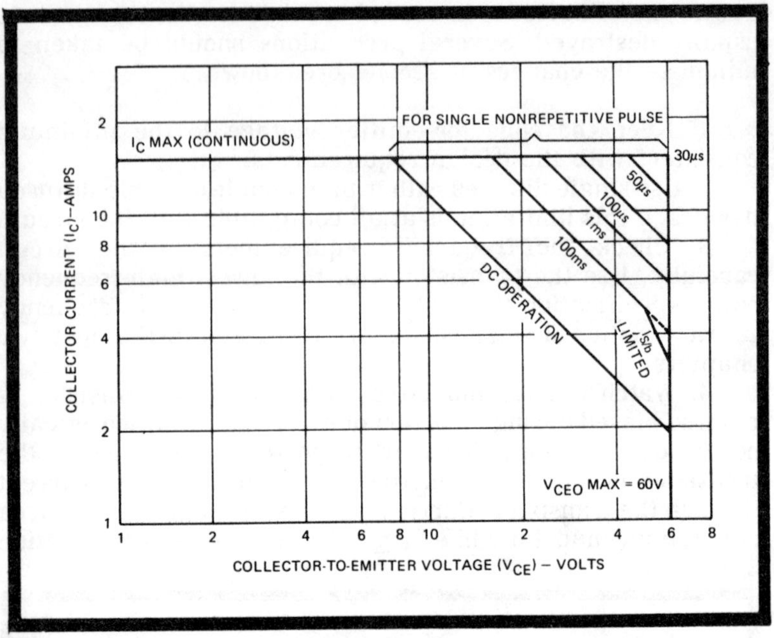

Fig. 8-3. Maximum safe operating conditions for a 2N3055. (Courtesy RCA)

to initiate second breakdown. In this case, the proper precaution is to keep the reverse base-emitter voltage as small as practical and make sure the resistor in the circuit is comparatively high in value.

Once transistors have been designed into an audio output circuit, they can be checked against safe operating area curves to ascertain if they are operating within the boundaries necessary to avoid second breakdown. These safe operating curves are provided by the manufacturer of the device. Curves of this type for the 2N3055 transistor are shown in Fig. 8-3.

A circuit should be established and tests performed to ascertain if the load line for a transistor is within the limits set by the curves. Thus, if there is a 100-millisecond pulse through the transistor, the load line for the transistor should be below the curve marked 100 ms. Similarly, if there is a 100-microsecond pulse in the collector circuit, the load line for the transistor should fall below the curve marked 100us. As the pulses get longer, the limiting curves get more severe. For intermediate size pulses not shown on the curves, the limits can be determined by interpolation. It is obvious that for a pulse of any duration, a safe load line will fall below the curve marked "DC operation." However, this severe load-line limit is not required should shorter pulses be reproduced.

The question may arise as to how the load line may vary with the duration time of the pulse. Previously we drew the DC and AC load lines and disregarded their dependence on frequency. It should be obvious that an inductive load line, due to a transformer or speaker load, is frequency sensitive and will vary in shape and size with the pulse width.

To test a circuit, sine waves rather than pulses may be used. Here, the duration time indicated on the curve can be considered safe if the load falls below the limit curve of ½f seconds, where f is the frequency of the sine wave.

The circuit may also be tested by using the actual haphazard signal, such as music or speech, normally present at the output of an audio amplifier. In this case, the continuously varying load lines should fall below the DC operation curve in Fig. 8-3. This is a severe test and a better indicator of transistor safety in operation than a test performed with any other type of signal.

To perform the test, the transistor in the output circuit can be connected to a calibrated scope as shown in Fig. 8-4. The collector current flows through R_E. The voltage across R_E is then proportional to I_C, the collector current. This voltage is plotted on the vertical axis of the scope in terms of collector current. At the same time, the collector-emitter voltage is applied to the horizontal input of the calibrated scope. The V_{CE}-I_C plot on the scope is the load line of the transistor. Vary the frequency fed to the amplifier or change the type of load, and the load line plot on the scope will be altered. To attain a safe operating condition, no point on the load line is to be above the limits set by the curves in Fig. 8-3 at any frequency of operation for any load and at any power output level. Identical test arrangements can be made for transistors in pushpull and output-transformerless circuits; I_C and V_{CE} are measured across one output device at a time.

THERMAL CIRCUITS AND LIMITS

The primary source of heat in a bipolar transistor is the power dissipated at the collector junction. The contribution of heat generated at the base-emitter junction to the overall temperature of the device is relatively negligible. The maximum operating junction temperature for silicon transistors is usually specified at 200 degrees C, while for germanium devices it is limited to 100 degrees C.

The heat within a transistor must be removed from the junction as it is being generated if the transistor is not to

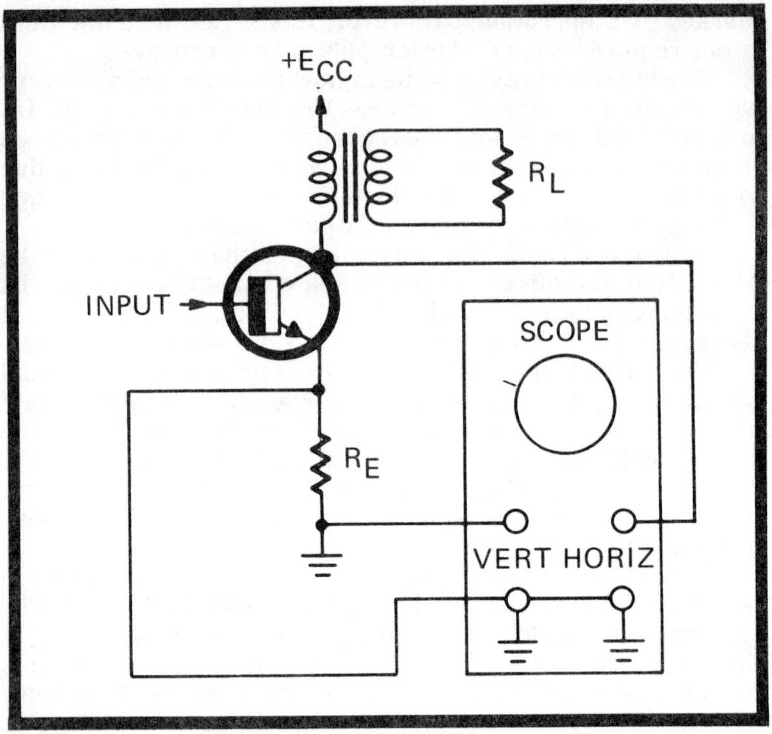

Fig. 8-4. Setup to check if a transistor is operating within the limits set by the safe operating curves.

become too hot and break down. There is a thermal resistance, θ (units are degrees Celsius per watt), which resists the flow of heat from the junction. The thermal resistance between the junction and the case of the transistor is θ_{JC}. The case is cooled by the surrounding air which is normally at a lower temperature than the case. There is a thermal resistance between the case and the air, assigned the symbol θ_{CA}.

Because the surface area of the case is small, very little air can flow over it within a specified period of time. Therefore, the case is usually mounted on a large metal surface or sink. It has a large area exposed to the ambient temperature. However, there is a thermal resistance between the case and the sink, θ_{CS}, and between the heat sink and the air, θ_{SA}. It is obvious that θ_{CA} equals $\theta_{CS} + \theta_{SA}$, but it should be noted that θ_{CA} is smaller when a heat sink is present that when only the case is exposed to the air. The thermal resistance between the case and sink, θ_{CS}, can be relatively sizable, but may be reduced if silicone grease is used between the case and sink to assure good contact between the surfaces.

Electronic circuits frequently require that the case (usually connected to the collector of the transistor) be insulated from the grounded heat sink. A mica insulator is usually placed between the case and heat sink to fulfill this function. The thermal resistance between the case and heat sink is increased when the mica is used, but this is unavoidable. Once again, the situation can be alleviated by applying silicone grease to both sides of the washer. Typically, θ_{CS} when a plain mica washer is used is about 0.8 degrees C per W, while the addition of silicone grease will reduce θ_{CS} to approximately 0.4 degrees C per W. Compare this with the thermal resistance when no washer is used. Here, θ_{CS} can be 0.2 degrees C per W without the silicone grease and 0.1 degree C per W with the use of the grease between the contacting surfaces.

Thermal resistance can be put into a thermal equivalent circuit and treated as an ordinary resistor in an electric circuit. The equivalent to the electric current is the power, P_{diss}, dissipated by the semiconductor. P_{diss} flows through the thermal circuit as current flows through an electric circuit.

When current flows through an electric circuit, a voltage is developed across a resistor in the circuit. Heat, measured in C degrees is developed across a thermal resistor when power flows through the thermal circuit. We can thus write an equation relating thermal resistance, power and temperature:

$$P_{diss} = \frac{T_J - T_A}{\theta_{JA}} = \frac{T_J - T_A}{\theta_{JC} + \theta_{CS} + \theta_{SA}} \qquad (8\text{-}1)$$

where T_J is the temperature at the junction and T_A is the ambient temperature. This is an important equation to remember. A schematic diagram presenting the information described by the equation is shown in Fig. 8-5. T_{JA}, T_{CA} and T_{SA} are temperatures from the junction, case and heat sink with respect to the ambient. Temperatures are developed across thermal resistors shown in the drawing. The temperatures represent the heat rise above the ambient, which is treated as if it were the circuit ground.

Using Equation 8-1, the temperature at the heat sink is T_S equals $\theta_{SA}P_{diss} + T_A$. Similarly, if the case temperature is $T_{CA} + T_A$, the junction temperature is $T_{JA} + T_A$ equals $T_{CA} + T_A + \theta_{JC}P_{diss}$. It is evident that the junction temperature is directly related to the case temperature. It is further

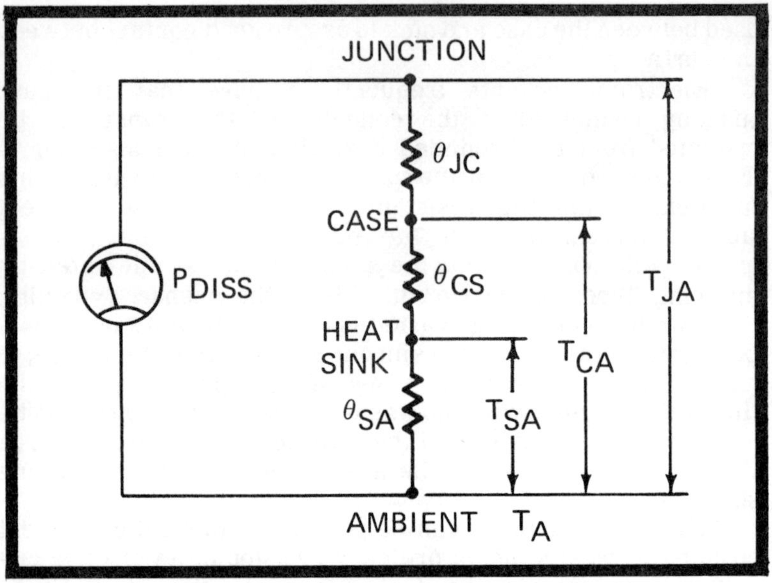

Fig. 8-5. Thermal equivalent circuit.

evident that if the junction temperature is limited to some maximum such as 200 degrees C, the maximum allowable case temperature will decrease as the power dissipated by the semiconductor rises. For example, if θ_{JC} is 1.5 degrees C per W, as it is for the 2N3055, and 50 watts is dissipated by the transistor, the maximum allowable case temperature is 200 degrees C - (1.5 degrees C per W) (50 watts) or 125 degrees C.

A curve can be drawn relating the maximum allowable power dissipation by the device to the temperature of the case. Since the maximum rated dissipation of the transistor is 115 watts, the curve for the 2N3055 is as shown in Fig. 7-2. Notice that at 125 degrees C the maximum dissipation as noted on the curve is 50 watts, substantiating our calculation from the data.

If the transistor temperature rises faster than its ability to dispose of the excess heat, the device will obviously overheat. The elevated temperature will force the current flow to increase. This is due primarily to I_{CBO}, the collector-base leakage current. In turn more heat is produced. The cycle will repeat itself, building up additional heat until the transistor is destroyed. This is known as thermal runaway. To avoid thermal runaway, the following inequality must be satisfied.

$$\theta_{JA} < \frac{T_{JA}}{[I_C + SI_{CBO}'] [E_{CC} - I_C (R_C + R_E)]} \qquad (8\text{-}2)$$

where:

T_{JA} is the junction temperature less the ambient temperature.

I_C is the idling collector current.

S is the stability factor discussed in Chapter 2 and equal to ΔI_C divided by ΔI_{CBO}.

I_{CBO}' is the maximum leakage current (collector-to-base with the emitter open) for the maximum temperature at which the semiconductor device can be used. As you recall, for conservative estimates, it doubles for every 6-degree C rise in the junction temperature of silicon devices and for every 10-degree C rise in the junction temperature of germanium transistors.

E_{CC} is the collector supply voltage.

R_C is the DC resistance in the collector circuit.

R_E is the DC resistance in the emitter circuit.

Heat sinks used to help maintain safe temperatures are rated by their thermal resistance. They are best used vertically with all surfaces exposed to the air in a type of chimney effect. Once θ_{SA} has been calculated, the size of a nonstandard heat sink may be estimated from Fig. 8-6.

THERMAL FATIGUE

It is well known that if a piece of sheet metal were bent back and forth along one line, it will eventually break. Some types of metal can be flexed more often than other types before cracking. So it is with transistors. They can be brought up to a specific high temperature and then cooled. This can be repeated only a specific number of times before the transistor will break down. The graph drawn in Fig. 8-7 shows the number of these heating and cooling cycles that a 2N6099 transistor can endure before breaking down. This **thermal-cycling rating** is related to both the power dissipated by the transistor and to the change of the case temperature during each thermal cycle.

At an average, audio amplifiers can be assumed to go through 1000 of these thermal cycles each year. If you expect the equipment to last five years (another average figure), the transistors must last for at least 5000 cycles. Equipment designed and built by the hobbyist can be made to last through many more cycles, if desired.

In the practical situation, the 2N6099 may be required to dissipate 20 watts. If the minimum number of thermal cycles should be 5000, the case temperature should not change by more than 50 degrees C. This can be determined from the

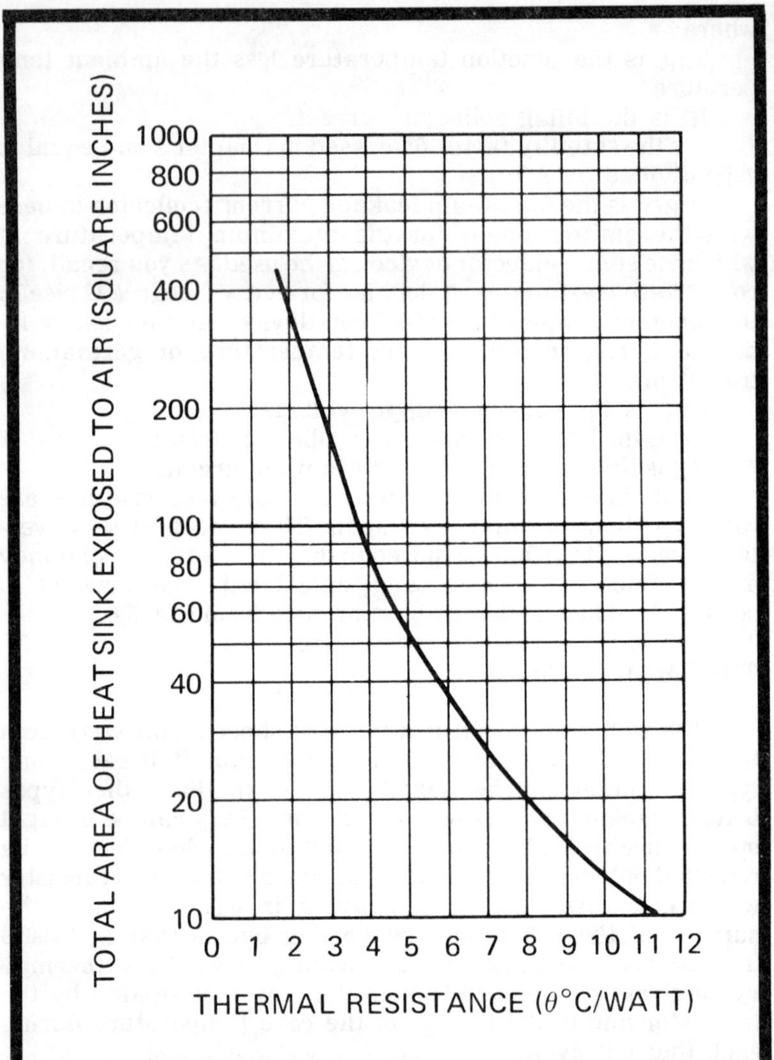

Fig. 8-6. Approximate thermal resistance of a heat sink made of ⅛-inch thick bright aluminum.

curves in Fig. 8-7. Used at an ambient temperature of 20 degrees C, the design should be such as to limit the case temperature to 70 degrees C when power is applied to the amplifier.

A Class AB amplifier dissipates different amounts of power when operating at different volume levels. As an example, assume that one transistor of a pushpull pair dissipates 4 watts when operating at low volume levels and 20 watts when producing loud music. Furthermore, assume that it is anticipated that the transistor in the amplifier will be dissipating 4 watts for 10,000 cycles and that the case temperature will increase 30 degrees C. At the 20-watt level, the case temperature may change by 80 degrees C, but it is required to go through only 2000 cycles. The problem is to determine if the device is being used within its thermal-cycling rating.

At the 4-watt level, the number of cycles permitted with a temperature change of 30 degrees C is 17,500, as determined from the curves. Since only 10,000 cycles are required, 10,000 divided by 17,500 or 0.57 of the maximum permissible number of cycles are used here. Similarly, when operating at the 20-watt level with a temperature change of 80 degrees C, 3000 cycles are permitted. Only 2000 divided by 3000 or 0.67 of the permissible number of cycles for this temperature change are used here. The sum of 0.57 and 0.67 is 1.24. The transistor is used within its thermal-cycling rating if the sum of the two decimals is equal to or less than 1. Obviously, operation of the above circuit is beyond the rating of the device.

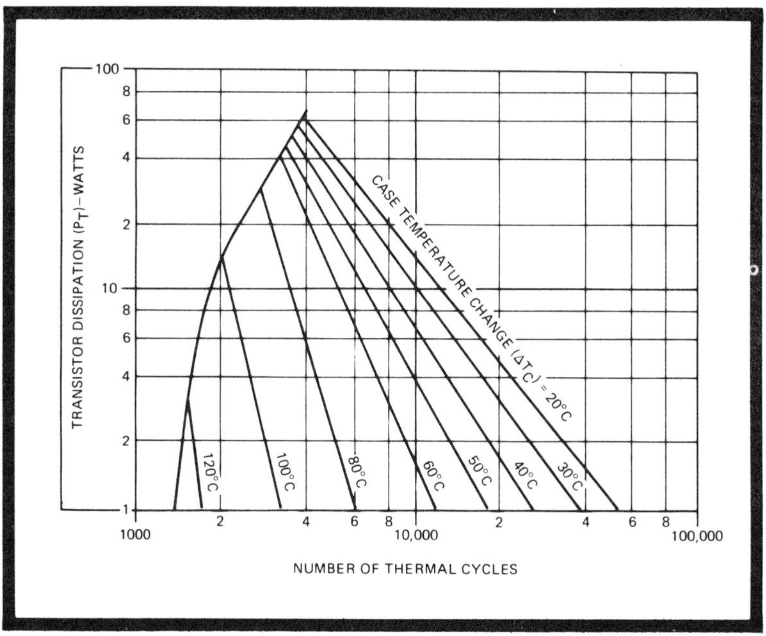

Fig. 8-7. Thermal-cycle curves for an RCA 2N6099 transistor. (Courtesy RCA)

If the number of cycles at the 4-watt level is fixed so that the decimal is 0.57, the number should not exceed 1 - 0.57 or 0.43 at the 20-watt level. Since 3,000 cycles are permitted when used exclusively at ΔT_C equals 80 degrees C, the number, N, of actual permitted cycles at the 20-watt level, can be calculated from N divided by 3000, is 0.43. Hence, N equals 1290 cycles.

Multistage Amplifiers

Chapter 9

The gain of an individual transistor circuit is limited primarily by device parameters, by the load it must feed, by the impedance of the input signal, and by the bias and bias stabilization components. Both JFET and bipolar devices are affected by all these factors, but to differing degrees. While the impedance of the input signal will probably have a severe effect on the gain of a circuit using a bipolar transistor as the input device, it will be relatively insignificant when facing the high input impedance of an FET.

Three arrangements are commonly used to couple several transistors into circuits designed to provide more gain than is possible with one device. All methods use the signal derived from one stage of amplification to drive a second stage of gain. The output from the second transistor may, in turn, drive a third device, and so on. The overall gain is the product of the gains of each individual amplifier circuit, remembering, of course, that the gain of each circuit is affected by the circuit and components preceding and following the one being analyzed.

Of the three types of coupling circuits, capacitive coupling is probably most frequently used. Here, a capacitor conducts the signal from the output of one stage to the input of the next, while all DC voltages are isolated within the individual stages. Although adequate in very many applications, capacitive coupling components discriminate against DC and the low frequencies of the signal being reproduced. There is also a frequency-dependent phase shift which many audio purists find objectionable.

The disadvantages are overcome by omitting the capacitor and using direct coupling. But circuits of this type are not free of drawbacks. Drain or collector current buildup due to instability in one stage is multiplied by the gain of succeeding stages. The current may build up to such proportions as to upset the quiescent conditions in the final stage of amplification. Furthermore, drift due to thermal causes can upset the balance of pushpull output stages, allowing DC current to flow through loudspeaker loads.

Transformer coupling, although frequently used due to convenience and circuit matching advantages, has many of the drawbacks exhibited by the capacitor coupling method. The characteristics of this arrangement are the first to be detailed here.

TRANSFORMER COUPLING

As you recall from Chapter 7, a transformer consists of two or more windings placed in a common magnetic field. The signal appearing in one winding is coupled to the second or whatever number of additional windings may exist. Transformers with a single tapped winding (autotransformers) are seldom used in audio applications because there is no DC isolation possible between coupled stages.

The impedance or resistance placed across the secondary winding is reflected into the primary winding. To determine the exact effect, multiply the impedance of the secondary load by the square of the turns ratio of the two windings. In Fig. 9-1, R_L is the load placed across the secondary winding of the transformer and the R_S is the DC resistance of that winding. With N_P turns in the primary winding and N_S in the secondary, the resistance reflected into the primary, R_L' due to the resistance in the secondary, is:

$$R_L' = \left(\frac{N_P}{N_S}\right)^2 (R_L + R_S) \tag{9-1}$$

A signal fed to the primary of the transformer sees R_L' in series with the DC resistance of the primary winding, R_P. R_P as well as R_S, the DC resistance of the secondary winding, are wasteful losses in the transformer. Good transformers are designed so that the resistances in the windings are negligible compared to the actual load placed across the winding. Consequently, R_S is made much smaller than R_L. In Equation 9-1, R_S is usually assumed to be zero.

A 25 percent loss factor is usually included in designs to compensate or account for all inefficiencies of the transformer. The primary factor affecting efficiency is the copper used for the windings. The percent copper efficiency can be readily determined from the equation:

$$\% \text{ Efficiency} = \frac{\text{Power output}}{\text{Power input}} = \frac{100}{1 + \left(\dfrac{R_P + n^2 R_S}{n^2 R_L}\right)} \% \tag{9-2}$$

where n is equal to N_P divided by N_S, the turns ratio of the transformer. Being directly proportional to the voltage ratio, the turns ratio can be determined experimentally by placing a known 400-Hz voltage, V_P, across the primary and reading the voltage, V_S, across the secondary winding with an accurate meter. The ratio V_P:V_S equals n.

The frequency range of a circuit may be limited by the presence of a transformer in that circuit. The circuit in Fig. 9-2A can be used to measure the frequency response of a transformer. A curve similar to that in Fig. 9-2B can be derived from the measurements. It shows the gain of a transformer circuit over a range of frequencies. The low-frequency rolloff is due to the incremental inductance of the primary winding of the transformer, while the high-frequency rolloff is due to the leakage inductance. The jagged characteristics of the curve at the high end are caused by various winding and stray capacitances.

The basic low-frequency equivalent of a circuit using a transformer is a resistor in series with an inductor, as shown in Fig. 9-3 . If the secondary of the transformer is open, R may simply be the sum of the primary winding resistance, R_P, and the collector or drain resistance of the transistor, usually r_d. Load the secondary with R_L and R_{in} becomes R_P + r in parallel with the value of the load resistor when reflected into the primary. Reflected into the primary winding, the resistance becomes R_L' equals n^2R_L. L in the circuit is not simply the inductance of the primary winding of the transformer, L_P. It is the incremental inductance, or the inductance of the primary winding when DC is present in the winding along with the audio signal.

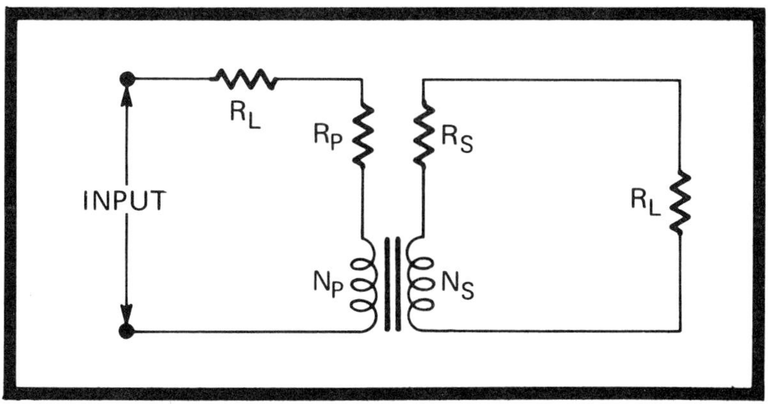

Fig. 9-1. Output load, R_L, is reflected into the primary winding of the transformer as R_L'.

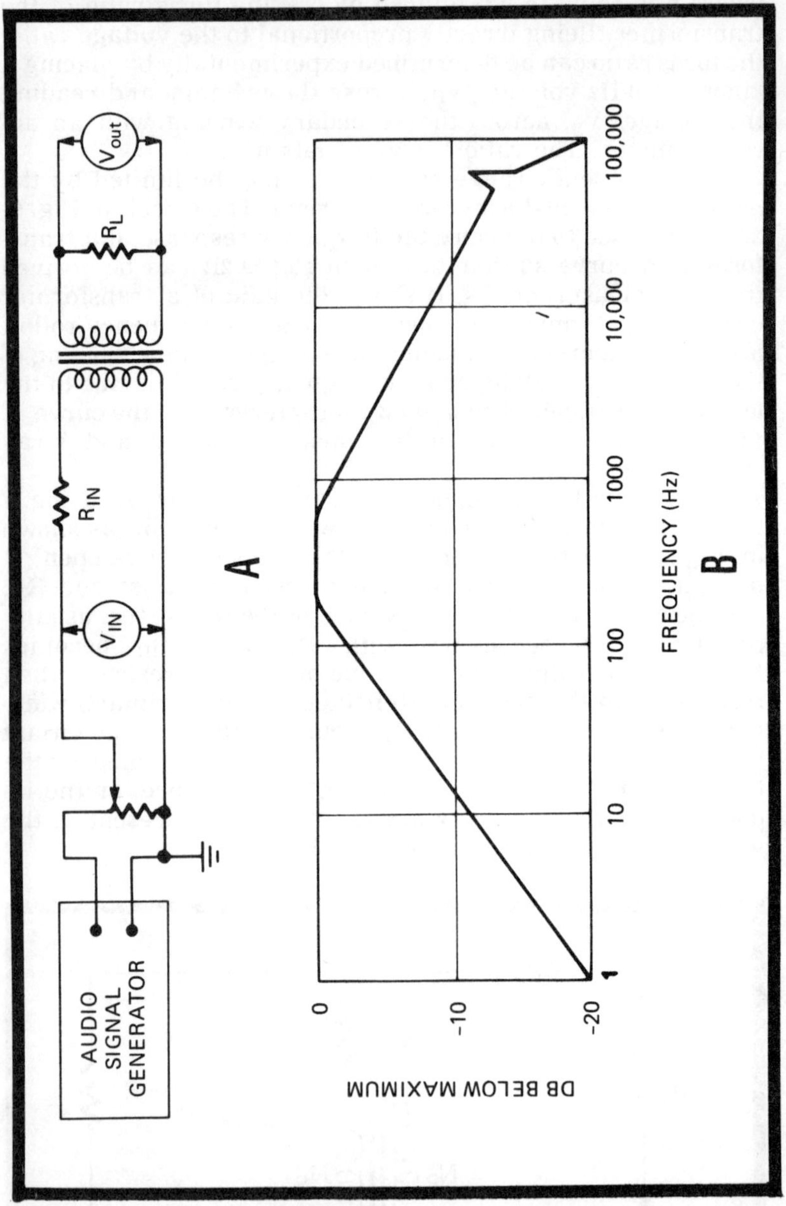

Fig. 9-2. Frequency characteristic of a transformer and the measuring procedure. The test setup (A) measures the frequency response of an output transformer. R_{IN} is the impedance or resistance of the circuit feeding the primary of the transformer. B shows a possible frequency response curve of a transformer. The ratio of the output to the input, expressed in db, is measured and plotted on the curve.

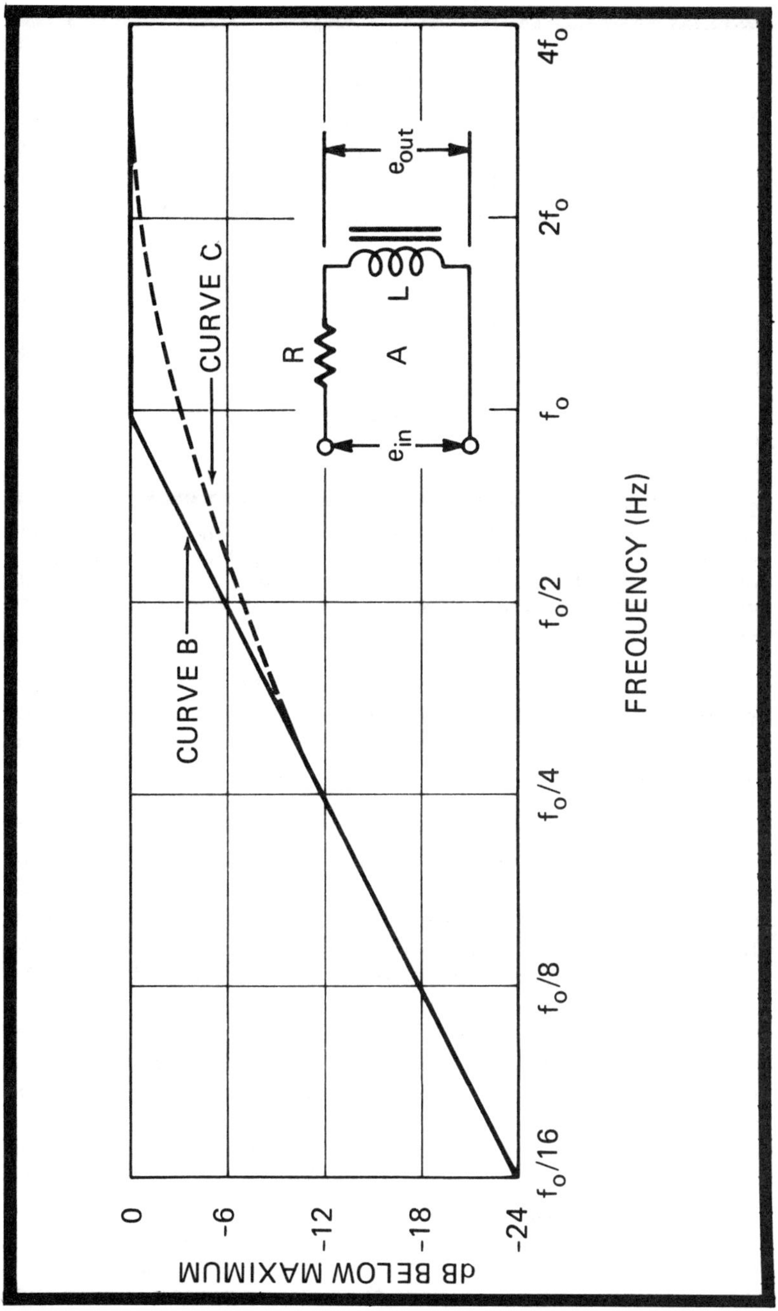

Fig. 9-3. High-pass R-L filter circuit (A). Also shown is the frequency response curve (B) and an approximation (C) to the curve at B.

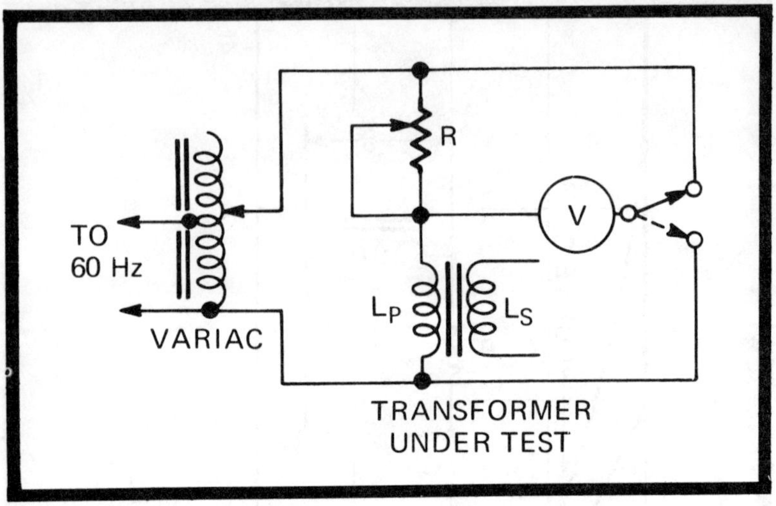

Fig. 9-4. Circuit used to determine the inductance in the primary winding of a transformer. It is determined by equating the resistance of the potentiometer to the impedance of the transformer. The resistance is equal to the impedance when the voltages across both components are equal.

The incremental inductance should be determined for a specific direct current through the winding while there is a 60-Hz AC voltage across the transformer. A bridge can be used for the measurement. The approximate inductance (not incremental) can be determined using a bridge or using the circuit in Fig. 9-4. Here, a 60-Hz AC voltage is fed from the AC lines across a series R_L circuit. A voltmeter is used to measure the voltage across R and L_P. R is adjusted until both voltages are equal. Then L_P equals R divided by 337. (Notice that when the voltages across both components are equal, the voltmeter will indicate more than one half the voltage of the supply due to phase shift in the RL_P circuit.)

A similar procedure can be used to measure the incremental inductance. Use the circuit in Fig. 9-4, with the addition of a DC supply placed across L_P. This supply must be connected across the winding through a large resistor so that the DC circuit will have no effect on the measurement of the incremental inductance, other than that due to the DC current flowing through L_P. Adjust the DC voltage so that the DC current in the primary of the transformer will be equal to the current expected to flow in the actual transistor circuit. Proceed as before to compare the voltage across R with that across L_P. When the voltages are equal, the incremental inductance is R divided by 377.

The low-frequency response characteristics of the circuit in Fig. 9-3A are shown in Fig. 9-3B, with an approximation of the exact curve drawn in Fig. 9-3C. Frequency f_0 is known as the corner frequency and can be determined from:

$$f_o = \frac{R}{6.28L} \qquad (9\text{-}3$$

At f_0, the gain has decreased 3 db from its mid-frequency value. Other important approximate points on the curve are at $2f_0$ where the gain has dropped 1 db from that at mid-frequency, at $f_0/2$ where the rolloff is 7 db, and the loss of gain at $f_0/4$ is 12 db. Below $f_0/4$, the gain decreases at the rate of 6 db per octave. The straight-line approximation to the curve has a 0 db rolloff at f_0 and the gain decreases at the rate of 6 db per octave below the corner frequency. It should be noted that when the gain drops at the rate of 6 db per octave, it is identical to a 20 db per decade rolloff.

At the high frequencies, the equivalent circuit, as shown in Fig. 9-5A, is a resistor in series with an inductor. Here, R is the sum of the primary resistance of the transformer, R_P, the load resistor reflected from the secondary winding into the primary R_L', and the collector (or drain) resistance of the transistor, usually r_d. The leakage inductance, L_{leak}, is the L in the circuit.

Leakage inductance is a stray inductance considered to be in series with the primary winding of the transformer. It is measured on a bridge by applying a 1000-Hz signal at 1 volt across the primary of the transformer while the secondary winding is shorted.

Leakage inductance can also be measured using the circuit in Fig. 9-6. The secondary winding is shorted and a variable resistor is placed in series with the primary winding. About 1.4 volts at 1 kHz is impressed across the series combination formed by the resistor and the primary of the transformer. The variable resistor is adjusted until the AC voltmeter reads the same voltage across the transformer and across the resistor. At this resistance, L_{leak} equals R divided by 6.28×10^3.

The actual high-frequency response curve for the circuit is shown in Fig. 9-5B and the approximation is shown in Fig. 9-5C. As was the case for the low-frequency characteristic, the rolloff of the approximate curve is at the rate of 6 db per octave or 20 db per decade, starting at f_0. f_0 is once again defined by Equation 9-3. At f_0, the actual gain is 3 db down from the gain at the mid-frequency. Gain is 1 db down at $f_0/2$, 7 db down

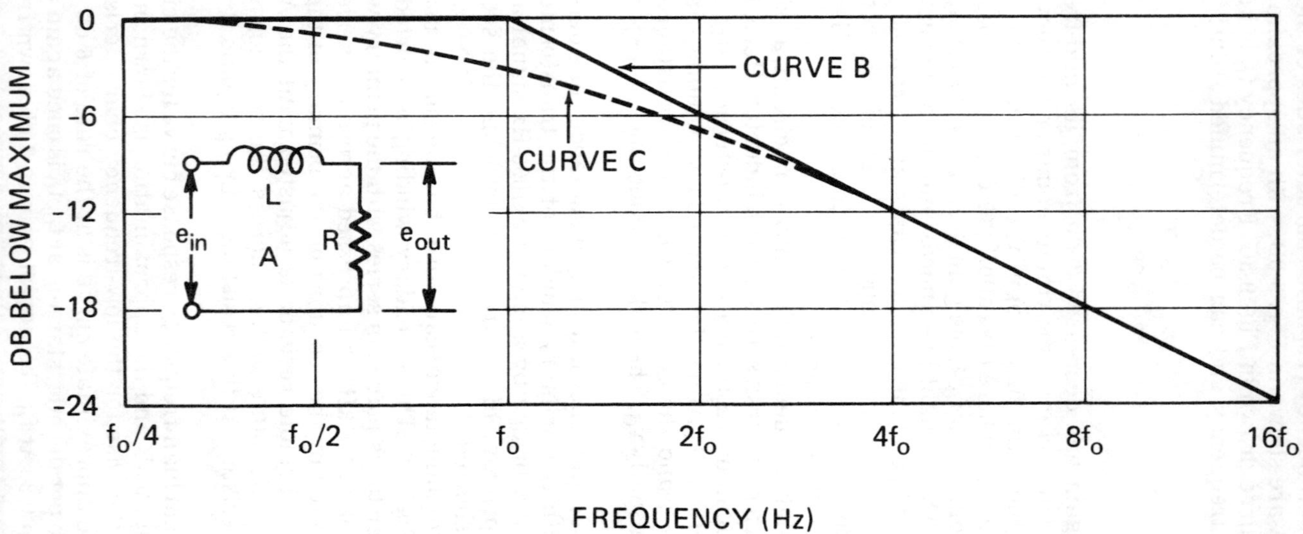

Fig. 9-5. Low-pass R-L filter circuit (A). The curve at B represents the frequency response of the filter and C is an approximation to the curve at B.

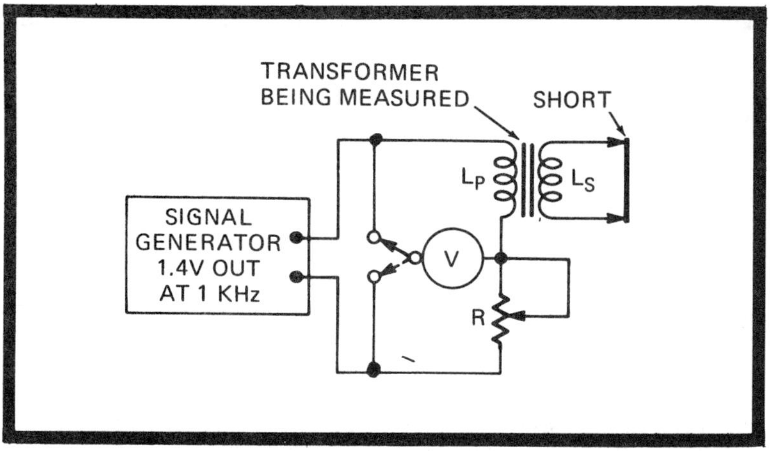

Fig. 9-6. Setup to measure leakage inductance. The secondary winding is shorted and potentiometer R is adjusted until the voltage across the primary winding of the transformer is identical to that across R.

at $2f_0$, 12 db down at $4f_0$ and it rolls off at the rate of 6 db per octave thereafter.

Due to various stray capacities, the high-frequency rolloff is not as smooth as depicted by the ideal situation described. There are various peaks in the response which cause bumps in the curve. Rolloff can proceed at the rate of 12 db per octave. The winding capacity responsible for the peaks can be found using the circuit in Fig. 9-7. Here, a few thousand ohms resistor, R, is placed in series with the primary winding of the transformer while the secondary is shorted. The frequency of the signal generator is varied until there is a peak voltage across the primary winding as read on the meter, V. Frequency f_1 is one of the frequencies at which a bump will

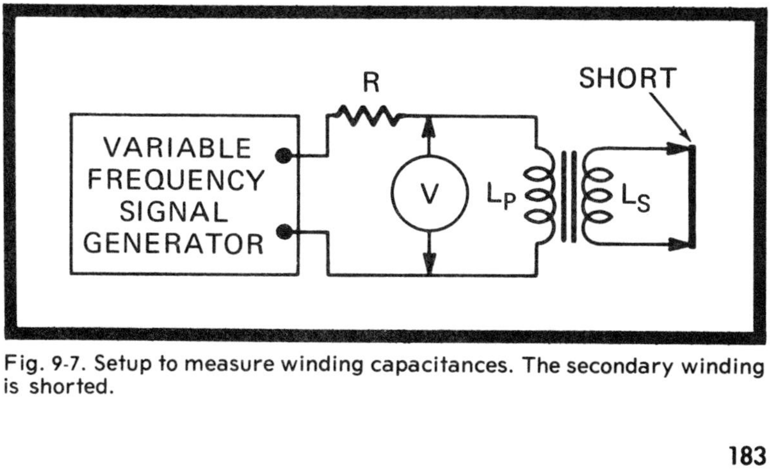

Fig. 9-7. Setup to measure winding capacitances. The secondary winding is shorted.

occur on the response curve. The winding capacity is C equals 1 divided by $(6.28f_1)^2 L_{leak}$. Several peaks may be noted for the various capacities which are characteristic of transformers. Some peaks will be more pronounced than others.

Using Coupling Transformers

In Chapter 7, we indicated that transformers are used to convert the low output impedance of a speaker load across the secondary to a high impedance in the primary winding. This relatively high impedance is to serve as the load for a transistor. In a similar manner, a transformer can be used to couple a high-impedance input source to the low-impedance at the input of a bipolar transistor. This impedance conversion is not generally required when feeding the high-input impedance of a JFET.

The audio input to an amplifier may originate at a microphone. Because of long lines, it is frequently desirable that a balanced input be used, as shown in Fig. 9-8. While a low-impedance long line is less susceptible to hum and noise pickup than a high-impedance line, the use of a transformer with the centertap connected to ground is an important asset in cancelling the in-phase hum and noise induced in the two "hot" leads of the connecting cable.

The signal fed through a low-impedance line may be applied directly to the input of an amplifier, or it may first be transformed so that it will appear as a higher impedance. Voltage gain is added as a fringe benefit. The transformer must be of the proper quality so as not to adversely affect the characteristics of the signal.

Interstage transformer coupling is commonly used in amplifiers. Although these components limit bandwidth and the flexibility of putting large amounts of feedback around the overall circuit due to stability problems (discussed in the next chapter) many excellent designs still do include driver transformers. However, they are found primarily in public address and low-price high fidelity equipment.

Let us now turn our attention to the circuit in Fig. 9-9, which uses a considerable amount of transformer coupling, and see just what the circuit component values and arrangements can be. More transformers than are normally used in any one design are shown here. The paper design should be checked in a laboratory for accuracy and reproducibility. On an actual chassis, care must be exercised to avoid magnetic coupling between transformers. Returns to B- and ground should be made in the proper sequence (see

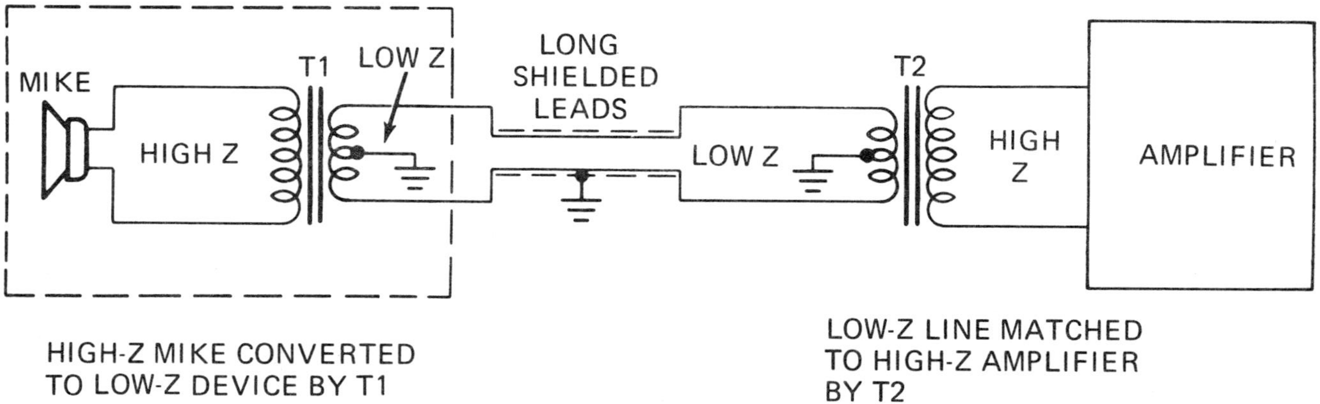

Fig. 9-8. Circuit showing a microphone connected in a balanced-line input circuit.

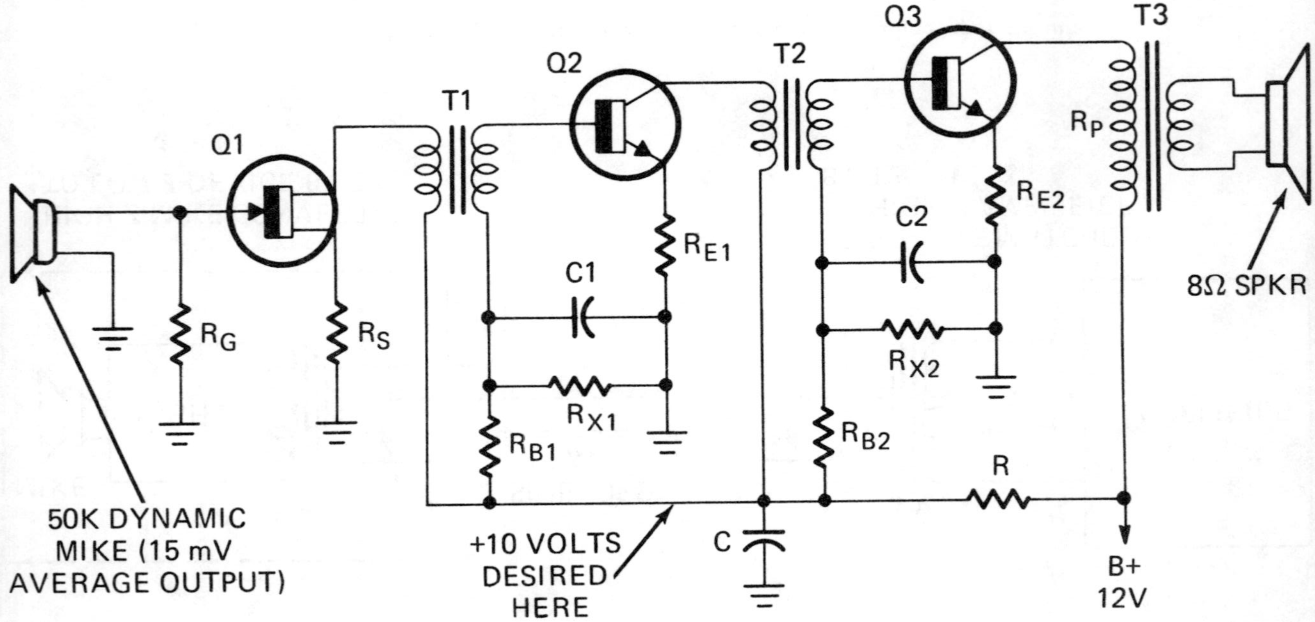

Fig. 9-9. Transformer-coupled amplifier used in the text design problem.

Chapter 12) to avoid coupling between stages due to resistance in the leads.

A 50,000-ohm dynamic microphone with 15 mv average output is the signal source. The signal is amplified by three stages of gain consisting of an N-channel JFET and two NPN bipolar devices. Each stage is transformer-coupled to the preceding and succeeding transistor. The required output is 1 watt delivered to an 8-ohm speaker. A 12-volt automobile battery is the source of power used directly as the B+ supply for the output stage, and applied through a decoupling RC network to feed the first two transistors. It is assumed that the first two devices require 10 volts B+. This voltage is present at the junction of R and C.

Starting with the final stage, let us first assume the output transformer is 75 percent efficient so that for 1 watt output, 1 divided by 0.75 or 1.33 watts is required at the primary of the transformer. Sufficient leeway can be built into the design if it is assumed that about 3 volts is lost across the output transistor due to leakage current and saturation voltage. About 2 volts of this can be attributed to saturation voltage. Adding this to power loss across R_{E2}, we can expect an additional 25 percent loss of power at Q3. The total power the output stage must be capable of delivering becomes 1.33+(0.25 x 1.33) or 1.66

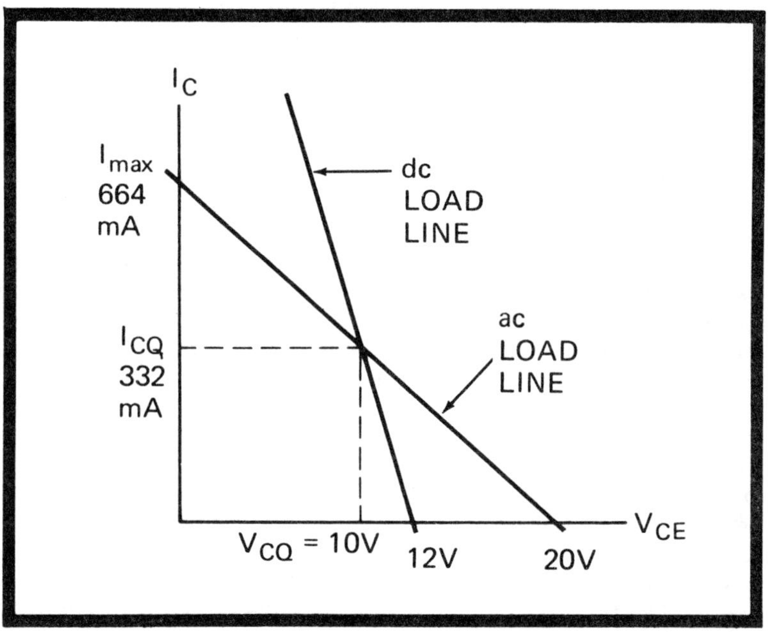

Fig. 9-10. Load lines used in the design of the output stage in Fig. 9-9.

watts. We must now determine if the relatively inexpensive 2N3053 transistor in a TO-5 package can deliver this power.

Since 1.66 watts is required across the primary winding, the transistor in the Class A circuit must dissipate 3.32 watts or twice the power the device must deliver. If we assume 2 volts will be developed across the DC resistance in the emitter circuit and the primary winding of transformer T3, 10 volts will be across the transistor in the quiescent state. The quiescent collector current is 3.32 watts divided by 10 volts or 332 ma. The peak current, I_{max} equals 2 x 332 ma or 664 ma, which is within the 700 ma rating of the 2N3053.

Draw AC and DC load lines as in Fig. 9-10. From the curve in the figure, the DC load resistance is (12 - 10) volts divided by 332 ma equals 6 ohms. The AC resistance is 20 volts divided by 664 ma or 30 ohms. The resistance that is to be reflected from the secondary of the transformer to the primary is 30 ohms - 6 ohms or 24 ohms, so that the impedance ratio of T3 is 24 ohms to 8 ohms, or 3:1, and the turns ratio is $(24 \text{ divided by } 8)^{1/2}$ equals 1.73:1. This assumes the DC resistance in the secondary winding is negligible when compared to the resistance of the 8-ohm load.

If we wish to design the transformer for a minimum of 25 percent efficiency, the maximum resistance of the primary winding can be determined (from Equation 9-2):

$$25 = \frac{100}{1 + \dfrac{R_P + 3R_S}{3R_L}} = \frac{100}{1 + \dfrac{R_P + 0}{24}} \qquad (9\text{-}4)$$

and Rp is 72 ohms. Since the DC design allows for a maximum of 6 ohms for the sum of the primary winding DC resistance and R_{E2}, the efficiency requirement will automatically be met in a proper design.

The voltage gain of an output stage is relatively unimportant when compared to its performance in delivering output power. Let us assume we would like a voltage gain of about 10 from this stage. Since the AC output impedance in the collector circuit is 24 ohms, the resistor in the emitter should be about one-tenth this value. Use a standard 2.7-ohm resistor for R_{E2}. This will also conform with another rule of thumb stating that about 1 volt should be across the emitter resistor of a power transistor when it is idling. Here the voltage is (2.7 ohms) (332 ma) or 0.9 volt, a value close enough to 1 volt. The DC resistance of the primary winding, R_P, can then be specified at less than 6 ohms - 2.7 ohms or 3.3 ohms.

1.33 watts must be delivered to the 24-ohm AC impedance across the primary of T3 if 1 watt is to be delivered to the load.

For this output, the RMS voltage across the primary of the transformer is V equals $(R_L'P)^{1/2}$ equals (24) $(1.33)^{1/2}$ equals 5.7 volts.

The emitter resistance, r_e, of the transistor is usually taken at 26 divided by I_E, where I_E is the quiescent emitter current expressed in ma. For a current of 332 ma, this figure becomes too small to be reasonable. We assume a minimum DC emitter resistance of 1 ohm for most transistors. The total AC resistance in the emitter circuit is r_e equals 1 ohm added to the R_{E2} equals 2.7 ohms, amounting to a total of 3.7 ohms. The calculated voltage gain of the output stage is the ratio of the 24 ohms at the primary of the output transformer to the 3.7 ohms just calculated, or 6.5. This differs from the approximate gain of 10 sought after above. More gain can be designed into the earlier stages. The voltage at the input to the transistor must be the 5.7 volts across the primary of the output transformer divided by the gain of 6.5, or 0.88 volt.

The output transformer usually represents the limiting factor on the frequency response. Assume our amplifier is to have an output from 20 Hz to 20 kHz, with the gain down 3 db

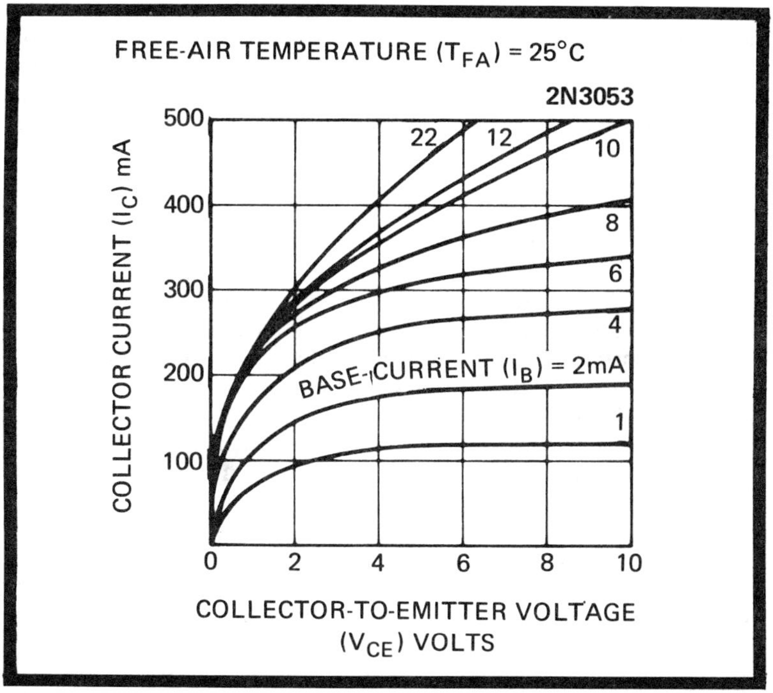

Fig. 9-11. Collector characteristic curves for the 2N3053 transistor. (Courtesy RCA)

from center at these extreme frequencies. To fulfill these goals, the minimum inductance and maximum leakage inductance tolerable in the transformer can be determined with the help of Figs. 9-3 and 9-5.

Since R in Fig. 9-3A is essentially R_L' equals 24 ohms (for it is R_L' in parallel with $r_d + R_p$), the incremental inductance of the primary winding of the transformer must be somewhat more than L equals R_L' divided by $6.28f_o$ from Equation 9-3. Letting f_o equal 20 Hz, L must then be more than 24 divided by 6.28(20) equals 0.19 henry.

At the high-frequency end of the band, from Fig. 9-5A, R equals $r_d + R_p + R_L$'. Substituting numbers, $R_p + R_L$' equals 3.3 ohms + 24 ohms or 27.3 ohms. The collector resistance, r_d, at 10 volts and 332 ma can be determined from the slope of the 2N3053 output characteristic curves in Fig. 9-11. It is about 180 ohms. R is thus 180 ohms + 27.3 ohms or 207.3 ohms. Using Equation 9-3, the maximum leakage current is L equals R divided by $6.28f_o$ equals 207.3 divided by 6.28(2 x 104) equals 1.65 millihenrys.

The bias circuit is the next item to be designed. R_{X2} should be specified at about 10 times the size of R_{E2} for stability purposes, or at less than 27 ohms. Since there is about 0.7 volt across the base-emitter junction and 0.9 volt across R_{E2}, the sum of these voltages, or 1.6 volts, is across R_{X2}. The current through R_{X2} is 1.6 volts divided by 27 ohms or 59 ma.

From the characteristic curves, it can be seen that about a 6 ma base current is required to get the 332 ma of current in the collector. The 59 ma flowing through R_{X2} and the 6 ma base current both flow through R_{B2} for a total of 65 ma. Since the circuit is designed for 10 volts at the junction of R and C, R_{B2} is (10 - 1.6) volts divided by 65 ma or 129 ohms.

The impedance across the secondary of T2 is $(R_{E2} + r_{e2})$ reflected into the base circuit. Since beta is 332 ma divided by 6 ma or 55, the reflected impedance is (2.7 + 1) x 55 or 204 ohms. Capacitor C2 must be a short circuit compared to the 204 ohms at 20 Hz (the lowest frequency to be amplified) in order to effectively bypass the bottom end of the transformer to ground. If it is made one one-hundredth of 204 ohms, or about 2 ohms at 20 Hz, the capacitor will be C2 equals 1 divided by (6.28) (20)(2) or 4000 mfd.

The 0.88 volt determined above and required at the input to Q3 must be developed across the 204 ohms at the input in the base circuit. Hence, e^2 divided by R equals $(0.88)^2$ divided by 204 equals 3.8 mw must be present at the output of T2.

Repeating the above procedure to design the circuit for Q2, 3.8 mw divided by 0.75 equals 5 mw must be present at the primary of T2 if the transformer is 75 percent efficient. Additional power was required previously from the large signal transistor due to the limitations set on the load line by the saturation voltage and leakage current. These boundaries can usually be disregarded when small signal voltage-amplifier-type devices are involved.

Since the transistor must be able to deliver 5 milliwatts, it must be capable of dissipating 10 milliwatts. A 2N3391 transistor can do this easily and, therefore, can be used here.

Curves drawn in Fig. 9-12 are for AC and DC loads. Using the 10 volts at the junction of R and C as the supply and assuming that 2 volts is developed across R_{E1} and the DC resistance of the primary of the transformer, the quiescent voltage across Q2 is 8 volts and the quiescent current is 10 mw divided by 8 volts or 1.25 ma. The DC load resistance is 2 volts divided by 1.25×10^{-3} amps or 1600 ohms. The AC load resistance is 16 volts divided by 2.5×10^{-3} amps or 6400 ohms. The AC impedance across the primary of T2 is 6400 - 1600 or 4800 ohms. We can easily let R_{E1} equal 470 ohms so that the gain of Q2 is 4800 divided by 470 or 10.2. The remaining DC resistance can be in the transformer winding.

The impedance ratio of the transformer is 4800:204, so the turns ratio and voltage ratio is about 4.9:1. Hence, the gain of the transformer is 1 divided by 4.9. Multiplying this by the gain of Q2, the overall gain of this part of the circuit is (10.2) (1 divided by 4.9) or 2.1. The signal voltage at the input to Q2 and

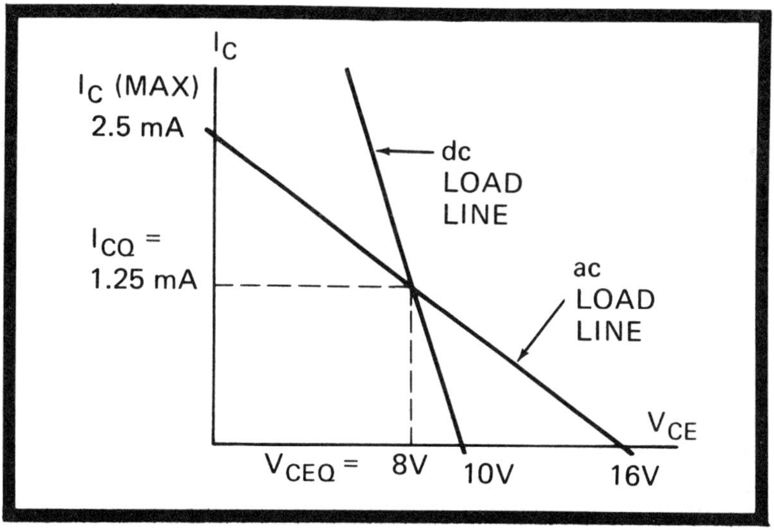

Fig. 9-12. Load lines used to design the proper quiescent condition of Q2.

across the secondary of T1 should thus be 0.88 volt divided by 2.1 or 0.42 volt. Since the emitter resistance, r_e, of Q2 is about 26 divided by I_E equals 26 divided by 1.25 or 20.7 ohms, and R_{E1} is 470 ohms, the total resistance in the emitter circuit is 470 ohms + 20.7 ohms or 490.7 ohms. From the specifications of the 2N3391, the average beta is 350, so the resistance of the emitter circuit when reflected into the base is (350)(490.7) or 1.74×10^5 ohms.

As for the bias circuit, make R_{X1} equal to 10 times the resistor in the emitter circuit or 4700 ohms. The voltage across this resistor is V_{BE} equals 0.7 volt plus (1.25 ma)(470 ohms) equals 0.59 volt for a total of 1.29 volts, so the current through R_{X1} is 1.29 divided by 4.7×10^3 or 0.27 ma. The average base current is the collector current divided by beta, or 1.25×10^{-3} divided by 350 or 0.0036 ma. Add the base current to the current through R_{X1} to determine the current through R_{B1}. It is 0.2736 ma. With 10 volts - 1.29 volts across R_{B1}, the resistor should be specified at about 8.71 volts divided by 0.2736×10^{-3} ma or 32,000 ohms.

Returning to the AC circuit, the impedance of C1 should be one-hundredth that across the secondary of the transformer, or 1.74×10^3 ohms at 20 Hz. C1 equals 1 divided by $(6.28)(20)(1.74 \times 10^3)$ or 4.5 mfd.

To deliver the required power, 0.42 volt must be developed across the 1.74×10^5 ohms at the base of Q2 and across the secondary of T2. The power required here is very small, $(0.42)^2$ divided by $1.74 \times 10^{-5} \approx 1$ milliwatt. This power can easily be delivered by Q1, the JFET. The efficiency of the transformer approaches 100 percent as R_L in Equation 9-2 becomes extremely large. We can now use the turns ratio of the transformer to supply gain and not be excessively concerned about losses and frequency characteristics of the component.

Select a 2N4303 N-channel JFET. Draw the AC and DC load lines for the 2N4303 JFET as shown in Fig. 9-13. Since the transformer winding can now have a large resistance without radically affecting performance, assume there is 6 volts across the transistor in the quiescent state. This leaves 4 volts to be dropped across the DC resistance of the primary winding of the transformer and across R_S, the resistor in the source lead. For low noise, which is an important consideration in the first stage of amplification, assume the quiescent drain current, I_{DO}, is near I_{DSS}. The I_{DSS} for the 2N4303 is specified to be between 4 and 10 ma. Make the quiescent current, I_{DQ}, equal to 3 ma, so the AC load line can theoretically swing from 0 to 6 ma and the AC impedance in the drain circuit is 12 volts divided by 6 ma or 2000 ohms.

The DC resistance—the sum of the resistance in the source circuit and the resistance of the primary winding of transformer T1—is (10 — 6) volts divided by 3 ma equals 1,333 ohms; therefore, the impedance R_L' reflected into the primary of the transformer is 2000 ohms — 1333 ohms or 667 ohms. The impedance ratio of the transformer is 667:1.74 x 10^5 or 0.0038. The turns ratio is $(0.0038)^{1/2}$ or 0.062, so the voltage gain of the transformer is 1 divided by 0.062 or 16. Thus, 0.42 volt divided by 16 equals 0.026 volt (26 millivolts) is required at the drain of the 2N4303 if the amplifier is to deliver 1 watt across the 8-ohm load at the output. (As you recall, the 0.42 volt is the signal voltage required across the secondary of transformer T1.)

The microphone delivers 15 mv. The gain of the JFET must be at least 26 mv divided by 15 mv or 1.73. Using a gain of three to compensate for variations in the actual semiconductor device and microphone, the source resistor, R_S, can be calculated from the equation:

$$A_V = \frac{g_m R_D}{1 + g_m R_S} = 3 \qquad (9\text{-}5)$$

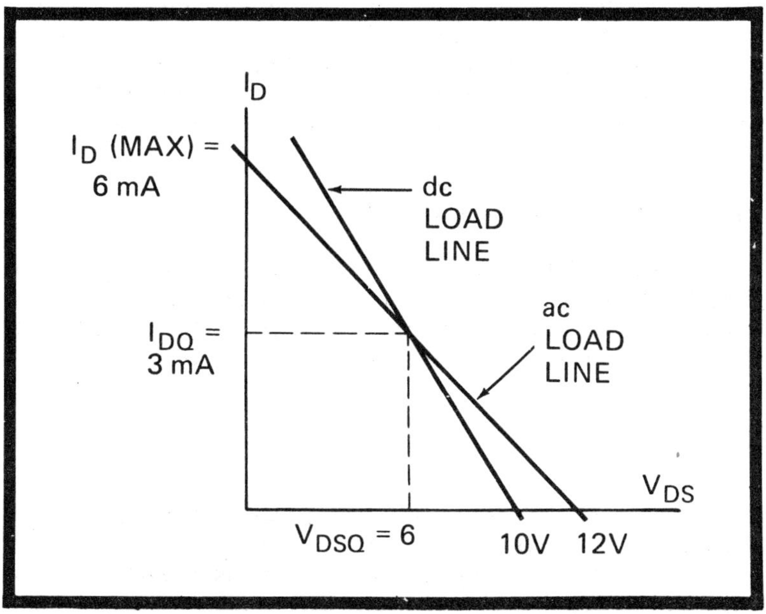

Fig. 9-13. Curves used to design a circuit around a JFET stage.

where A_V is the voltage gain of the JFET, g_m is the transconductance at the quiescent current (3 ma), R_D is the resistance in the drain circuit, and R_S is the source resistor.

The minimum transconductance at I_{DSS}, when the gate-to-source voltage is zero, is g_{mo} equals 2000 micromhos. Using the maximum value of I_{DSS} equals 10 ma and an idling current, I_D equals 3 ma, we can find the minimum transconductance of the circuit at the operating point from the equation:

$$g_m = g_{mo}\left(\frac{I_D}{I_{DSS}}\right)^{1/2} = 2000\left(\frac{3}{10}\right)^{1/2} = 1100 \qquad (9\text{-}6)$$

Substituting this into Equation 9-5:

$$3 = \frac{1100\,(667)}{1 + 1100\,R_S} \qquad (9\text{-}7)$$

Solving the equation, R_S equals 223 ohms.

If conditions were such that the circuit could not supply sufficient gain, a "stepup" transformer can be used between the microphone and the input to the JFET for this purpose.

It must be noted that the microphone is not loaded with its characteristic impedance. If it were, the voltage delivered to the load at the input of the JFET would be half the voltage delivered when the microphone is feeding the extremely high input impedance at the transistor. The load on the microphone would become a factor if the input transistor were of the bipolar variety.

RESISTOR-CAPACITOR COUPLING

The limits inherent in R-C or resistor-capacitor coupled circuits are similar to those of transformer-coupled circuits. Capacitive reactance is inversely related to the frequency of the voltage across the capacitor. In the formula:

$$X_C = \frac{1}{6.28fC} \qquad (9\text{-}8)$$

X_C is the reactance of the capacitor in ohms, f is the frequency of the applied signal in Hertz, and C is the capacity in farads. Include the capacitor in a circuit with a resistor and you have a frequency-dependent R-C filter. These filters are part and parcel of the R-C coupled transistor circuit.

R-C Filters

Two basic filter circuits using one resistor and one capacitor in each are shown in Fig. 9-14. The high-pass filter in Fig. 9-14A discriminates against the low frequencies and freely passes the upper end of the spectrum. The exact and approximate frequency characteristic curves for this filter are shown in Figs. 19-15A and 19-15B, respectively. Notice the 6 db per octave or 20 db per decade rolloff which makes these curves identical to those applied to the R-L filters. The corresponding curves for the low-pass filter of Fig. 9-14B are shown in Figs. 9-15C and 9-15D. The corner frequency, f_0, can be determined from the equation:

$$f_o = \frac{1}{6.28RC} \tag{9-9}$$

where R is the resistance in ohms and C is the capacitance in farads.

Two or more such filters can be connected into the circuit to increase the rate of the rolloff beyond 6 db per octave. At any particular frequency, the reduction in gain below maximum is equal to the sum of the reduction in gain of each individual filter section at the frequency in question. Five such sums are performed graphically in Fig. 9-16.

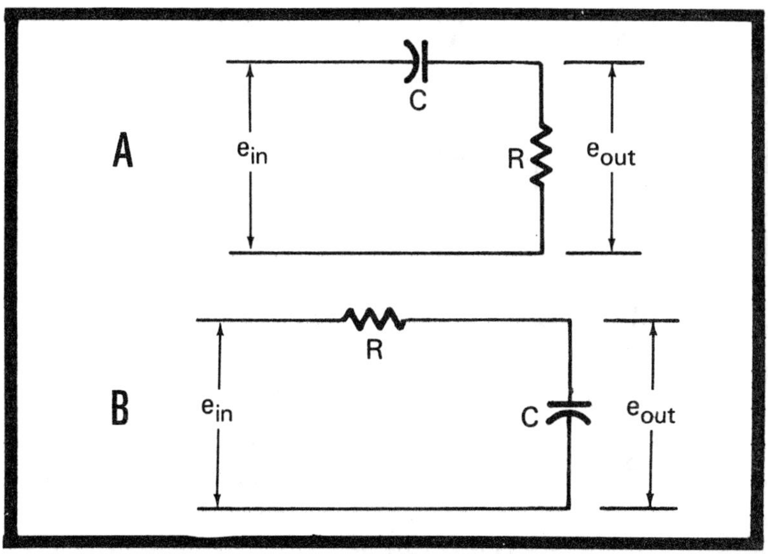

Fig. 9-14. Basic R-C filters: High-pass filter, A, and low-pass filter, B.

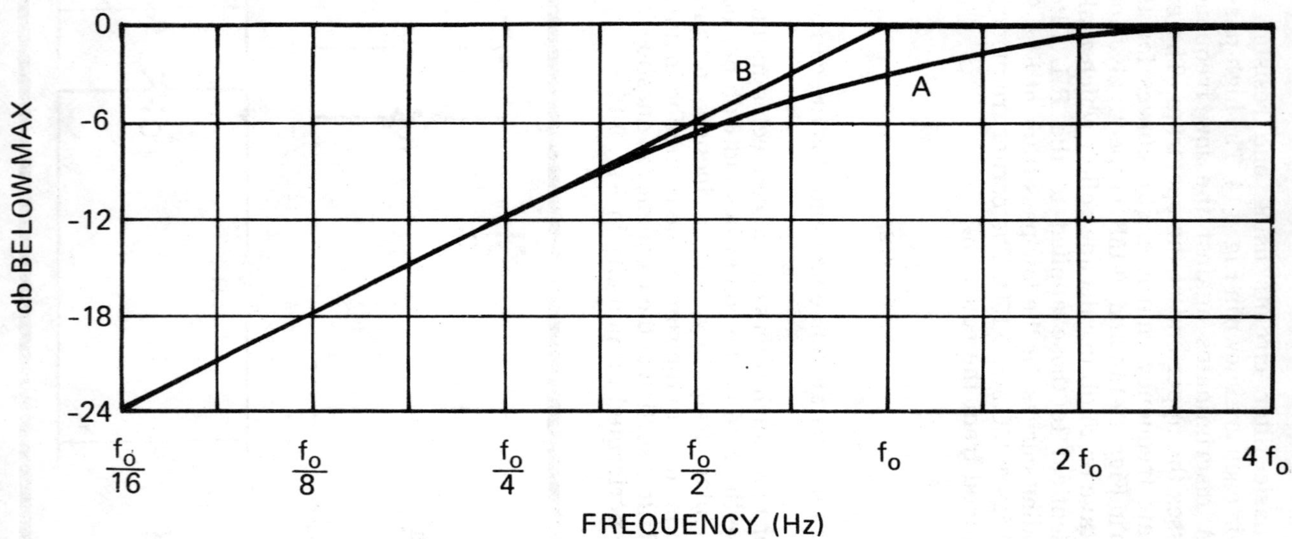

0
-6
-12
-18
-24
db BELOW MAX
A
B
$\frac{f_o}{16}$
$\frac{f_o}{8}$
$\frac{f_o}{4}$
$\frac{f_o}{2}$
f_o
$2 f_o$
$4 f_o$
FREQUENCY (Hz)

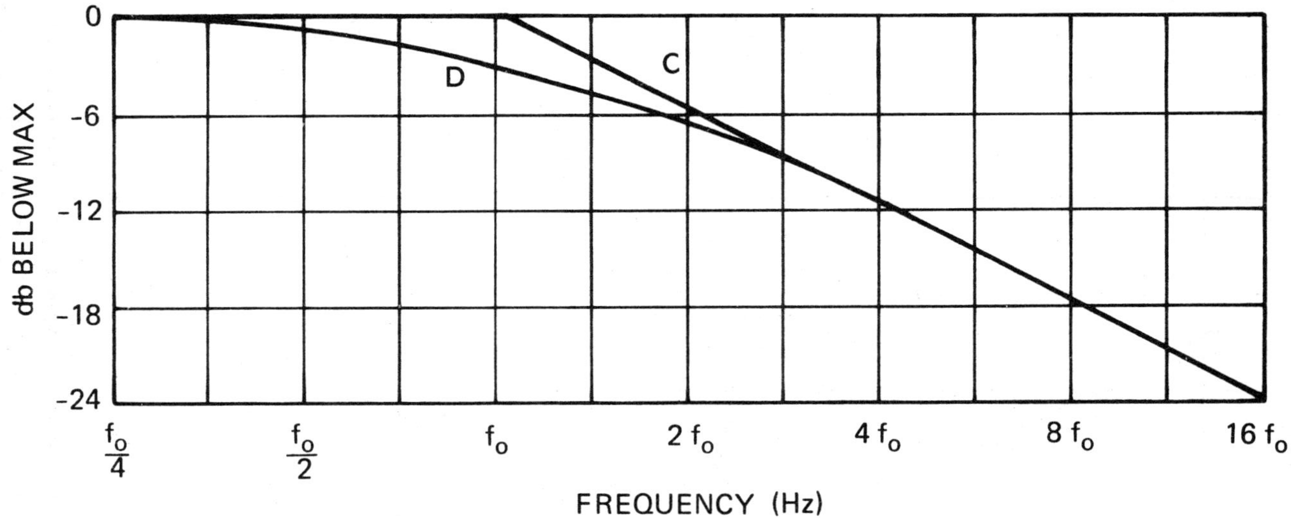

Fig. 9-15. Curves showing the frequency characteristics of the R-C filters in Fig. 9-14. The exact frequency response of the high-pass filter in Fig. 9-14A is illustrated by Curve A. Curve B is an approximate curve for the filter in Fig. 9-14A. Curve D represents the exact frequency response of the low-pass filter in Fig. 9-14B. An approximate curve for filter in Fig. 9-14B is shown at C.

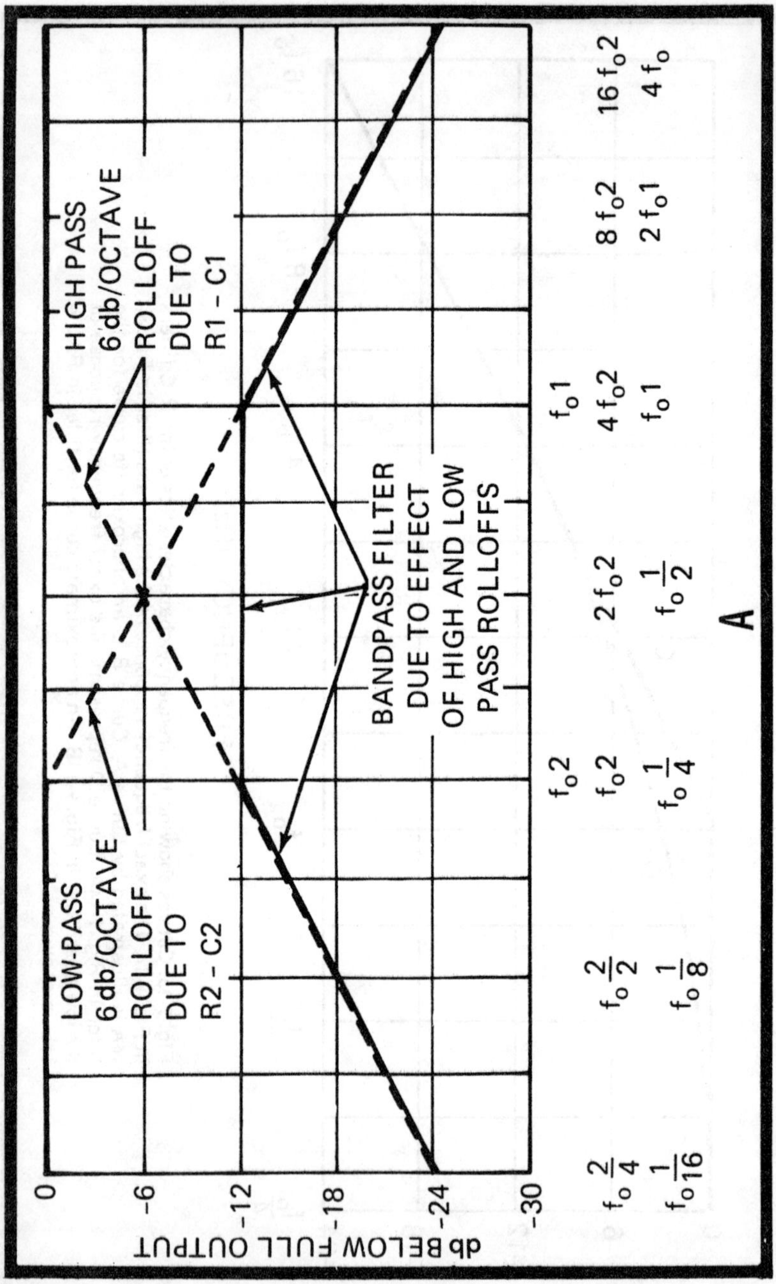

Fig. 9-16A. Two low-pass filters connected in one circuit. f_{o1} is the corner frequency of the R1-C1 filter and f_{o2} is the corner frequency of the R2-C2 filter section.

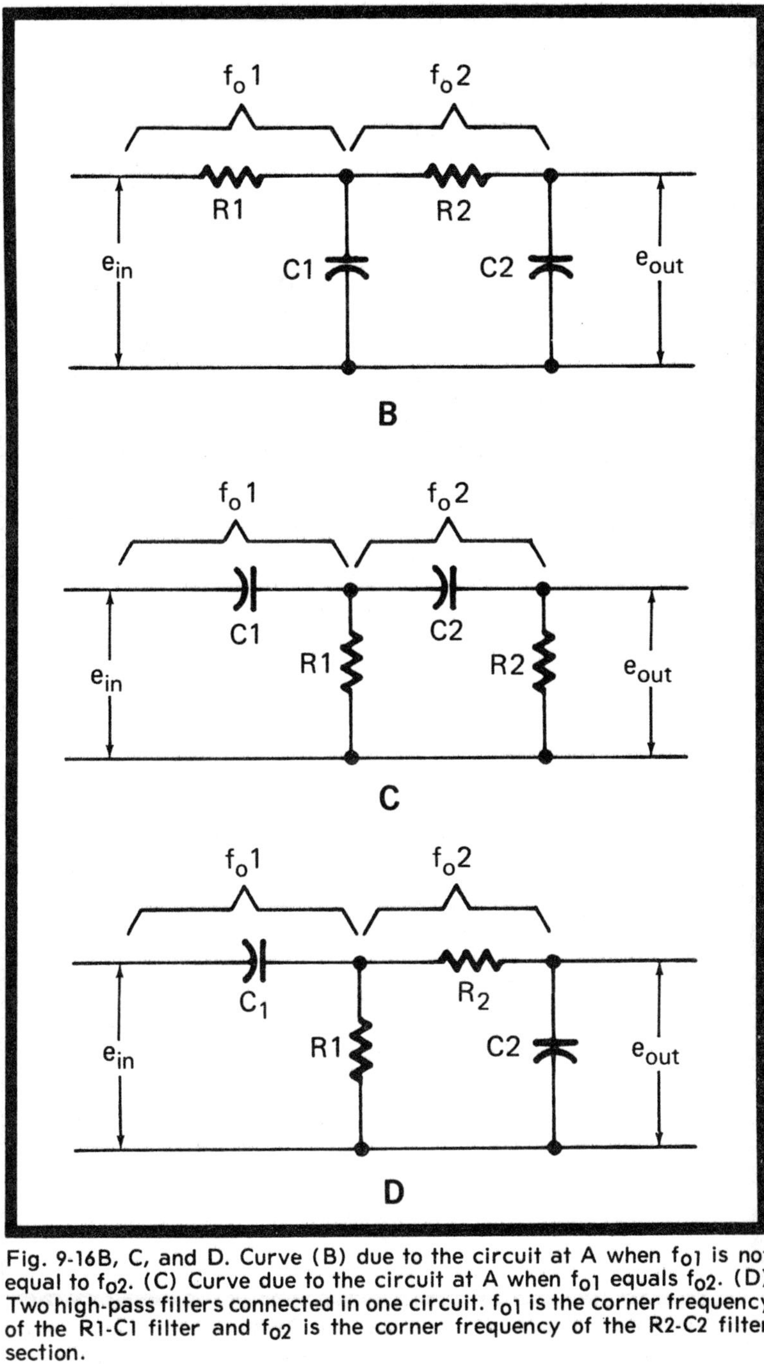

Fig. 9-16B, C, and D. Curve (B) due to the circuit at A when f_{o1} is not equal to f_{o2}. (C) Curve due to the circuit at A when f_{o1} equals f_{o2}. (D) Two high-pass filters connected in one circuit. f_{o1} is the corner frequency of the R1-C1 filter and f_{o2} is the corner frequency of the R2-C2 filter section.

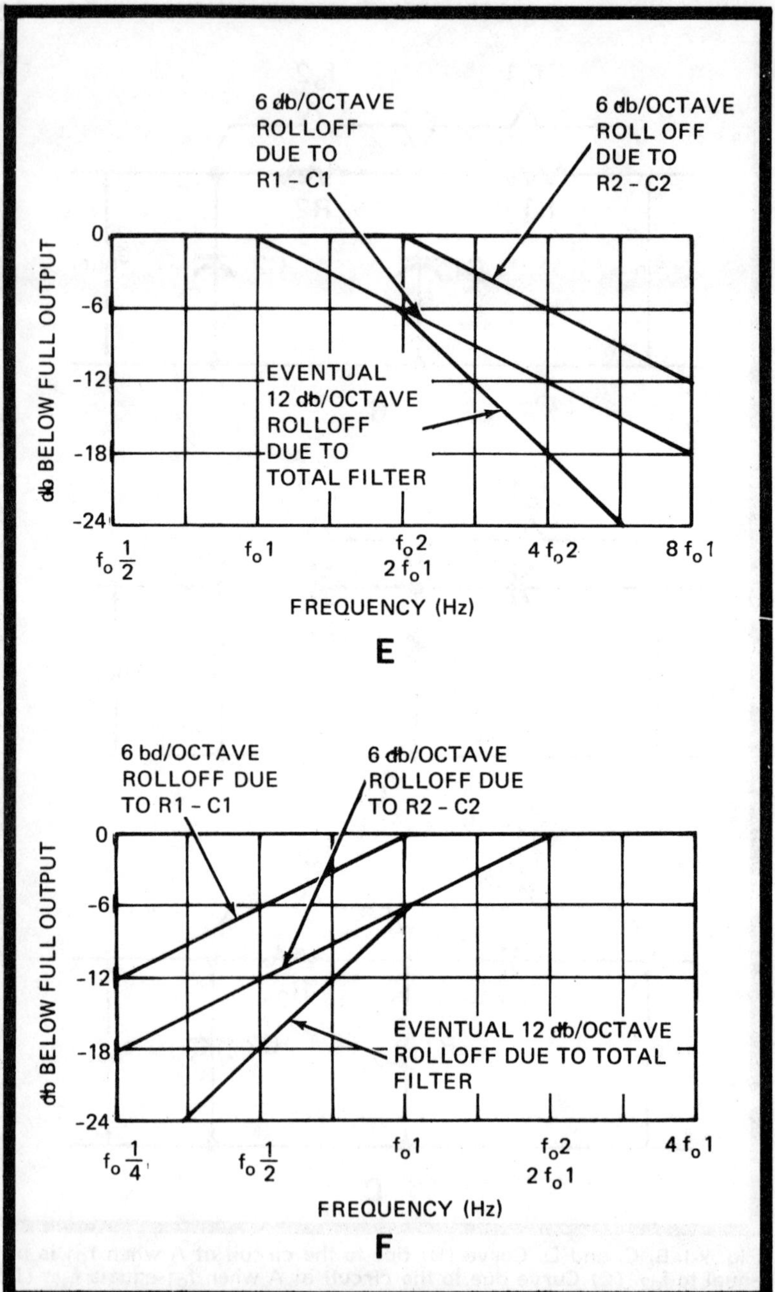

Fig. 9-16E and F. Curve (E) of the circuit at D when f_{o1} is not equal to f_{o2}. (F) Curve of the circuit at D when f_{o1} equals f_{o2}.

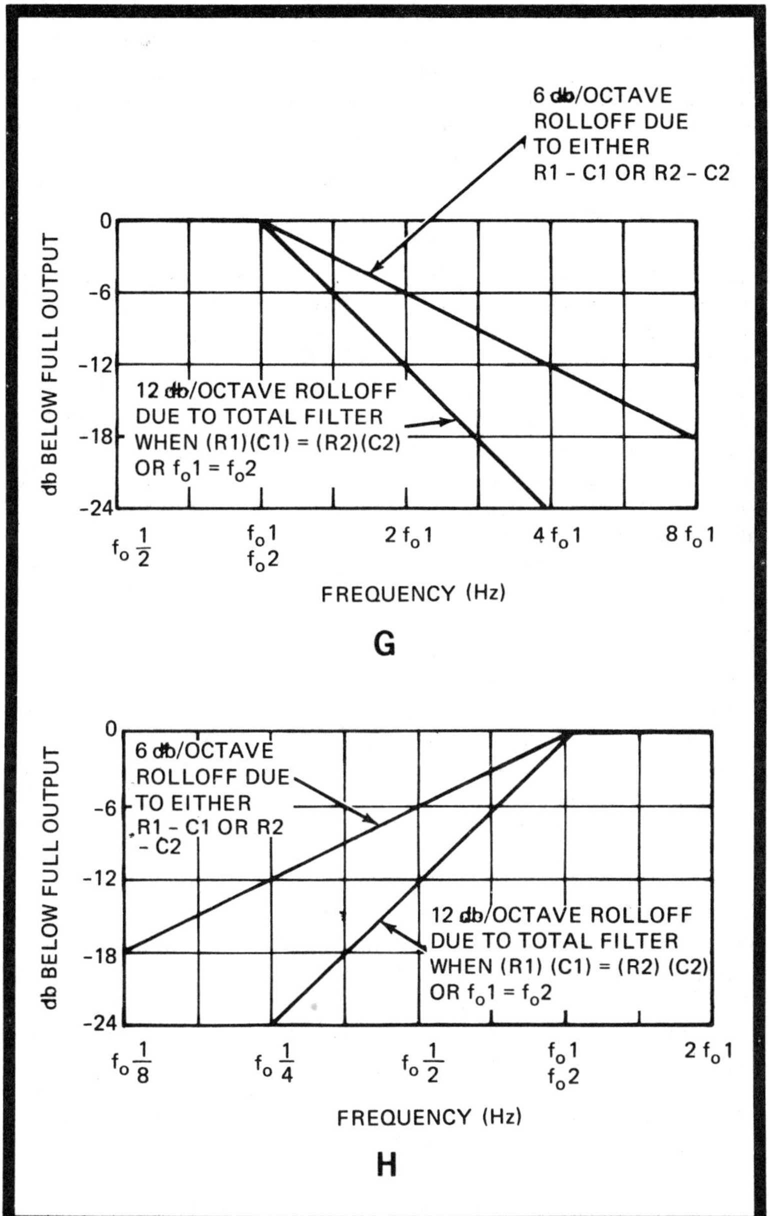

Fig. 9-16G and H. One low-pass filter (G) consisting of R2-C2 and one high-pass filter consisting of R1-C1 in one circuit. The combination is a bandpass filter, fo1 is the corner frequency of the R1-C1 filter while f_{o2} is the corner frequency of the R2-C2 filter section. (H) Bandpass filter response of the circuit at G.

The circuit in Fig. 9-16A consists of two low-pass 6 db per octave filters; one is composed of R1-C1 and the second is R2-C2. The corner frequency of the R1-C1 combination is f_{o1} and that of the R2-C2 section is f_{o2}. Should f_{o2} be higher than f_{o1}, the approximate frequency characteristic curve of the total filter is the heavy line drawn in Fig. 9-16B.

In the drawing, the low-frequency end of each curve is flat with 0 db attenuation. At f_{o1}, one curve starts to roll off at the rate of 6 db per octave and beginning at f_{o2} the second curve rolls off at the same rate. The heavy line is the sum of the two curves. Up to f_{o1}, neither R-C combination attenuates the signal, so the heavy curve is flat with 0 db attenuation. Between f_{o1} and f_{o2} the rolloff is affected only by R1-C1, so that this portion of the curve rolls off only at the rate of 6 db per octave. From f_{o2} on up to an ideally infinite frequency, the slope due to the rolloff characteristic of the R2-C2 section must be added to the slope due to R1-C1. Since both 6 db per octave circuits are involved, the heavy curve is characterized by a 12 db per octave rolloff.

Another way of plotting the effect of the total filter is to add the curves of each section vectorially, point by point. Up to f_{o1}, the attenuation of both curves is zero, so 0 + 0 is 0 db. At $2f_{o1}$, or at f_{o2}, the attenuation of one curve is 6 db and that of the second curve is still zero. The point on the heavy line curve is thus 6 db + 0 db equals 6 db. Continuing this, at $4f_{o1}$ equals $2f_{o2}$, one curve shows an attenuation of 12 db while the other is 6 db down. Since the total is 18 db, this is another point on the heavy line curve.

All points can be connected by a straight line, with the resulting heavy-line curve being the approximation to the complete filter. The characteristics are altered somewhat from the ideal due to the effect of the loading of one filter section on the other. The procedure outlined is more accurate when the two filter sections are isolated from each other. This isolation can be accomplished by placing a high-impedance transistor stage between the two R-C sections. Should f_{o1} and f_{o2} coincide, the rolloff is 12 db per octave, beginning at the coincident corner frequency. This is shown in the plot in Fig. 9-16C. Similar plots for the high-pass filter in Fig. 9-16D are made in Figs. 9-16E and 9-16F.

Characteristic curves of the bandpass filter in Fig. 9-16G are drawn in Fig. 9-16H. The vector additions producing the sum curve defining the response of the total filter are performed as before.

An equivalent to the input circuit of a transistor can take the form shown in Fig. 9-17. While R can be the Thevenin equivalent resistance of the output of a preceding circuit, R_p

may represent the input resistance of a transistor or may be the resistor placed at the input of a transistor circuit. C could be the input capacitance of a JFET or bipolar device. The actual and approximate attenuation curves of this circuit are shown in Figs. 9-15C and 9-15D. However, it is displaced from the 0 db maximum level by a number of db determined by the resistive components in the circuit.

The horizontal portions of this curve, from 0 Hz to f_o Hz, are not at 0 db, as was the case when the curve in Fig. 9-15 was drawn for the circuit in Fig. 9-14B. There is already some attenuation at the low frequencies. At these frequencies, the attenuation is determined by the resistive voltage divider formed by R and R_p, because up to f_o, C is effectively an open circuit. Up to f_o:

$$e_{out} = \left(\frac{R_p}{R + R_p}\right) e_{in} \tag{9-10}$$

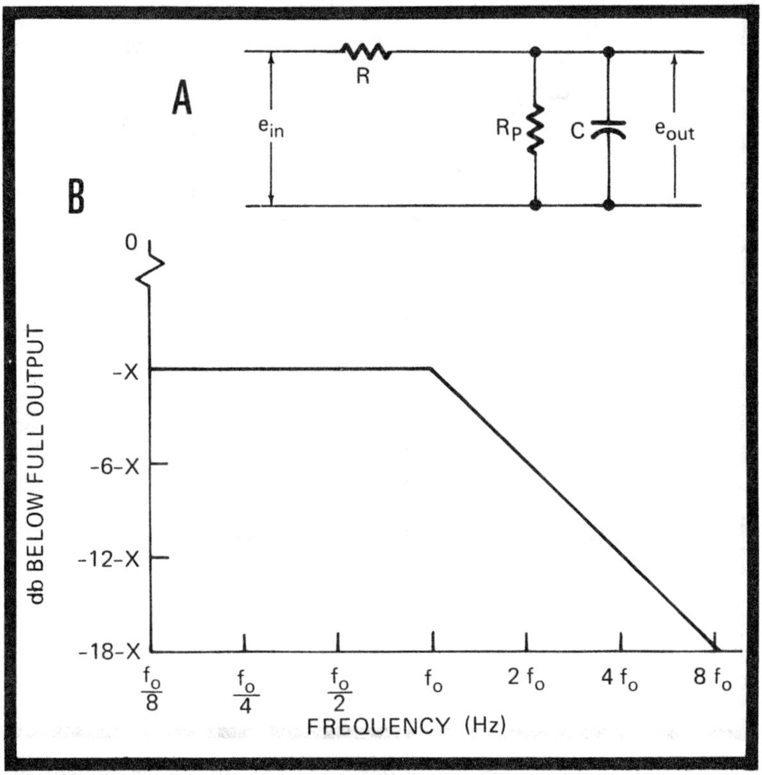

Fig. 9-17. Practical form (A) of the equivalent circuit at the input to a transistor. B is the frequency response curve for the circuit.

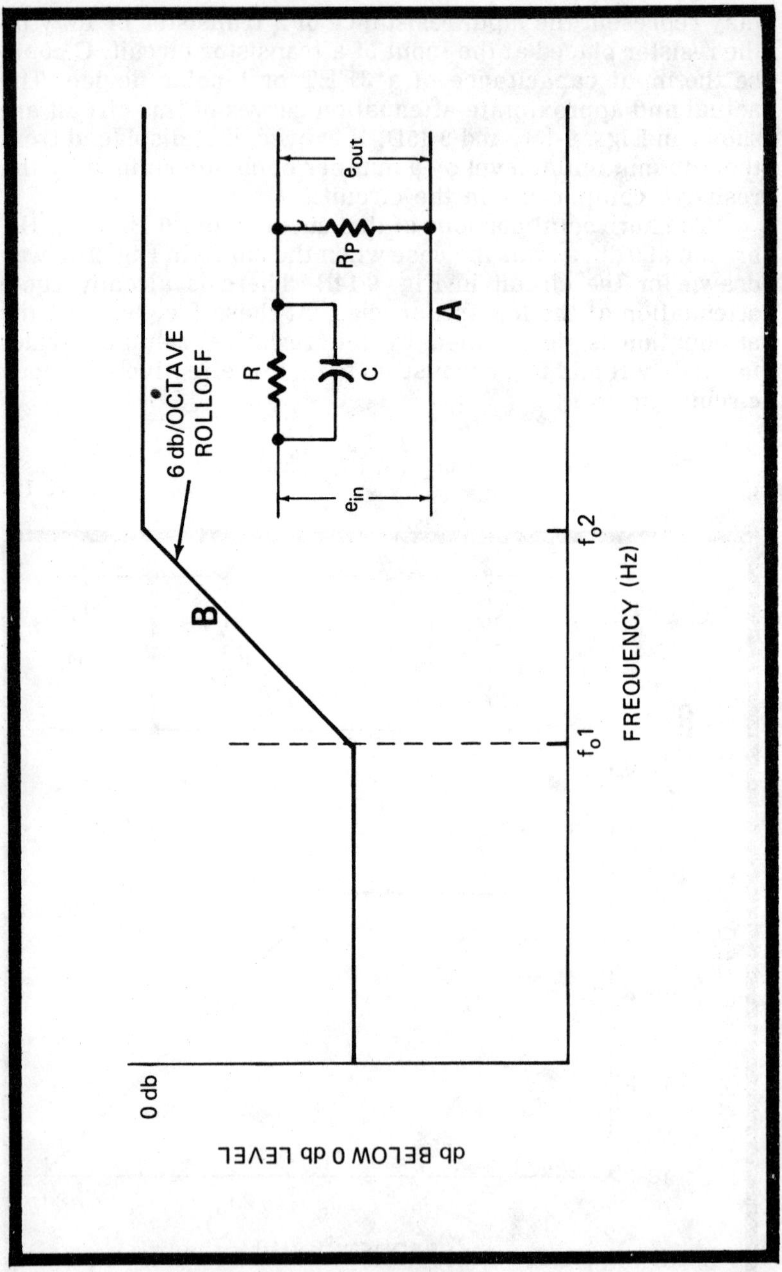

Fig. 9-18. Practical high-pass filter (A) used as the equivalent of a coupling circuit for a transistor amplifier. B is the frequency characteristic curve for the circuit.

so that the db attenuation at these frequencies can be determined from the equation:

$$\text{dB attenuation} = 20 \log_{10}\left(\frac{e_{out}}{e_{in}}\right) = 20 \log_{10}\left(\frac{R_p}{R + R_p}\right) \qquad (9\text{-}11)$$

The corner frequency can be determined by shorting the input voltage, e_{in}, while the output circuit at e_{out} is left open. Then note the effective resistance shunting the capacitor, C. For this circuit, R is in parallel with R_P across C. Using the symbol $R \| R_p$ to represent the parallel resistor combination:

$$f_o = \frac{1}{2\pi\,(R \| R_p)C} \qquad (9\text{-}12)$$

This equation is identical in form to Equation 9-9, where 2π equals 6.28.

A high-pass filter circuit frequently encountered is shown in Fig. 9-18A. In a practical circuit, R_p can be the input resistance of a transistor circuit, while C may be a coupling capacitor shunted by a resistor, R. R may be the resistance due to the leakage of an electrolytic coupling capacitor. The frequency characteristic curve for this circuit is also shown in Fig. 9-18.

As was the case with the circuit in Fig. 9-17, the output from 0 Hz to f_{o1} Hz is dependent only upon the resistive voltage divider formed by R and R_p and can be determined using Equation 9-11. The lower corner frequency can be calculated from the parallel RC circuit when e_{in} is open circuited and the output is shorted at e_{out}. It is simply equal to 1 divided by 6.28RC as stated in Equation 9-9. The curve from f_{o1} to f_{o2} has a 6 db per octave slope. Equation 9-12 can be used to determine f_{o2}. As before, e_{in} is shorted to determine the resistance across the capacitor, $R \| R_p$, while e_{out} remains open.

$$f_{o2} = 1/6.28\,(R \| R_p)C \qquad (9\text{-}13)$$

The curve in Fig. 9-18 is actually the sum of two curves, for there are two corner frequencies. The addition is performed in a manner similar to that described for Figs. 9-16 and detailed in Fig. 9-19.

Let us now see just what we did to determine the corner frequencies. One corner frequency, f_{o2}, is determined from the parallel R-C circuit when e_{in} is shorted and e_{out} is left open circuited. For the high-pass filter, the low frequency rolls off or decreases below f_{o2}. The second corner frequency, f_{o1}, is determined from the parallel R-C circuit when e_{in} is open circuited and e_{out} is short-circuited. In this case, the output

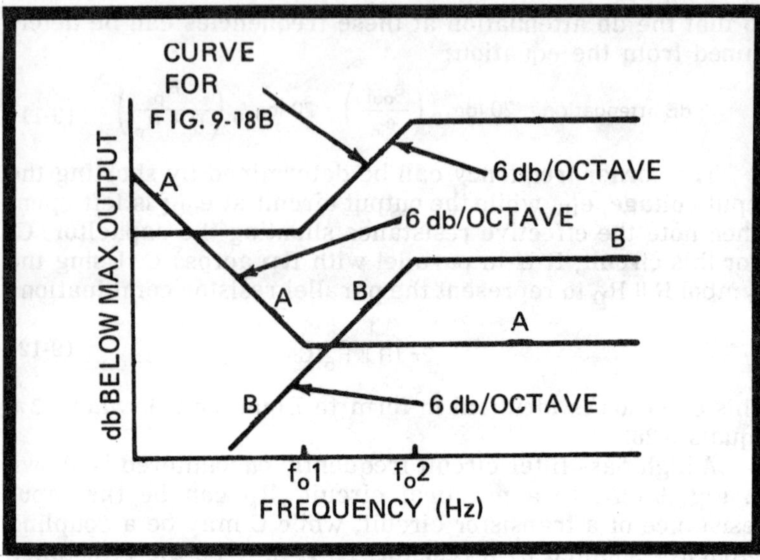

Fig. 9-19. Curves A and B are used to generate the sum curve in Fig. 9-18B. f_{o1} and f_{o2} must be determined for Curves A and B.

increases at the rate of 6 db per octave from the corner frequency.

One more coincidence is evident with respect to the low-pass filter in Fig. 9-17 and the representative curve in Fig. 9-15C. The corner frequency was determined by shorting e_{in} while e_{out} was left open. The curve above f_o rolled off or decreased at the rate of 6 db per octave. This rolloff was common to both the high- and low-pass filters when the corner frequency was determined after e_{in} is shorted while e_{out} is left untouched. We can now write two rules.

1. Determine one corner frequency by shorting the input while the output remains open. Substitute the equivalent resistance of all resistors in parallel with C into Equation 9-9. The curve will roll off or decrease at the rate of 6 db per octave from this frequency. The direction of the rolloff will depend upon whether the circuit is a high- or low-pass type of filter.

2. Determine a second corner frequency by opening the input circuit while shorting the components at e_{out}. Substitute the equivalent resistance of all resistors in parallel with C. The curve will rise at the rate of 6 db per octave from the frequency determined by using all resistance across C. If the parallel R-C combination is in series with the signal, the output will rise to 0 Hz as shown on curve A in Fig. 9-19. If a parallel R-C network is a component at the output terminals of

the filter, the curve will rise to ∞ Hz beginning at the corner frequency.

Let us apply these two rules to the circuit in Fig. 9-20A. Since C is across a portion of the output, it is essentially a low-pass filter circuit. Determine f_{o1} by shorting the input e_{in}. The equivalent resistance is $(R + R_S)$ in parallel with R_P, written $(R + R_S) \parallel R_P$. Thus, f_{o1}, the frequency at which the rolloff starts is 1 divided by 6.28 $(R + R_S) \parallel R_P$ C. Leaving e_{in} open and shorting the component at e_{out}, the second corner frequency, f_{o2} equals 1 divided by $(6.28)(R_P \parallel R)C$, can be determined. The individual curves and the sum of the curves are shown in Fig. 9-20B. The number of db the horizontal

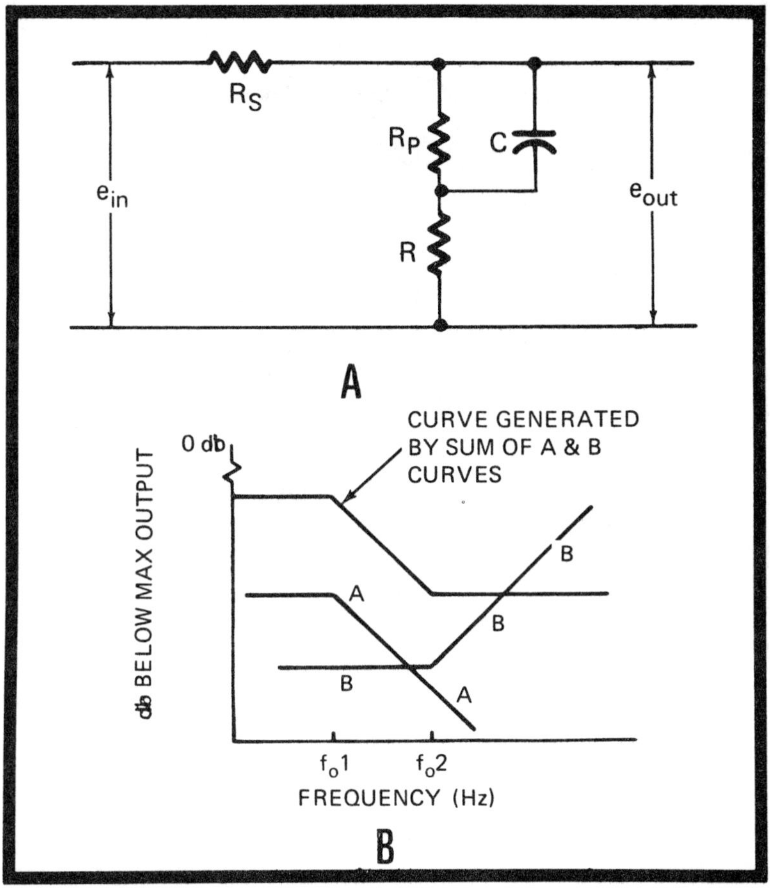

Fig. 9-20. Analysis of the circuit and the final curve generated by adding Curves A and B.

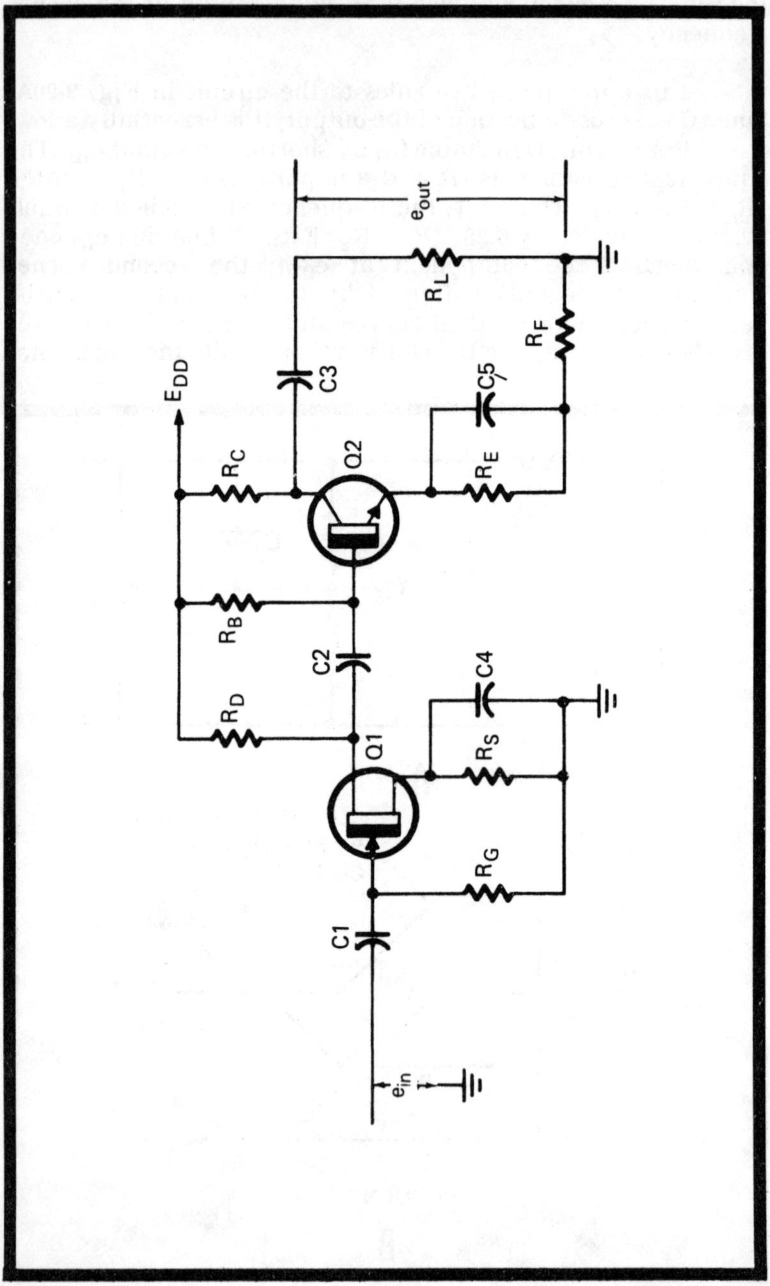

Fig. 9-21. Capacitor-coupled circuit using a JFET and a NPN bipolar transistor.

section of the curve is below the 0 db reference can be determined from the resistor components and is equal to:

$$\text{dB below zero} = 20 \log_{10} \left(\frac{R_p + R}{R_p + R + R_s} \right) \qquad (9\text{-}14)$$

Frequency Characteristics of an R-C Stage

The circuits discussed can be applied to practical R-C and direct-coupled stages. A possible circuit is shown in Fig. 9-21. The input signal is fed through capacitor C1 to the gate of a JFET. The voltage developed across R_D is coupled through capacitor C2 to the base of a bipolar transistor. The output of the 2-stage amplifier, appearing across R_C, is fed to load resistor R_L through capacitor C3.

Should e_{in} be a voltage source, the corner frequency for the circuit consisting of R_G and C1 is 1 divided by $6.28R_GC1$. The curve picturing the frequency characteristic of the signal across R_G is shown in Fig. 9-15A. The identical curve is used for the circuit involving R_D-C_2-R_{in} (assuming R_B is much larger than R_{in}; R_{in} is the input resistance of the bipolar transistor circuit) where the corner frequency is 1 divided by $6.28C2(R_D + R_{in})$ as well as for the circuit including R_C-C3-R_L with a corner frequency at 1 divided by $6.28C3(R_C + R_L)$.

Corner frequencies are also affected by the capacitor-resistor combination in the drain and emitter circuits. Due to these combinations, the gain rises at the rate of 6 db per octave from the corner frequency up to ∞ Hz. For the JFET the corner frequency is 1 divided by $6.28C_4R_S$, while for the bipolar device it is 1 divided by $6.28C5R_E$.

The unbypassed resistor, R_F, in the emitter, in conjunction with C5 and the balance of the circuit, comprise a circuit with the frequency characteristic shown in Fig. 9-15C. The corner frequency here is:

$$\frac{1}{6.28\,C5} \left(\frac{R_E + R_Q}{R_E R_Q} \right). \qquad (9\text{-}15)$$

where R_Q is $R_F + r_e +$ (resistance seen by the base circuit) divided by β . R_Q is the resistance seen by R_E looking back into the emitter circuit of the transistor.

Response curves are all approximate due to stray and component capacitance and resistance ignored in the calculations. All calculations should be checked in the laboratory.

The load lines for the input transistor in this circuit (the JFET) are drawn in Fig. 9-22. The DC load line involves all

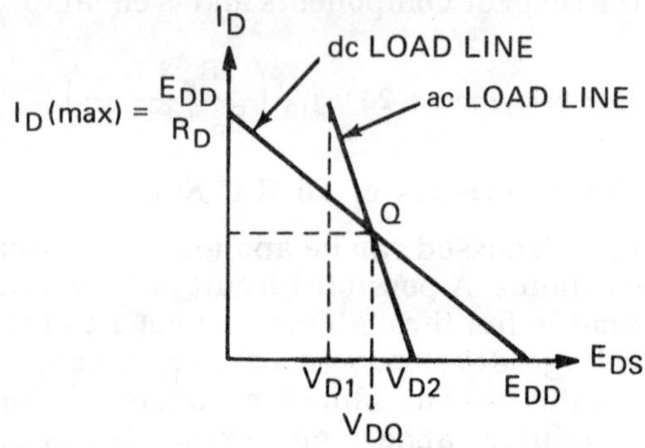

Fig. 9-22. Load lines of the input transistor. In Fig. 9-21, it is an FET. The construction also applies to a bipolar transistor circuit.

resistors in the drain circuit and is the sum of resistors R_D and R_S. If, for this discussion, R_S is negligible compared to R_D, the maximum drain current shown in the drawing is $I_{D(max)}$ equals E_{DD} divided by R_D. The quiescent current and voltage point is normally chosen at the center of the load line when the transistor is operated in Class A.

At the mid-frequencies, all capacitors are considered as short circuits. The mid-frequency AC load line, therefore, is determined by R_D shunted by R_B, while this combination is in turn shunted by the impedance looking into the base circuit of the bipolar transistor or $r_b + \beta\ (r_e + R_F)$. Assign R_{ac} as the symbol for the resistance shunting R_D when the AC load line is considered. The effective AC load line is then determined by R_{ac} in parallel with R_D. It passes through the quiescent point on the DC load line.

Notice that the possible voltage swing across the DC load resistance ranges from 0 to E_{DD}. Shunting resistor R_D with R_{ac}, as is required to determine the AC load line, the possible output voltage swing is limited from V_{D1} to V_{D2}. It is obvious that the larger R_{ac} is made with respect to R_D, the larger the voltage swing of the output signal can be. R_{ac} is normally designed to be at least equal to R_D, although it is most desirable that its resistance be more than ten times that of the resistor in the drain circuit.

Should the input transistor be of the bipolar variety, the identical considerations are involved when the resistor in the collector circuit is shunted by other resistors capacitively coupled to it to compose the AC load line. Regardless of the input transistor, R_{ac} at least partially depends on the loading the second device places on the output circuit of the first transistor. If the second transistor is a JFET, the gate resistor of the second amplifier stage must be considered. Should the second device be a bipolar transistor, R_{ac} is the sum of the base resistance, r_b, with beta times the emitter resistance and the unbypassed resistor R_F in the emitter circuit. All this is shunted by the components used to bias the device.

As in the case of all coupled circuits, the mid-frequency gain is the product of the AC gains of each stage when the gains are calculated using the proper AC load at the output. The proper AC load includes all components in, and the loading effect of, the succeeding stage.

Coupling Without Capacitors

Capacitor C2 in Fig. 9-21 is vital in keeping the DC from the first stage out of the input circuit of the next stage. Improper DC voltage appearing at the input of the second transistor will upset the quiescent bias conditions of the device.

Capacitors in coupling circuits introduce several disadvantages. They are frequency selective. They limit the swing of the output voltage. They may cause noise pulses to be produced at the output while the capacitors are charging. They may cause a delay in the establishment of the final stable bias conditions and a delay before the transistor begins to amplify. They may cause the transistor to conduct heavily while charging the capacitor, causing the semiconductor to break down. These drawbacks can be overcome by using direct-coupled circuits.

DIRECT-COUPLED CIRCUITS

Of all methods used to couple one transistor to another, the direct-coupled circuit is the most versatile. Its importance is further enhanced by its ability to accept considerable amounts of feedback (discussed in the next chapter) around the circuit without unduly subjecting the amplifier to low-frequency instability problems. However, quiescent bias point stability problems are much more stubborn than those encountered when using other methods of coupling.

Bias and Stability

A typical direct-coupled amplifier circuit using bipolar devices, is shown in Fig. 9-23. A signal, e_{in}, is fed to the base of

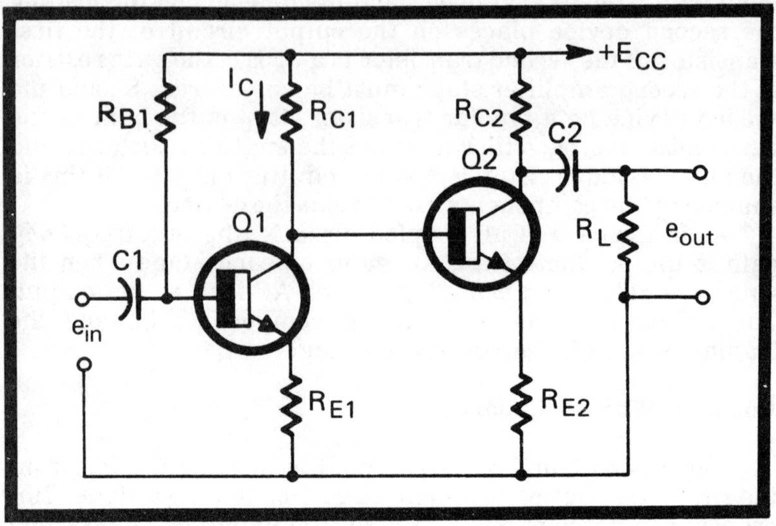

Fig. 9-23. Direct-coupled amplifier circuit.

Q1 through C1, amplified by the direct coupled stages, Q1 and Q2, and fed to the load through C2.

The quiescent base current through the first transistor is established by R_{B1} and R_{E1}, as well as by the DC beta, β_{1DC}, of Q1. As you may recall, the base current of Q1 is I_{B1} equals E_{CC} divided by $(R_{B1} + \beta_{1DC}R_{E1})$. The collector current is beta times the base current.

It is a more complex task to determine the base current, I_{B2}, of Q2. The circuit can be analyzed using Thevenin's equivalent methods, as follows. First, disconnect the collector circuit of Q1 from the base circuit of Q2. This means disconnecting the base lead of Q2 from the collector of Q1 as well as disconnecting the ground loads between R_{E1} and R_{E2}. Stage 1 is now isolated from amplifier stage 2. Find the voltage at the collector of Q2 under these conditions.

I_{B1} was determined above. I_{C1} is $\beta_{IDC}I_{BI}$. Only I_{C1} flows through R_{C1} when the base of Q2 is not in the circuit. The voltage at the collector, therefore, is $V_{C1}(TH)$ equals E_{CC} - $I_{C1}R_{C1}$. This is the Thevenin equivalent voltage of the circuit. The Thevenin resistance of the circuit can be found by connecting the supply, E_{CC}, to ground. Resistor R_{C1} is essentially in parallel with r_{d1}, the collector resistance of transistor Q1. In most cases, r_{d1} is much larger than R_{C1}, so that the Thevenin resistance is only R_{C1} equals $R_{C1}(TH)$.

The Thevenin equivalent circuit consists of $V_{C1}(TH)$ in series with $R_{C1}(TH)$. The base and ground leads of Q2 can now be connected to the Thevenin equivalent of the output circuit of Q1. The DC resistance at the base of Q2 is the DC beta, β_{2dc}, of Q2 multiplied by the resistors in the emitter circuit or $\beta_{2dc}R_{E2}$. In Fig. 9-24, the Thevenin equivalent is shown connected to the equivalent resistance in the base circuit. Ohm's Law can be applied to determine that the base current of Q2 is V_{C1} (TH) divided by (R_{C1}(TH) + $\beta_{2dc}R_{E2}$).

Voltage gain of the second stage is highly dependent upon the resistor in that emitter circuit. It is a simple matter to make R_{E2} small and get a large amount of gain. But since this resistor is an important factor in setting the base bias current, a small R_{E2} may produce improper bias conditions for Q2. The problem may be overcome by placing a parallel R-C combination in series with a small R_{E2}. Capacitor C is made so large as to make the impedance of the R-C combination completely negligible at the lowest frequency to be reproduced by the amplifier. While the gain remains a function only of R_{E2}, the base current is dependent upon β_{2dc} multiplied by the sum of R_{E2} and the R in the parallel resistor-capacitor circuit.

Direct-coupled circuits such as shown in Fig. 9-23 suffer from bias drift problems. In R-C coupled circuits, one stage is

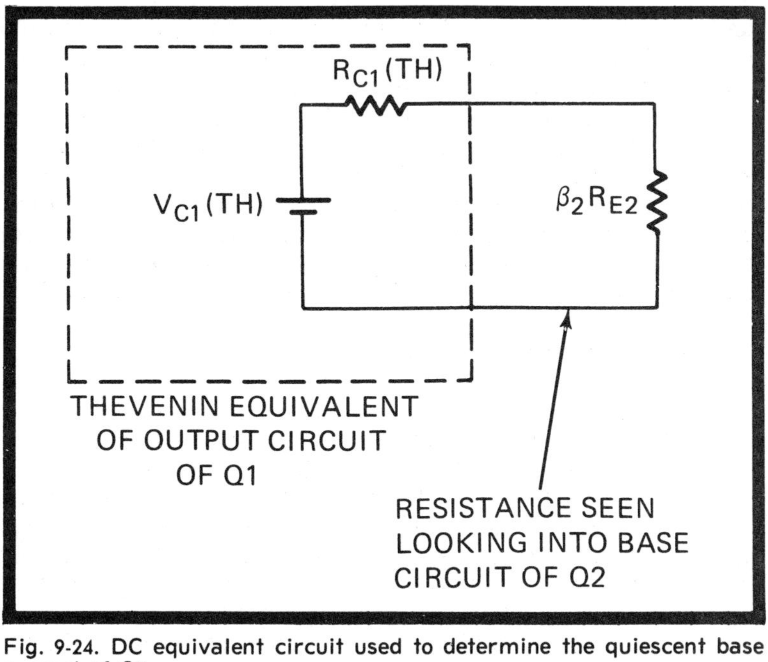

Fig. 9-24. DC equivalent circuit used to determine the quiescent base current of Q2.

isolated DC-wise from the other by the coupling capacitor. Each circuit in this type of amplifier must be stabilized within itself. As far as bias stability is concerned, one stage does not affect the preceding or succeeding ones.

As for DC-coupled circuits, any drift in collector current in one stage is passed on to the succeeding amplifier and multiplied in magnitude. The stability factor, S, for the overall circuit in Fig. 9-23 is:

$$S = S_1 \left(\frac{\beta_{2dc} R_{C1}}{\beta_{2dc} R_{E2} + R_{C1}} \right) \tag{9-16}$$

S_1 is the stability factor of the first transistor stage. This can be determined using methods described in Chapter 2.

Three or more stages can be coupled, thus extending the circuit in Fig. 9-23 to include more transistors. For a 3-stage amplifier, the stability factor is as stated in Equation 9-16, multiplied by:

$$\left(\frac{\beta_{3\,dc} R_{C2}}{\beta_{3\,dc} R_{E3} + R_{C2}} \right) \tag{9-17}$$

where the factors with the subscript 3 refer to components in and characteristics of the third stage. Additional similar multiplying factors are used in the equation for each additional stage of amplification added to the circuit.

Stability problems are also emphasized in direct-coupled JFET circuits. They must be designed to assure the maximum stability in each amplifier stage, with the primary emphasis being placed on the stability of the first stage of amplification. The stability problem may be alleviated by using a balanced circuit in a differential amplifier arrangement. Popular circuits of this type are used in high quality power amplifiers.

Bipolar Differential Amplifier

Two useful differential amplifier circuits are shown in Fig. 9-25. Although both circuits have many similar qualities, each has its own advantages and disadvantages.

Both differential amplifiers amplify the **difference** between the two signals at their inputs. The voltage, $e_{01} - e_{02}$, appearing between the two collectors in Fig. 9-25A is related linearly to the difference of the input signals, $e_{s1} - e_{s2}$. Voltage e_0 across R_C in Fig. 9-25B is similarly related to $e_{s1} - e_{s2}$.

Two identical signals, when applied in phase to the inputs of a differential amplifier, should produce zero volts at the output. This type of signal can be applied in series with R_E, as is normally the case with noise voltages rather than at e_{s1} and e_{s2}. It is referred to as common-mode operation. A figure of merit to determine how well this signal is rejected by the circuit is known as the common-mode rejection factor, CMR, and is defined by the equation:

$$CMR = \frac{A_v \text{ (Differential Mode)}}{A_v \text{ (Common Mode)}} \tag{9-18}$$

A_V (differential mode) is the voltage gain when the input voltages are equal but with opposing phase (applied at e_{s1} and e_{s2}). A_V (common mode) is defined as the differential voltage gain when the input signals are equal and in phase.

It is desirable that the CMR be very large. This is accomplished by making R_E large. Being common to the emitter circuits of the two transistors, in the differential mode two identical but out-of-phase audio signals flow through R_E. The gain is thus not affected by its presence or size. However, the quiescent emitter currents of Q1 and Q2, $I_{EQ1} + I_{EQ2}$, also flow through R_E. A large DC voltage is developed across R_E. This dictates that the $-E_{EE}$ supply be very large.

To overcome this drawback, R_E can be replaced by a circuit simulating a constant-current source. The desirable characteristic is that the AC impedance of this type of circuit is large, while the DC voltage drop across the semiconductor device in the circuit is small. A basic circuit to accomplish this is shown in Fig. 9-26.

The divider network in the base circuit is supplied by a fixed voltage. A reasonably stable voltage is developed across R_X. Consequently, the same stable voltage is also across the series circuit in parallel with this resistor, namely R_E and the Q base-emitter diode. Because the voltage across the base-emitter junction is relatively constant regardless of the current through the junction (over a limited current range), the voltage across R_E is constant. Using Ohm's Law, current through R_E is the voltage across that resistor divided by the size of the resistor. As this is also the emitter current of the transistor, it is just about equal to the collector current as well.

The AC output impedance of the circuit is equal to the ratio of the change in collector-to-emitter voltage to the change in collector current. Since the change in current is just about equal to zero (for it is constant), the impedance is high.

Improvements in this circuit are possible. R_X may be replaced by two series-connected forward-biased diodes

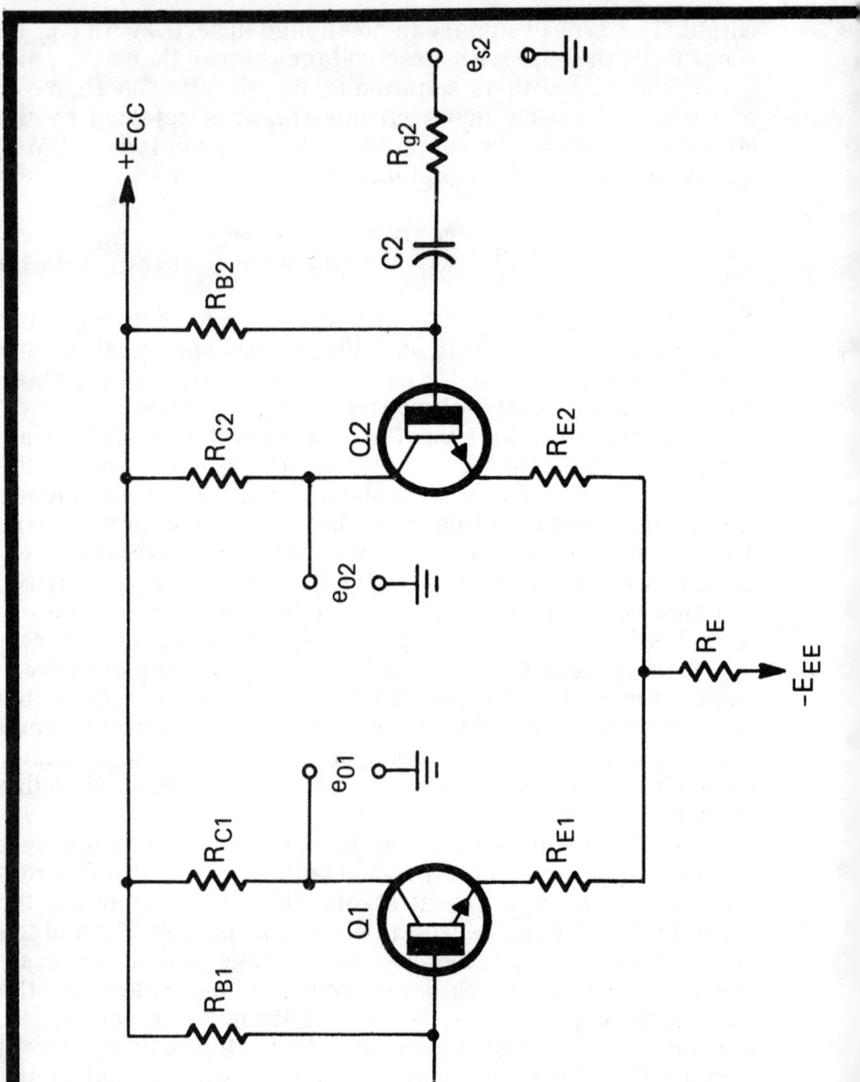
+ECC
RB2
RC2
Rg2
C2
es2
Q2
RE2
e02
RE
-EEE
e01
RC1
RE1
Q1
RB1
C1
Rg1
es1
A

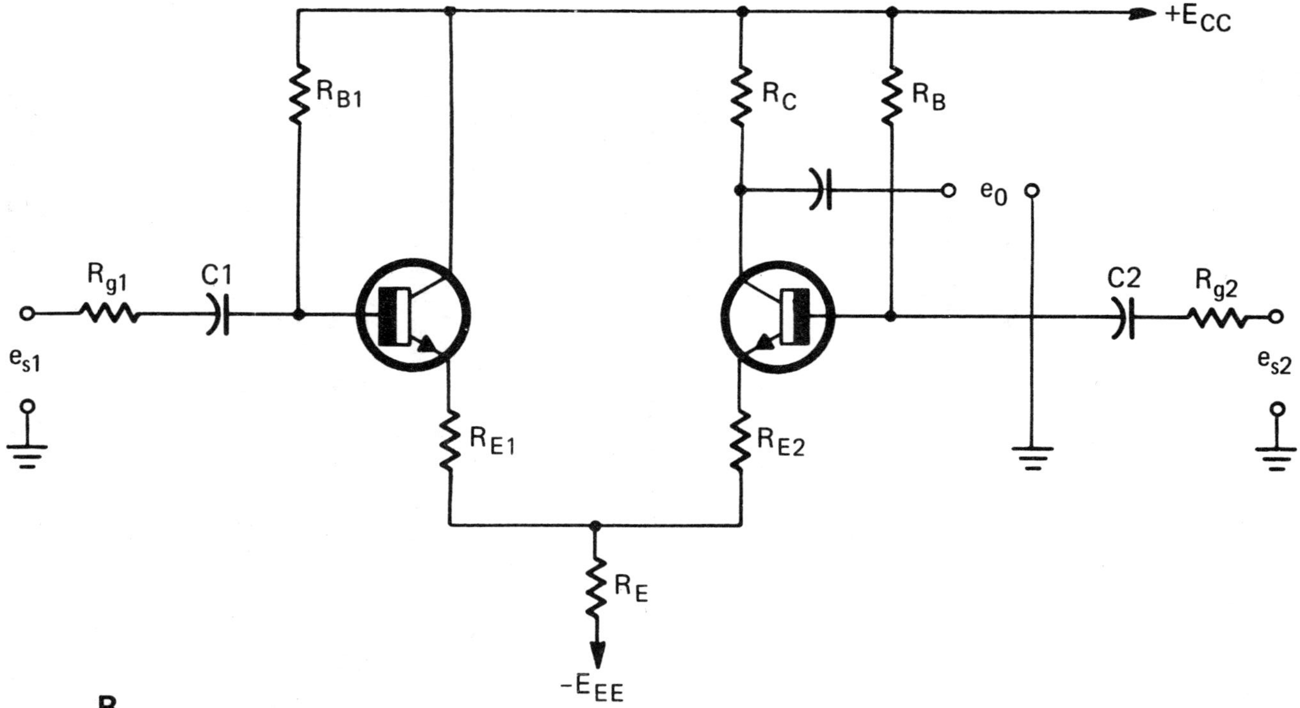

Fig. 9-25. Two versions of differential amplifiers: Balanced differential amplifier (A) and an unsymmetrical differential amplifier (B).

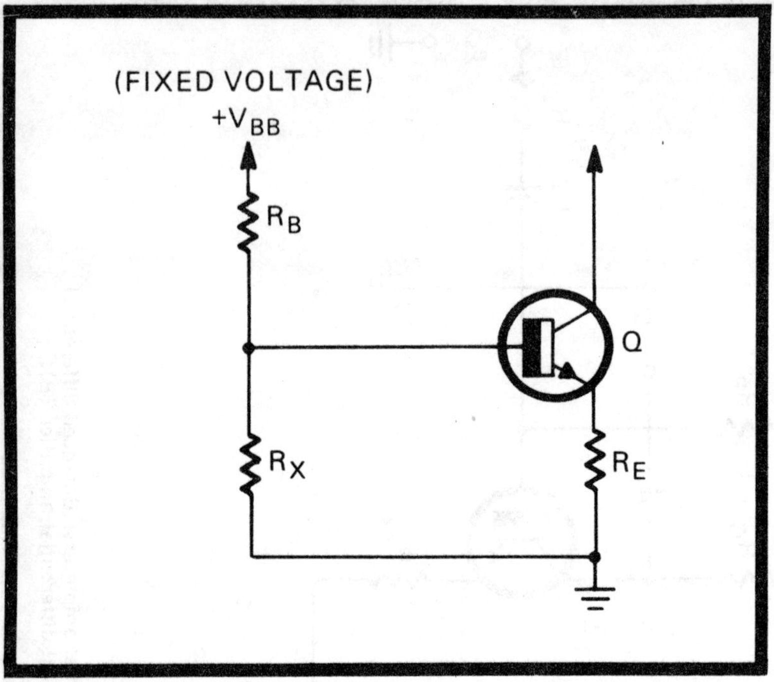

Fig. 9-26. Basic constant-current source circuit. A high impedance exists at the collector of Q.

where the cathodes of both devices are oriented toward ground. The diode current is held relatively constant by virtue of the fact that the magnitude of this current is determined by a current flowing through a large resistor, R_B. The voltages across the diodes vary little with changes in current through the devices.

Because the diodes are in parallel with the base-emitter junction and R_E, the voltage across one diode is identical (when matched properly) to the voltage across the base-emitter junction. The voltage across the second diode is identical with that across R_E. Since the voltage across a forward-biased diode is relatively constant with minor changes in diode current, the voltage across and hence the current through R_E are both stable. Since the change in emitter and hence the collector currents are zero, the requirement of establishing a constant current source has been met!

The quality of the circuit is further enhanced when R_X in Fig. 9-26 is replaced by a zener diode with the anode lead closest to ground. Because of its low impedance when in the breakdown region, the voltage across the zener can, within any reasonable boundaries of operation, be considered a constant.

The AC impedance seen at the collector of the circuit is the same as the output impedance of a common-emitter circuit. It is:

$$Z_{out} = \frac{r_d\left[\beta(R_E + r_e) + (R_G + r_b)\right]}{R_E + r_e + R_G + r_b} \qquad (9\text{-}19)$$

where r_d is the collector impedance of the device in the common-emitter mode of operation, equal to r_c divided by β; r_e is the emitter resistance equal to 26 divided by I_E when I_E is the collector current expressed in milliamperes; R_G is the resistance of R_X or the impedance of the semiconductors substituting for R_X; and r_b is the base resistance, usually assumed to be about 500 ohms.

FETs can be used effectively as constant-current sources. When operated in the pinch-off region, the drain current is relatively independent of the drain-source voltage. All that is required is to establish the gate-source bias, and the fixed drain current can be determined from the equation:

$$I_D = I_{DSS}\left(1 - \frac{|V_{GS}|}{|V_p|}\right)^2 \qquad (9\text{-}20)$$

discussed in Chapter 3. When the gate-source voltage is zero, the drain current is obviously constant at I_{DSS}.

More generally, the circuit in Fig. 9-27 should be considered. Here, the bias voltage is developed across R_S and the gate-source voltage is the drain (and source) current, I_D, multiplied by the resistor, R_S. Once the desired drain current has been established, R_S can be determined from the equation below. It was stated in Chapter 3 in another form.

$$R_S = \frac{V_p}{I_D}\left[1 - \left(\frac{I_D}{I_{DSS}}\right)^{1/2}\right] \qquad (9\text{-}21)$$

where I_D is the desired drain current, I_{DSS} is the drain current when V_{GS} equals 0, V_p is the pinch-off voltage of the FET in question. The impedance at the drain is equal to the common-drain output impedance of the device:

$$Z_{out} = r_{ds}(1 + g_m R_S) + R_S \qquad (9\text{-}22)$$

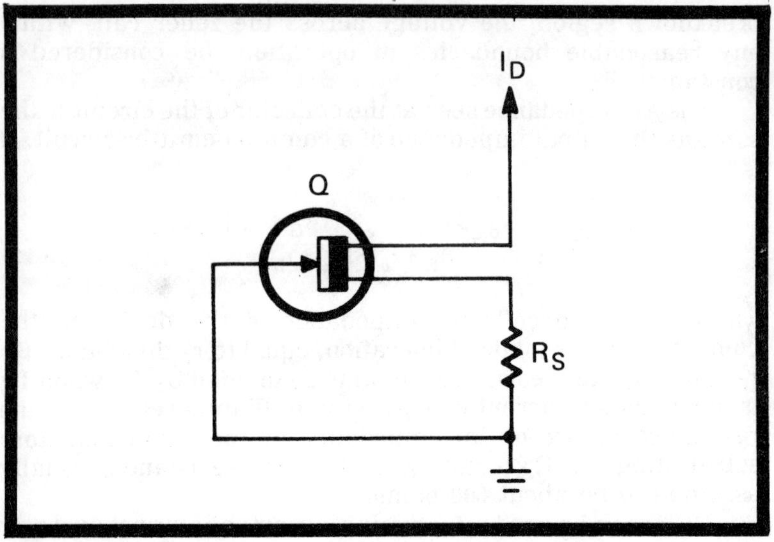

Fig. 9-27. FET constant-current circuit. A high impedance exists at the drain of Q.

where r_{ds} is the drain-to-source resistance of the FET, g_m is the transconductance at the point of operation.

Any of the constant-current circuits or simply a large resistor common to the two emitters can be used for the differential amplifiers in Fig. 9-25. The common-emitter circuit is designed to maintain constant the sum of the collector currents through both amplifier circuits, regardless of the input signal. The resistors in the individual emitters do cause degeneration of the signal, but are required to stabilize the difference in the relative emitter currents. In some cases, R_{E1} and R_{E2} are combined into a potentiometer or variable resistor with the wiper arm connected to R_E. It is adjusted for equal current to flow through both amplifiers.

The overall circuit may drift and become unbalanced due to several factors:

1. The leakage current and the base-emitter voltages of the two devices may change because the ambient temperature change at one device is not identical to the change at the second device. While the unsymmetrical circuit in Fig. 9-25B will alleviate the drift problem due to V_{BE} variations, the balanced circuit in Fig. 9-25A will also compensate for I_{CBO} variations.

2. Large temperature changes can cause a shift in the relative output voltages, because the two transistors may not

have identical V_{BE} and I_{CEO} characteristics even at normal room temperature.

3. One transistor may be heated more by the signal than the other device in the differential amplifier circuit.

The drift is never entirely eliminated, but for identical devices this drawback is negligible when the output is taken from between the two collectors. The effect can be minimized by using low-leakage silicon transistors. Further improvement can be realized if both devices are imbedded in one heat sink, or if the power dissipated due to the signal swing is small when compared to the quiescent power dissipated by each device.

The stability factor, S, of each individual section is identical to the S of one half the circuit when it is isolated from the second half. The S for specific circuits is discussed in Chapter 2. To determine the S from the equations discussed previously, do not substitute the large emitter resistor value into the equation. Instead, substitute $2R_E$, or two times the size of the resistor. All other components in the equation are defined as before.

As with all other amplifiers, the ultimate purpose of this amplifier is to provide gain. For the circuit in Fig. 9-25B, the voltage gain is:

$$A_v = \frac{\beta R_c}{2\left[(R_g + r_b) + \beta (R_{E1} + r_e)\right]} \qquad (9\text{-}23)$$

when the output is taken across R_C from the collector to ground. Should the output be taken between the collectors in the circuit in Fig. 9-25A, the voltage gain is:

$$A_v = \frac{\beta R_c}{R_g + r_b + \beta (R_{E1} + r_e)} \qquad (9\text{-}24)$$

When the output is taken from each collector to ground, the voltage gain is one-half the A_V in Equation 9-24.

As far as the input impedance of the circuit is concerned, it is normally double that of each section taken individually (R_E is not involved in this calculation) while the output impedance is essentially equal to the collector resistor. It may be reduced by adding emitter followers at the output.

In audio applications, the input signal is normally fed to the base of Q1 while a feedback signal is fed to the base of Q2. Two out-of-phase outputs appear at the collectors of Q1 and Q2 in Fig. 9-25A to feed a pushpull output circuit. If there is in-

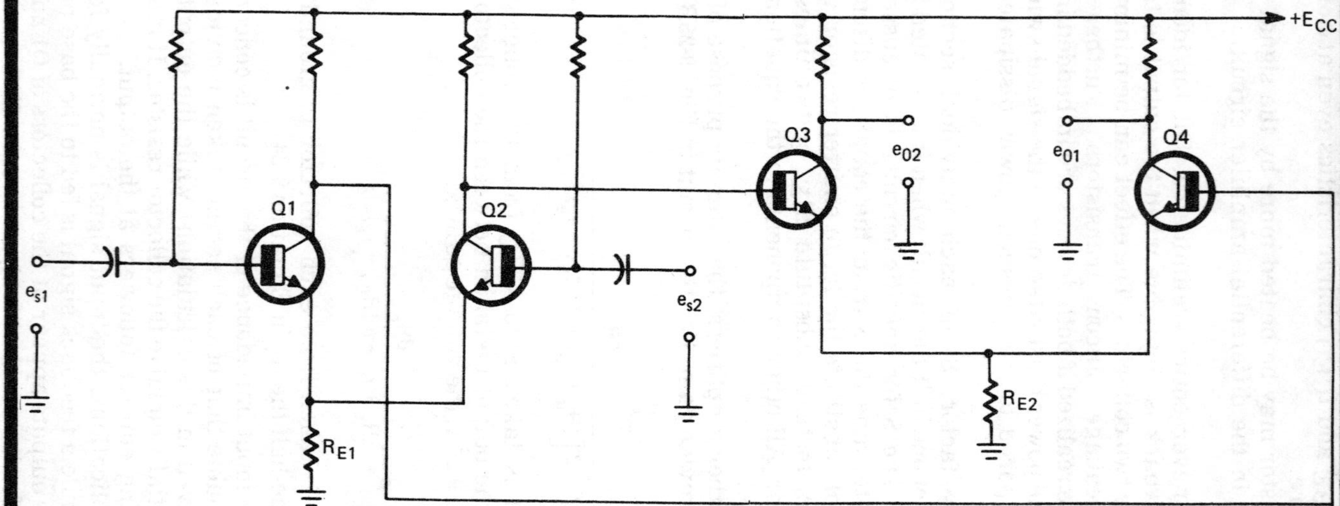

Fig. 9-28. Cascaded differential amplifier circuit using two differential amplifier stages coupled to each other.

sufficient gain, additional differential amplifier stages may be added as shown in Fig. 9-28. Should the circuit in Fig. 9-25B be used, additional phase inversion circuitry must be added to feed a pushpull output stage.

We will now look at the considerations involved in the design of a balanced differential amplifier commonly used in audio circuits. See the circuit in Fig. 9-29. Q1 and Q2 are components in the differential amplifier while Q3 is a constant-current source. In a balanced circuit, Q1 is matched to Q2 for beta, leakage and the V_{BE} characteristic. The equivalent resistors in the circuit are all made equal. As an example, R_{C1} equals R_{C2}.

Once the desired quiescent current, I_{CQ}, through each device has been determined, the current flowing through Q3 must be established at $2I_{CQ}$. Assuming that the voltage, V_D, across one diode in the base circuit of Q3 is equal to that across R_E (since the voltage across the second diode is equal to V_{BE} of Q3), R_E is made equal to V_D divided by $2I_{CQ}$. The current through the diode is usually made equal to the collector current of the constant-current device, so that R_B will be

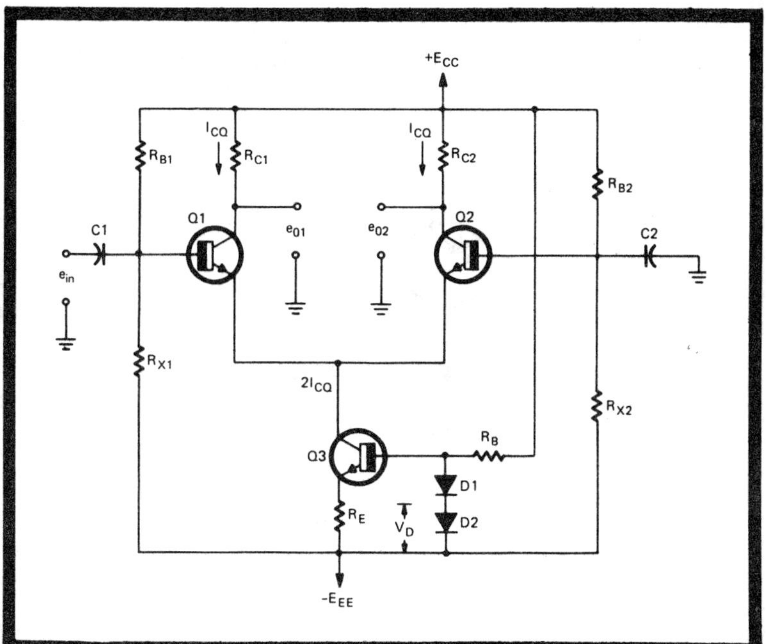

Fig. 9-29. A practical differential amplifier circuit.

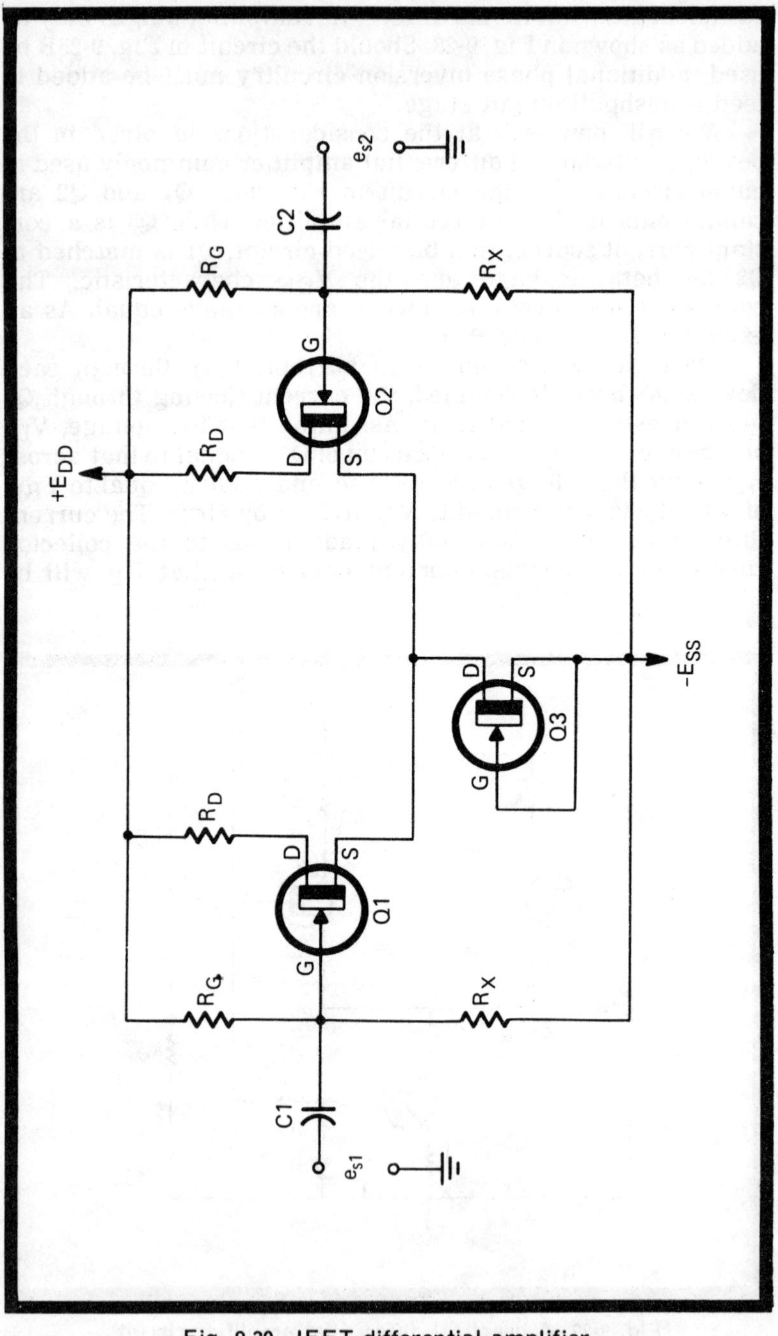

Fig. 9-30. JFET differential amplifier.

approximately (E_{CC} - E_{EE}) divided by $2I_{CQ}$. (Notice that E_{EE} is negative so that (E_{CC} - E_{EE}) is the sum of two voltages.) This assumes the usual case when the voltage across the diodes is negligible compared to a voltage equal to the sum of the two supply voltages. The AC impedance at the collector of Q3 can be determined from Equation 9-13 if we let R_G equal the impedance of the diodes.

The DC quiescent voltages at the collectors of Q1 and Q2 are usually equal to E_{CC} divided by 2 so that R_{C1} and R_{C2} can each be made equal to E_{CC} divided by $2I_{CQ}$.

The voltage at the emitters of Q1 and Q2 can be placed at ground or zero potential. The current through the base resistors may be equated to I_{CQ} so that R_B equals (E_{CC} - V_{BE}) divided by I_{CQ}, where V_{BE} is the base-emitter voltage of Q1 and Q2. R_X is then equal to about (E_{EE}-V_{BE}) divided by I_{CQ}. (Since E_{EE} is negative, (E_{EE} - V_{BE}) is the sum of two voltages. Although this sum is negative, R_X is positive as the direction of the current flow is reversed and hence negative in this equation.) This assumes the base current to be negligible when compared to the currents in the base resistors.

The gain of the circuit, the ratio of (e_{o1} - e_{o2}) to e_{in}, can be determined from Equation 9-24. Here, R_{E1} and R_g equals 0 while r_e equals 26 divided by I_{CQ}, when I_{CQ} is expressed in milliamperes. The gain of each individual section, e_{o1} to e_{in}, or e_{o2} to e_{in}, is half the gain calculated from the equation.

FET Differential Amplifier

A circuit using a constant-current source is shown in Fig. 9-30. Q3 may be replaced with a large resistor to keep the sum of the drain currents through Q1 and Q2 constant, but it requires a large -E_{SS} supply: As for the bipolar circuit, the total current through the FETs, Q1 and Q2, is established by Q3 (here it is the I_{DSS} of Q3) while the AC impedance in the source leads of Q1 and Q2 can be calculated from Equation 9-22.

The voltage gain or the ratio of the signal developed between the two drains to the input signal, is $R_D g_m$, where g_m is the transconductance of either device. The gain from the input to the voltage developed across one R_D resistor is $R_D g_m$ divided by 2. For FET circuits, the common-mode rejection factor, CMR, is equal to $2g_m r_{ds}$, where r_{ds} is the drain-to-source resistance of the device.

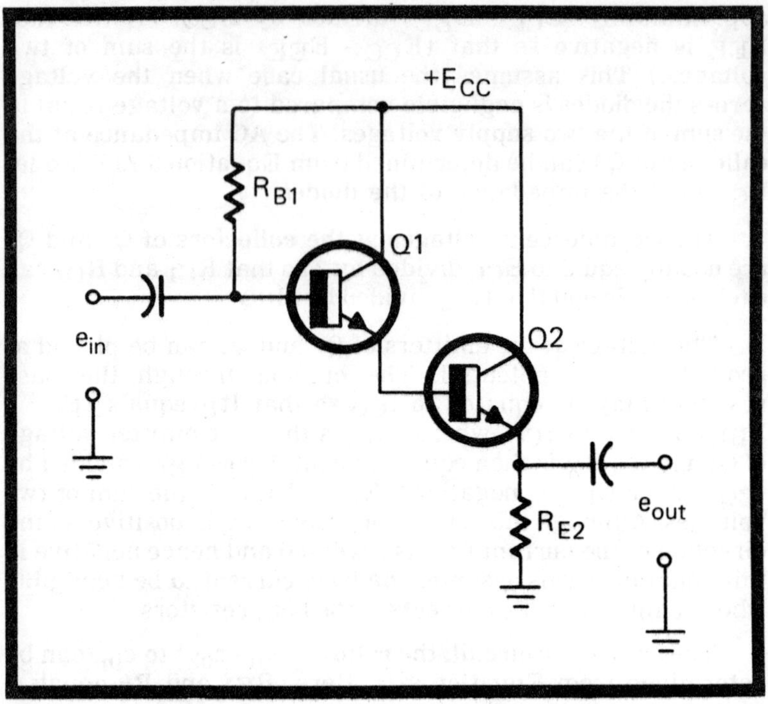

Fig. 9-31. Basic Darlington circuit.

Darlington Circuit

Different DC-coupled transistor arrangements are used to isolate the output circuit from the input or to feed a signal to a circuit with a high input impedance so that the intelligence will emerge from an output with a low impedance. One such circuit is the FET source follower discussed in Chapter 5. The bipolar counterpart, the emitter follower, was likewise detailed in a previous chapter.

As you know, the emitter follower has a voltage gain of somewhat less than 1 and a current gain equal to beta. Take two emitter followers and connect them in series. Known as the Darlington circuit, it is shown in Fig. 9-31. The voltage gain remains at somewhat less than 1, while the current and power gains are the product of the betas of the two transistors.

We must pause here to recall that the beta to be used in determining impedances and the current gain is not merely the AC beta as stated on specification sheets. It must always be modified to include the effect of r_d, the collector-emitter resistance of the device. As you know, the effective beta is the transistor beta multiplied by r_d divided by $(r_d + R_L)$, where R_L is the load resistor regardless of whether it is placed in the collector circuit or the emitter circuit. Although not frequently applied, errors are compounded when this factor is ignored. It is negligible, however, when R_L is much smaller than r_d. The effective beta should be used when evaluating the gain and various impedances of the Darlington circuit if accuracy is an important factor in a specific problem.

There are many variations of the Darlington circuit. For one very common modification, a small resistor is placed from the base of Q2 to ground. The function of this resistor is to increase the collector-emitter breakdown voltage point of the second device. Don't make this resistor too small because it may cause the input impedance of the circuit to be reduced below an optimum level as well as divert excess signal from the base of Q2, reducing the current gain.

A second and useful variation is shown in Fig. 9-32. The circuit is designed so that diode D1 has the same forward characteristics as the base-emitter diode of Q2. Also, let R_{E1}

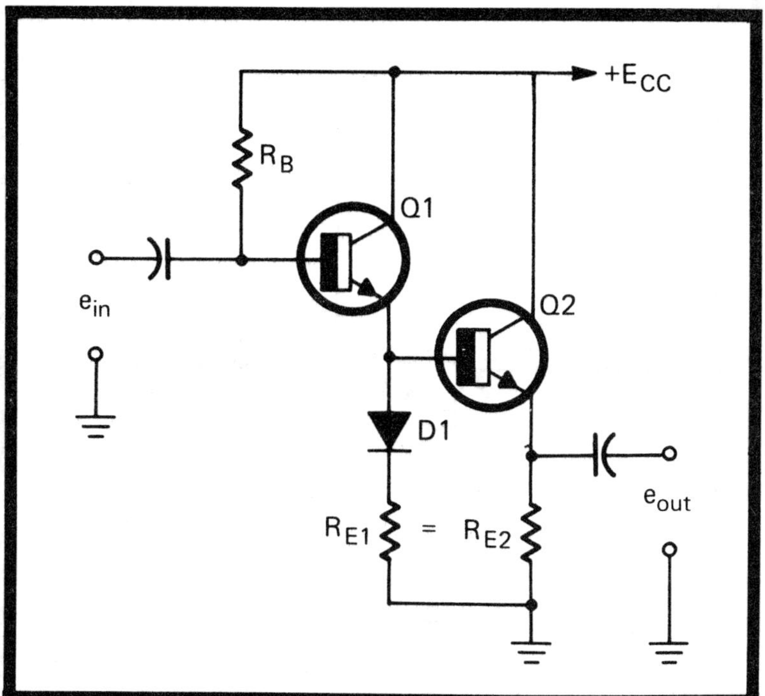

Fig. 9-32. Darlington circuit arranged to make the emitter current through Q1 equal to the emitter current through Q2.

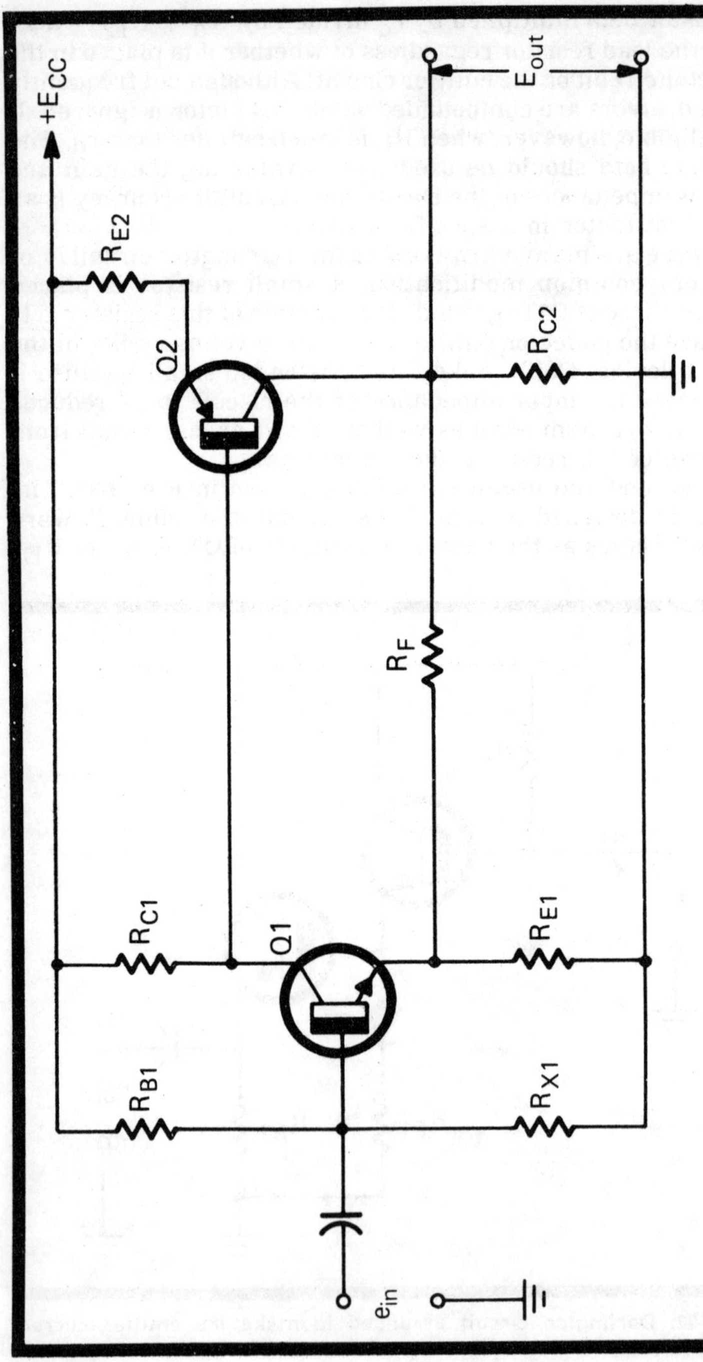

Fig. 9-33. Basic complementary circuit.

equal R_{E2}. Since the D1-R_{E1} circuit is in parallel or across the series combination of the base-emitter diode of Q2 and R_{E2}, the voltages across the two combinations are equal. Since the voltages across D1 and the base-emitter junction of Q2 are equal, the voltage and, hence, the currents through R_{E1} and R_{E2} (as well as through Q2) are identical. Hence, the collector current through Q2 is equal to that through Q1 if the base current flowing through Q2 is considered negligible in comparison. (This circuit is used in Chapter 11 to describe a transformerless power amplifier.)

Complementary Circuits

Another circuit with a total beta equal to the product of the betas of two transistors is shown in Fig. 9-33. An NPN transistor is used at the input, and the output is a PNP device. (An identical circuit can be drawn with the role of the NPN and PNP devices reversed.) The input signal is fed to the base of Q1, and the amplified output appears across R_{C1}. It is in turn fed to the input of Q2. The output signal from the circuit, appearing across R_{C2}, is fed back (feedback is discussed in the next chapter) to the emitter of Q1 through R_F. The various equations describing this circuit are:

$$A_v = \frac{R_F + R_{E1}}{R_{E1}} = R_F/R_{E1} + 1 \tag{9-25}$$

$$A_i = \beta_1\beta_2 \tag{9-26}$$

$$G = A_vA_i \tag{9-27}$$

$$R_{in} = \frac{R_{B1}R_X}{R_{B1} + R_X} \tag{9-28}$$

$$R_{out} = \frac{R_{C2}R_F}{\beta_2R_{C2} + R_F} \tag{9-29}$$

Different components can be omitted from the circuit and the appropriate substitutions made in the equations. Some of the variations are shown in Fig. 9-34.

For Fig. 9-34A, R_{X1} equals ∞, R_{E1} equals 0, R_F equals ∞ and R_{E2} equals 0. Substituting these into Equations 9-25 through 9-29, A_V equals 1, G equals A_i, R_{in} equals R_{B1} and R_{out} equals R_{C2}. The circuit in Fig. 9-34B is identical to the one in Fig. 9-34A, with the exception that R_{C1} has been

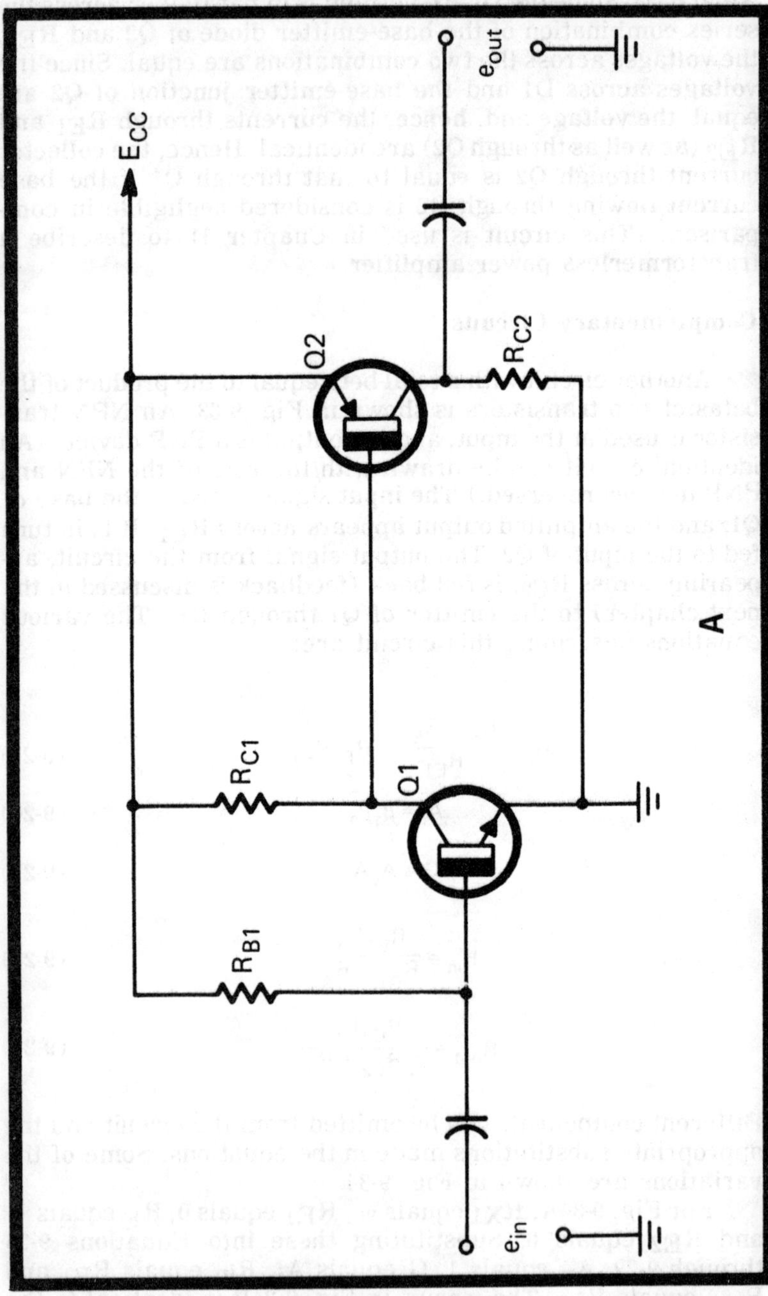

Fig. 9-34A. Variation of the circuit in Fig. 9-33. R_{X1} is ∞, R_E is 0, R_F is ∞, and R_{E2} is 0.

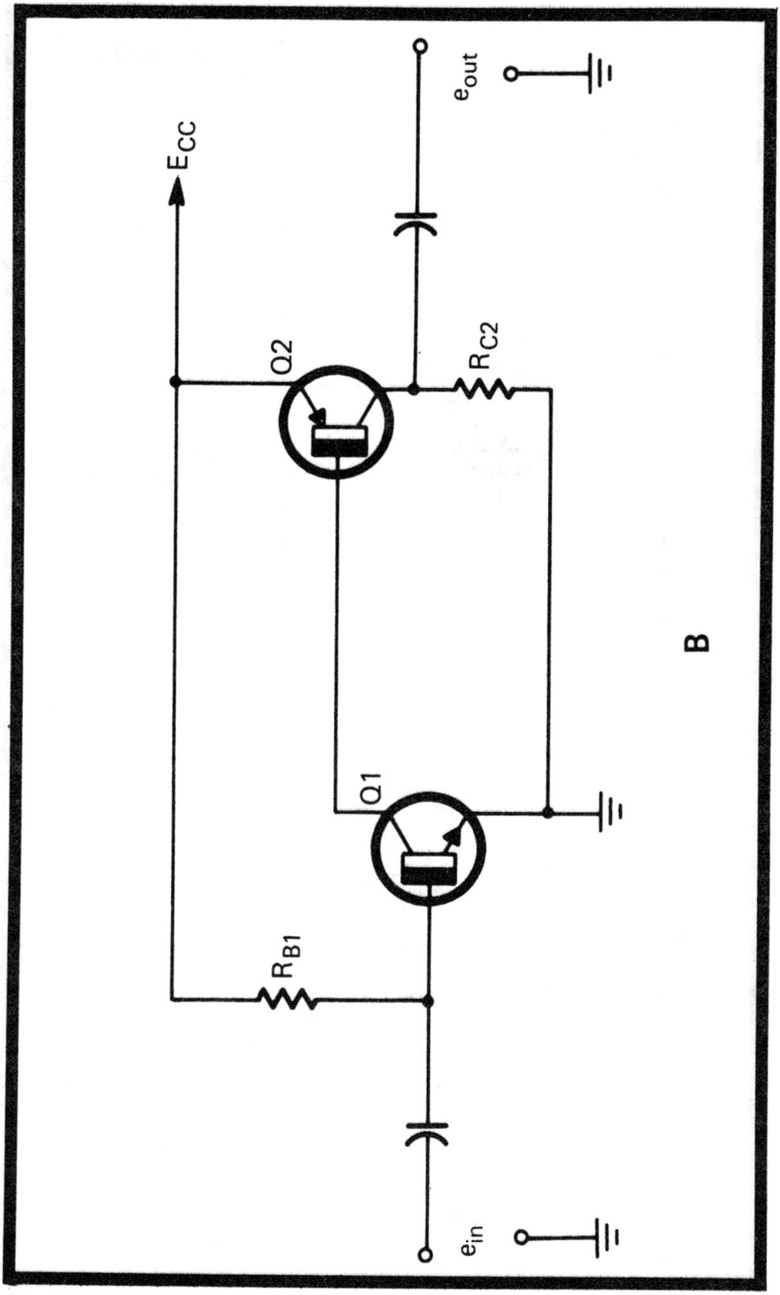

Fig. 9-34B. Variation of the circuit in Fig. 9-33. Circuit at A with R_{C1} equal to ∞ .

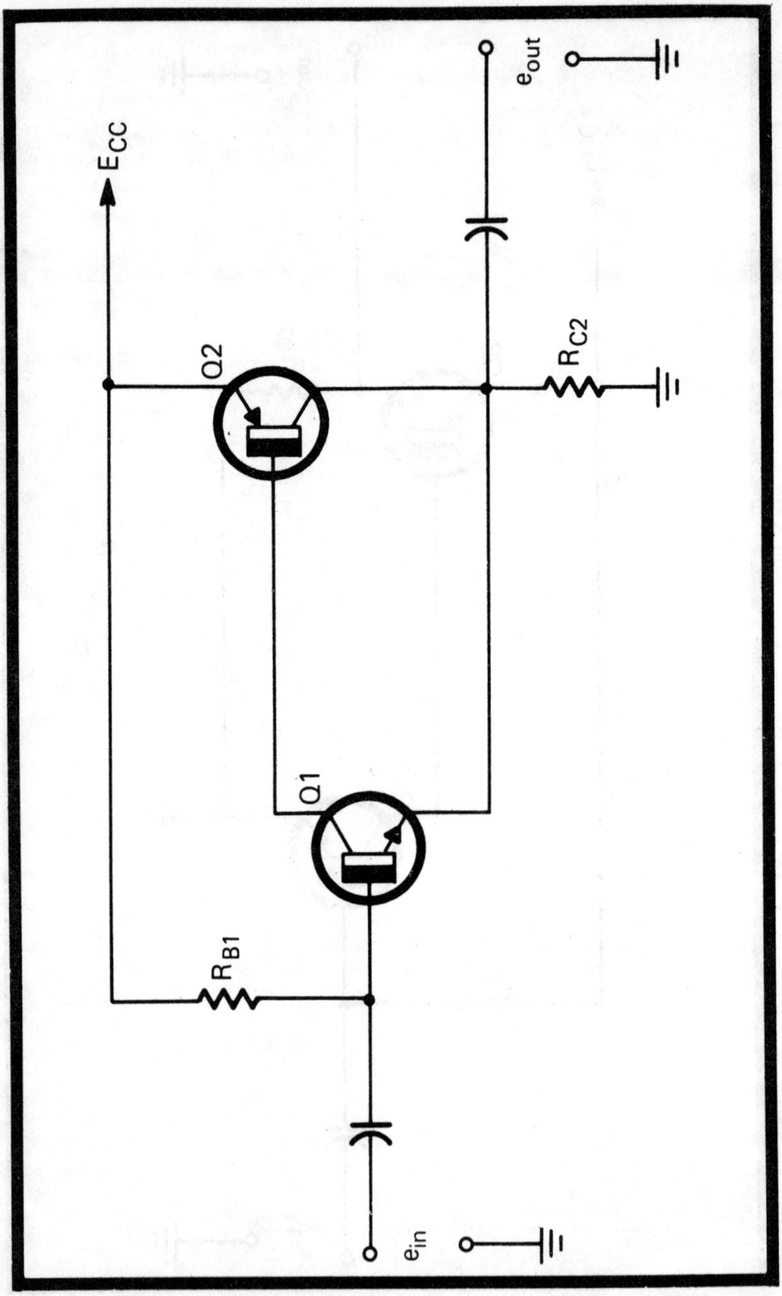

Fig. 9-34C. Circuit in Fig. 9-33 in which R_{X1} is ∞, R_{C1} is ∞, R_{E1} is ∞, R_F is 0 and R_{E2} is 0.

omitted, making it infinite. The impedance and gain characteristics of both circuits are identical.

As for Fig. 9-34C, R_{X1} equals ∞, R_{E1} equals ∞, R_{E2} equals 0, R_F equals 0 and R_{C1} equals ∞. Substituting this information into Equations 9-25 through 9-29, the characteristics of the circuit are A_V equals 1, G equals A_i, R_{in} equals R_{B1} and R_{out} equals 0. Notice the big advantage of this circuit over the ones in Figs. 9-34A and 9-34B. The output impedance here is very low and equal to zero. Since R_F equals 0, there is an extremely large amount of feedback. Distortion is thus extremely low. The circuit in Fig. 9-34C is well worth remembering.

Other complementary circuits are frequently used as the power output stages in pushpull amplifiers. A commonly used one is drawn in Fig. 9-35. The input signal, e_{in}, is fed to the

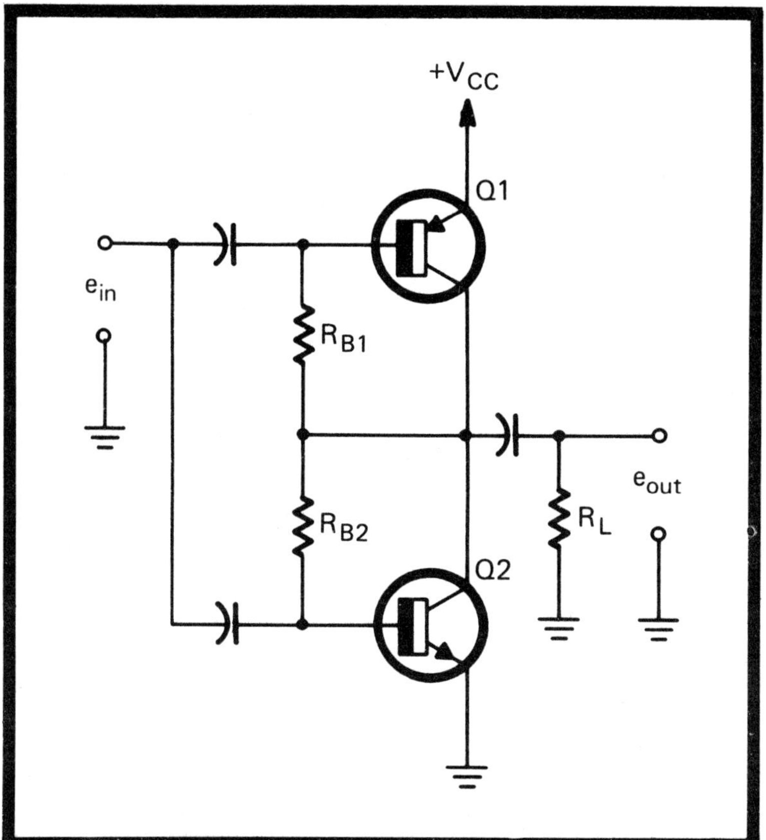

Fig. 9-35. Complementary pushpull output circuit.

bases of the two transistors. The positive half cycles are amplified by Q2 and the negative halves are amplified by Q1. The composite signal is recreated as e_{out}. As in all direct-coupled circuits, here, too, bias stabilization is difficult due to DC drift with temperature. DC feedback is an important factor in alleviating the problem.

Audio Feedback

Chapter 10

In a circuit using bipolar transistors the voltage gain can vary considerably with the beta, emitter resistance and collector resistance of the devices. As for the JFET circuit, the voltage gain is directly dependent upon the transconductance. Consequently, a design may be limited to a few devices only. The circuit will then operate properly with only a narrow group of transistors categorized by a specific range of characteristics.

It would be ideal if the gains and impedances in a circuit were independent of device parameters. Goals of this type can be approached when the designer applies judicial amounts of feedback around his circuit or around the individual device.

A system including feedback uses some method of sensing the output voltage, output current, or both, and feeding the derived information back to an earlier stage or to the input of a single stage. For audio applications, the phase is arranged so that the gain of the overall circuit is usually reduced when feedback is applied (negative feedback). Because the amount of gain is a function of the feedback, the gain of a circuit tends to remain relatively constant regardless of the amplification in the forward circuit before feedback is applied. (The forward circuit amplification is the gain of an amplifier from the input to the output before any feedback loop is closed.)

Types of feedback can be specified in several ways. If the output voltage is sensed, it is referred to as voltage feedback. Obviously, current feedback involves the sensing of the current at the output and feeding back a voltage proportional to that current. Compound feedback involves a circuit using both voltage and current feedback.

Feedback can also be classified as to how it is applied. Series feedback, as the name implies, is applied in series with the input signal. It then follows that parallel feedback is applied in parallel with the input. The different types of feedback and different methods of application lead to an overall circuit with differing characteristics. The various attributes are important in the many distinctive audio applications.

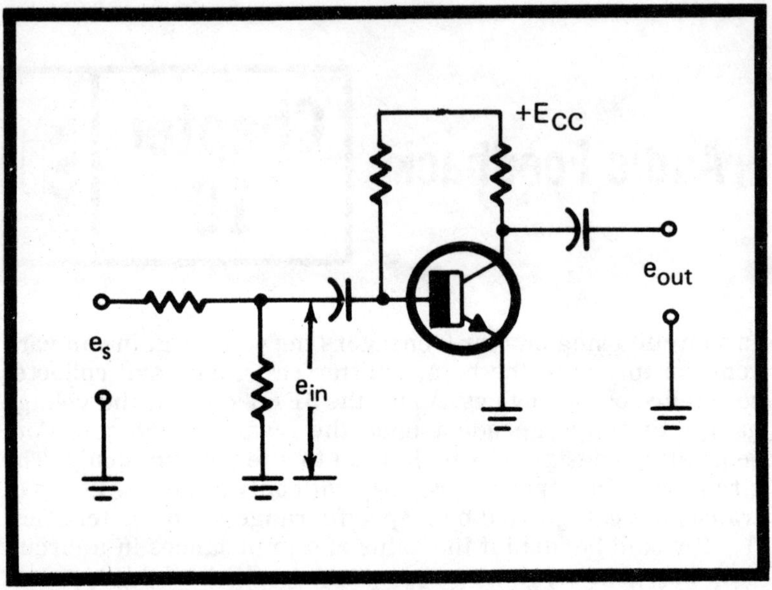

Fig. 10-1. Circuit used to define e_S and e_{in}.

THE FEEDBACK EQUATION

Let us start by defining two voltages, e_S and e_{in}. Choose the convention where e_S is the input signal to an amplifying system. The signal may feed passive components at the input to the amplifier, in which case, e_{in} is the input signal at the input to the amplifier proper, after the signal has passed through the passive components (see Fig. 10-1). It will include e_S and the effect of passive components as well as all other signals mixed with or added to e_S. If there are no other signals besides e_S feeding the amplifying system, and there are no passive components between e_S and the input to the amplifier, then e_{in} equals e_S. In this case, the voltage gain, A_V, of an entire amplifier circuit is:

$$A_v = \frac{e_{out}}{e_s} = \frac{e_{out}}{e_{in}} \tag{10-1}$$

where e_{out} is the voltage at the output of the amplifying system.

Let us now take a fraction, B, of the output voltage, e_{out}, and feed it, B e_{out}, back so that it is out of phase with the input signal. (See Fig. 10-2.) The difference between these two signals is now feeding the input to the amplifier to establish

e_{in}. In this case e_{in} is no longer equal to e_s, but because of Be_{out} it is:

$$e_{in} = e_s + Be_{out} \tag{10-2}$$

Substituting this into Equation 10-1, the voltage gain of the entire circuit with feedback becomes:

$$A_{vf} = \frac{e_{out}}{e_{in} - Be_{out}} \tag{10-3}$$

Divide the numerator and denominator by e_{in} and let the ratio e_{out} divided by e_{in} equal A_V. Equation 10-1 becomes:

$$A_{vf} = \frac{A_v}{1 - BA_v} \tag{10-4}$$

This is the standard often quoted feedback equation.

For negative feedback, BA_V is negative, so the feedback factor, $1\text{-}BA_V$, is greater than 1. This implies a 180-degree phase shift between the input and output. Should there be 0 degrees phase shift, BA_V is positive. When BA_V is equal to +1, the denominator is zero and the gain is infinite—an oscillating condition. This can occur at very high frequencies, just above the audio spectrum.

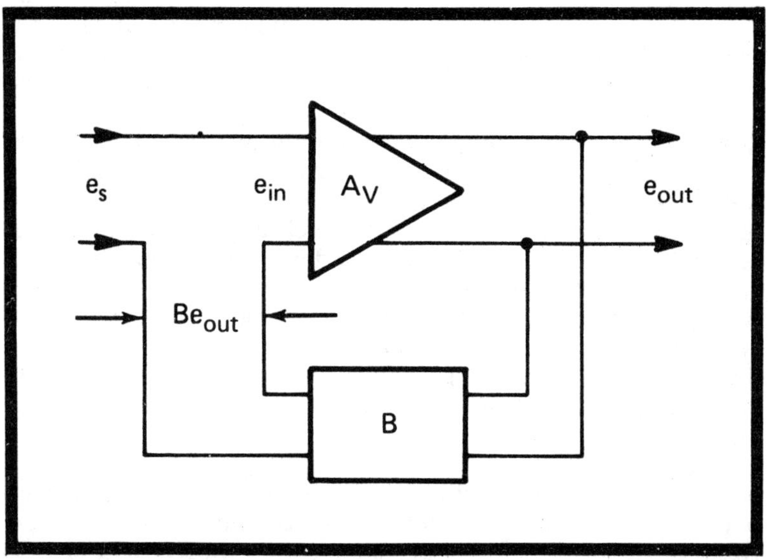

Fig. 10-2. Block diagram of a voltage feedback amplifier.

MARGIN OF STABILITY

It is obviously undesirable that an amplifier be unstable at any frequency or band of frequencies, even if the resulting oscillation is outside the audible band. High-frequency oscillation has been known to burn out the tweeters of loudspeaker systems. Power used to produce such oscillation is wasteful and can be put to more useful purposes. Oscillation at high frequencies can also cause the reproduced program material over the entire band to sound distorted.

The tendency of an amplifier to oscillate can be determined in several different ways. An analysis can involve a Nyquist plot. This is a graph of the real against the imaginary components of the BA_V factor in Equation 10-4. This curve is plotted at all frequencies that can be amplified, even if they are outside the audio band. Should the resulting curve touch or encircle the point -1, +j0 on the graph, the amplifier will be unstable and oscillate. This method, while accurate, requires quite a bit of calculation.

A simple method, best applied experimentally, can be explained using the voltage feedback circuit in Fig. 10-3A as an example and the revision of the circuit in Fig. 10-3B. In the revised circuit the feedback loop is opened. Two types of feedback are used in the circuit in Fig. 10-3A. Disregarding the connection between R_F and the emitter of Q1, the first stage of amplification has an unbypassed resistor, R_{E1}, in the emitter. The signal current flowing through this resistor develops a signal voltage across R_{E1}. This voltage, eR_{E1}, opposes e_s so that the signal feeding the amplifier, e_{in}, is e_s - eR_{E1}. Since the current through the resistor determines the voltage developed across it, this is an example of current feedback.

The signal at R_{C1} is amplified by Q2. The output voltage is fed back through R_F. A portion of this output voltage is developed across R_{E1} due to the voltage divider action of resistors R_F and R_{E1}. The voltage developed across R_{E1} due to this network depends upon the size and type of output voltage. This, the second type of feedback in the circuit, is a voltage feedback arrangement. B for this circuit is R_{E1} divided by R_F. Since the gain in the forward circuit is usually large, the voltage gain with feedback is approximately R_F divided by R_{E1}.

The procedure used to experimentally check the stability of an amplifier starts with the disabling of the voltage feedback circuit. The connection between R_F and R_{E1} is broken. A resistor, R_{EF}, equal to R_{E1} is connected from R_F to ground so that the output circuit is not upset. This is shown in Fig. 10-3B.

The voltage fed back to the input is e_f. A curve (Fig. 10-3C) is drawn to show how the voltage varies in db with frequency. This is accomplished by connecting an audio sine wave signal generator at e_s in Fig. 10-3B. Keep e_s constant while varying the frequency of the signal. The output is measured across e_f. The ratio 20(log e_f divided by e_s) is the ratio in db of the voltage gain at a specific frequency. Note this ratio at all frequencies. Using the highest number as a 0 db reference, plot the number of db the output varies below the 0 db point over the frequency range of the amplifier. You should get a curve similar to that shown in Fig. 10-3C.

Now determine the amount of voltage feedback present in the circuit. Feed a signal at e_s. The frequency used must have 0 db attenuation as plotted in Fig. 10-3C. Place a meter at e_{out} for the circuit in Fig. 10-3B. Call this output voltage e_{out1}. Then reconstruct the circuit as in Fig. 10-3A and read e_{out}, referring to this latter reading as e_{out2}. The voltage feedback in db can be determined from the equation db (feedback) equals (20)log(e_{out1} divided by e_{out2}).

Draw a horizontal line at the db(feedback) level below 0 db. Note where this line crosses the curve previously drawn. If the rolloff of the curve is slower than at the rate of 12 db per octave at these points, the amplifier will be stable. It is most desirable that the design be for a slower rolloff than 9 db per octave to provide a reasonable margin of stability for the overall circuit.

The curve and feedback can, of course, be calculated from the circuit, but this can be a very tedious and approximate operation. Some factors affecting the curve may not be considered in the analysis or are so complex as to make paper calculations difficult if not impossible.

One of the major factors that must be accounted for is stray lead capacity. Another factor, most important when dealing with large power transistors, is the beta cutoff frequency, written $f_{\alpha e}$ or f_β. At this frequency, the beta is 3 db less (about 0.7 of the size) than it is at the low frequency normally specified on data sheets. Beta rolls off at the rate of 6 db per octave, following the high-frequency rolloff curves discussed in Chapter 9 with reference to capacitive and inductive filters. It is redrawn in Fig. 10-4 showing the beta variations.

Another important frequency on the curve is f_T, the frequency at which the beta is equal to 1 (beta has rolled off to 0 db). This is approximately equal to the alpha cutoff frequency, written f_{hbo}, $f_{\alpha b}$, or f_β, at which alpha has rolled off 3 db from its low-frequency value. Similar to the beta

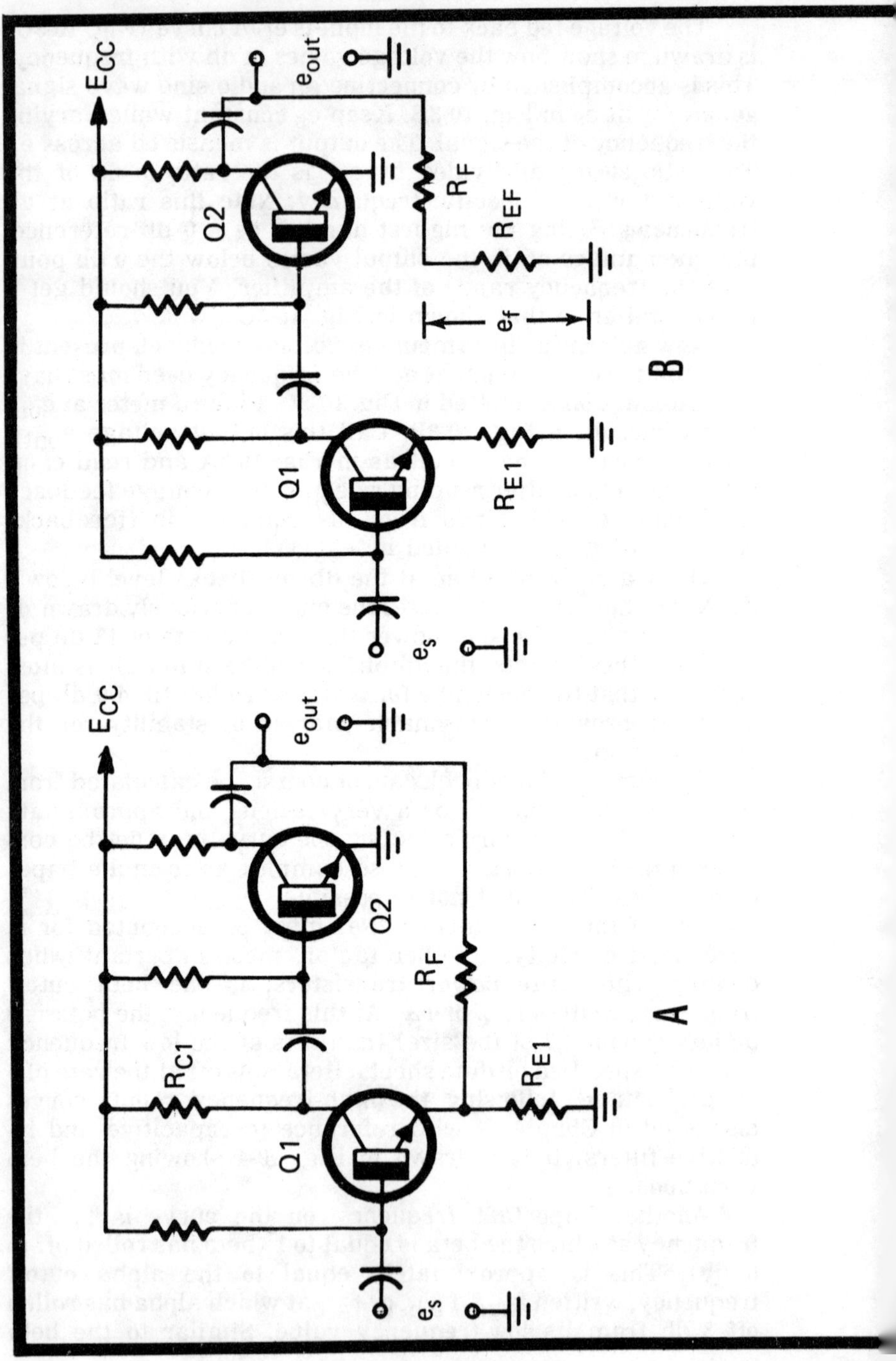
E_CC
e_out
Q1
Q2
R_C1
R_E1
R_F
R_EF
e_f
e_s
A
B

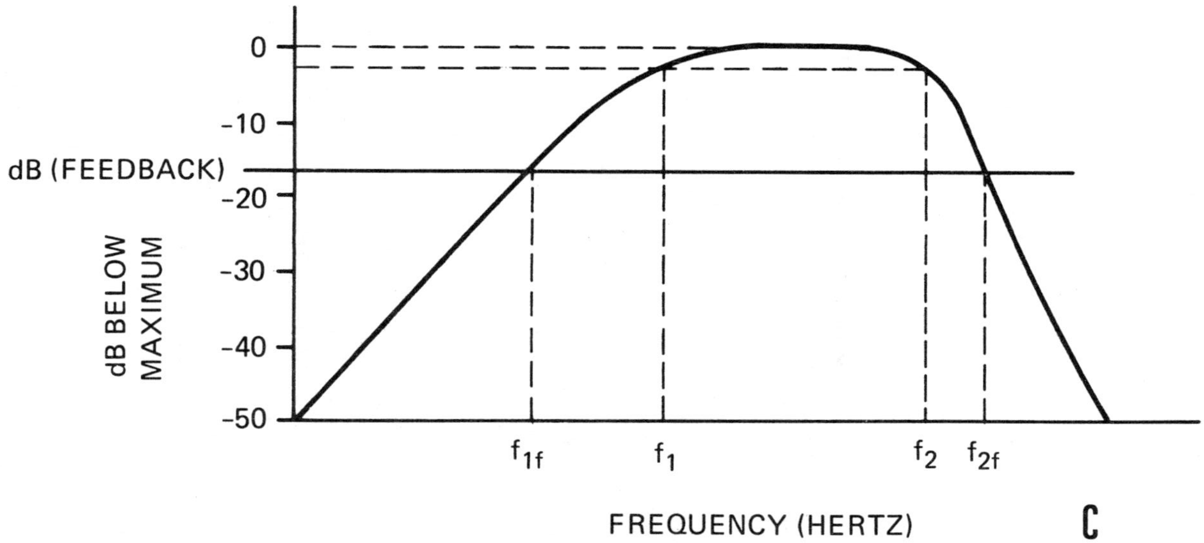

Fig. 10-3. Procedure to make a frequency response plot of a forward amplifier circuit. Circuit A is a 2-stage amplifier with feedback. Circuit B has been modified to enable you to measure the response curve of the forward circuit. C is the response curve of the forward circuit of the amplifier.

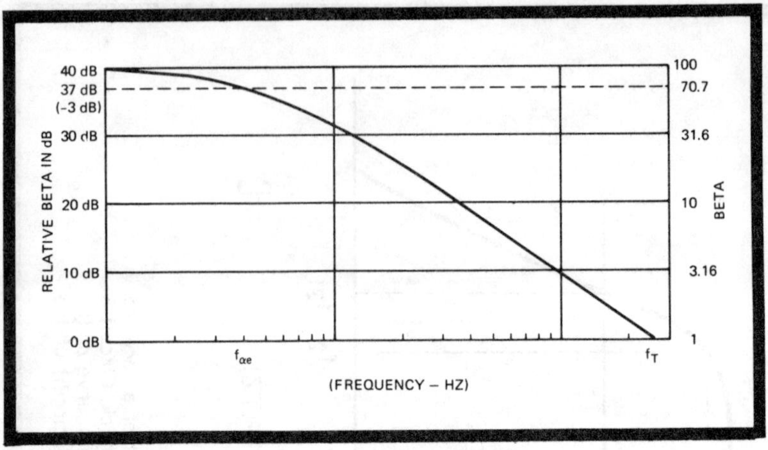

Fig. 10-4. Curve showing the variation of beta with frequency. For this particular transistor, it is assumed that beta is equal to 100 at low frequencies.

curve, the curve plotting alpha against frequency exhibits the 6 db per octave characteristic shown in Fig. 10-4 for the beta gain.

Several interesting relationships can be evolved. On the rolloff portion of the beta curve, the product of the frequency and the beta at any frequency remains constant. Hence, f_T can be found by simply setting the beta equal to 1 and equating f_T to the product of the beta and frequency at a known point on the curve. There is an approximate relationship between f_α and f_β, namely f_α equals β_0 f_β, where β_0 is the beta specified at a low frequency.

As for the JFET, there is input and output capacity. Here, as for the bipolar device, the Miller effect can be important. Basically, the Miller effect states that the input capacity of a device is the actual input capacitance added to the product of the output capacity and the voltage gain of the device. For the JFET, the input capacity is $C_{iss} + A_V C_{rss}$, where C_{iss} is the sum of the gate-to-source capacitance, C_{gs}, and the gate-to-drain capacitance, C_{gd}. C_{rss} is approximately equal to C_{gd}. These are discussed in Chapters 5 and 6.

As for the bipolar device, the input capacitance resulting from the Miller effect is $C_{b'e} + A_V C_{b'c}$. In this relationship, b' represents a point inside the transistor that is not accessible by a lead or probe. $C_{b'e}$ is the capacitance from that point to the emitter lead and $C_{b'c}$ is the capacitance from that point to the collector. Using more easily recognizable symbols, the input capacitance is approximately equal to $C_{ob} + A_V$ divided by 6.28 $r_e f_\alpha$. Here, C_{ob} is the collector-to-base capacitance in the common-base equivalent circuit and r_e is the AC emitter resistance.

However analyzed, feedback has two glaring disadvantages. There is, of course, a loss in gain. There is also the tendency towards instability. However, the many benefits far outshine these drawbacks.

CHARACTERISTICS OF CIRCUITS USING VOLTAGE FEEDBACK

There are several consequences of voltage feedback when placed around an audio circuit. Should the feedback be applied in series with the input signal, the input impedance increases to:

$$Z_{if} = Z_i (1 - BA_v) \qquad (10\text{-}5)$$

where Z_i is the input impedance without feedback and Z_{if} is the input impedance with feedback. In Fig. 10-3A, the portion of the output voltage developed across R_{E1} due to the voltage divider action of R_F and R_{E1}, when placed in series with e_s, is an example of series voltage feedback. Here, a portion of the output is fed back across R_{E1} and subtracted from e_s.

In all voltage feedback circuits, the output impedance with feedback, Z_{of}, is less than without feedback, Z_o. They are related by the equation:

$$Z_{of} = Z_o/(1 - BA_v) \qquad (10\text{-}6)$$

Another consequence of feedback is the extension of the bandwidth of the overall circuit. This can be illustrated with the help of Fig. 10-3C. Without feedback, the bandwidth is from f_1 to f_2. Feedback extends this bandwidth so that it is from f_{1f} to f_{2f}, where the db(feedback) line crosses the plot of the frequency response curve of the amplifer's forward circuit. The frequency f_{1f} can be determined from the equation:

$$f_{1f} = f_1/(1 - BA_v) \qquad (10\text{-}7)$$

while the upper frequency of the band, f_{2f}, can be found from:

$$f_{2f} = f_2 (1 - BA_v) \qquad (10\text{-}8)$$

The voltage gain of an amplifier varies with the various parameters and components. This variation is reduced con-

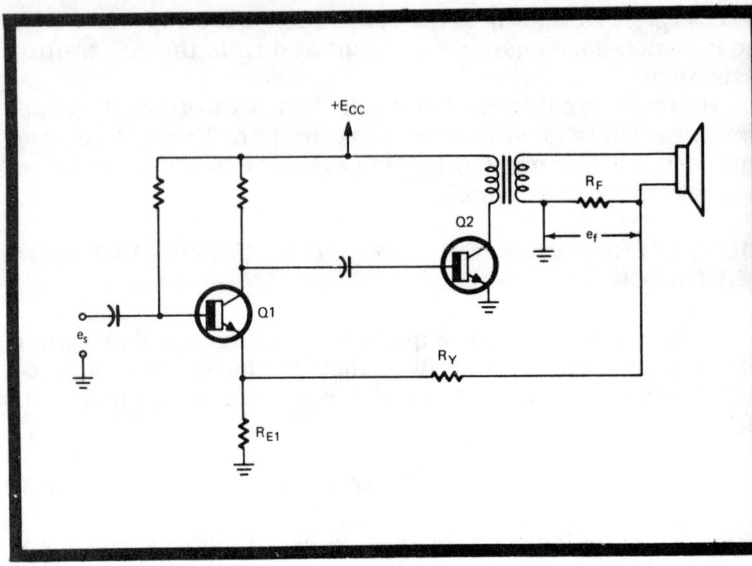

Fig. 10-5. Series-applied current feedback circuit.

siderably when feedback is applied. The amount the gain changes with feedback, ΔA_{vf}, is related to the amount the gain changes when there is no feedback, ΔA_v, by the equation:

$$\frac{\Delta A_{vf}}{\Delta A_v} = \frac{A_f}{A_v}\left(\frac{1}{1 - BA_v}\right) \tag{10-9}$$

Distortion is similarly reduced by a factor of 1 divided by $(1\text{-}BA_V)$. Noise is reduced by the same factor only if it is repetitive and the feedback loop encompasses the source of the noise.

CHARACTERISTICS OF CURRENT FEEDBACK

The circuit in Fig. 10-5 is that of a single-ended audio amplifier feeding a loudspeaker. A small resistor, R_F, is placed in series with the speaker. The current that flows through the speaker also flows through R_F. A voltage is developed across R_F due to the current flowing through the resistor. The voltage is proportional to the output current. A portion of this voltage is applied through a resistor to R_{E1}. This voltage bucks e_s, reducing the voltage gain of the overall circuit. The gain is thus related to the output current. The voltage fed back is related to the resistors R_F, R_Y and R_{E1}, and the load resistance of the loudspeaker, R_L. The voltage gain with feedback is:

$$A_{vf} = \frac{A_v}{1 - \delta A_v} \qquad (10\text{-}10)$$

The equation is very similar to that for the voltage feedback. Here, γ is BR_F divided by R_L, where B is the fraction previously defined for voltage feedback. The ratio R_F divided by R_L can easily be determined by reconsidering Equation 10-2. Here, e_{in} to the amplifier proper is $e_s + Be_{out}$. But in Fig. 10-5, the sensed signal fed back is not from the output voltage, but is that appearing across R_F. If this voltage is e_f, e_{in} becomes $e_s + Be_f$. Since e_f is R_F (IL equals $R_F(e_{out}$ divided by R_L), e_{in} becomes $e_s + Be_{out}$ (R_F divided by R_L). The $B(R_F$ divided by $R_L)$ in the relationship is the γ in Equation 10-10 and the rest of the equation reverts to the form of Equation 10-4. B is, of course, R_{E1} divided by $(R_Y + R_{E1})$, which is approximately equal to R_{E1} divided by R_Y.

Another example of a series-applied current feedback circuit is shown in Fig. 10-6. The output current is sensed across R_E. The voltage developed across R_E bucks the input

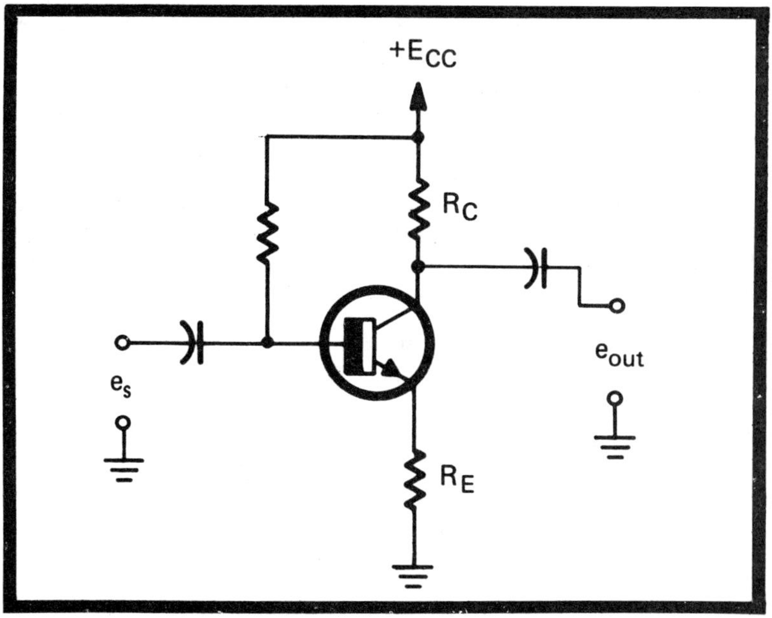

Fig. 10-6. Series-applied current feedback circuit in one stage of am plification.

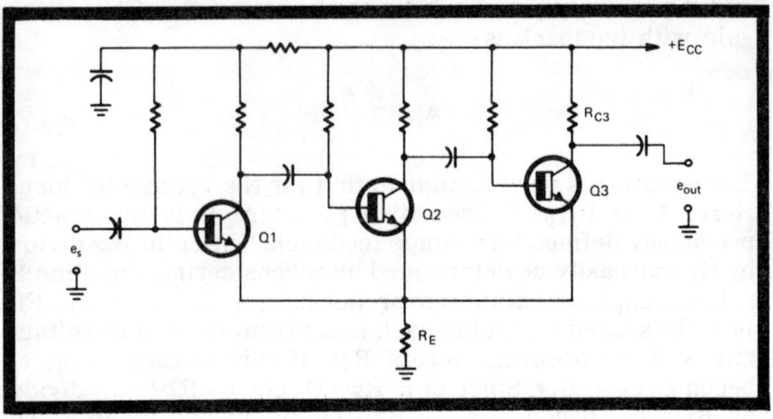

Fig. 10-7. Series-applied current feedback direct-coupled to three stages of amplification.

signal, e_S. Here, γ equals R_E divided by R_C while the voltage gain with feedback is R_C divided by R_E when γA_V is very large.

A similar situation exists if a common-emitter resistor is used for several stages of amplification, as shown in Fig. 10-7. Here, γ equals R_E divided by R_{C3} and the voltage gain despite the large amount of amplification in the forward loop of the amplifier is R_{C3} divided by R_E.

The above examples all deal with series-applied current feedback. Shunt-applied current feedback is shown in Fig. 10-8. This circuit can be considered as a voltage feedback or current feedback arrangement. As a voltage feedback circuit, the portion of e_{out} fed back to the input is related to the voltage divider formed by the parallel combination of the input impedance of the transistor, Z_i, and R_S. This combination, $R_S || Z_i$, is in series with R_B, so B in Equation 10-4 becomes:

$$\left(\frac{R_S \parallel Z_i}{R_S \parallel Z_i + R_B}\right) \tag{10-11}$$

This factor has little effect on the voltage gain of the circuit.

On the other hand, the current gain is dependent upon R_C, R_B and $R_S || Z_i$. The portion of the current fed back from the output is R_C divided by $R_B + (R_S || Z_i)$. Letting this portion of the output current be B_i, the current gain with feedback becomes:

$$A_{if} = \frac{A_i}{1 - B_i A_i} \tag{10-12}$$

where A_{if} is the current gain with feedback and A_i is the current gain without feedback. If R_B is much larger than $(R_S || Z_i)$, B_i is R_C divided by R_B. Using Equation 10-12, the current with feedback is R_B divided by R_C when $B_i A_i$ is very large.

Few of the circuit characteristics due to current feedback differ from those using voltage feedback. In the case of series-applied current feedback, only the output impedance is different. While it is reduced in the case of voltage feedback, it increases by a factor of $(1 - \gamma A_v)$ in current feedback circuits. As for shunt-applied current feedback, the input impedance is reduced by a factor of 1 divided by $(1 - B_i A_i)$ rather than being characterized by the increase inherent in the voltage feedback arrangements.

POSITIVE FEEDBACK

The circuit in Fig. 10-9A is an example of positive feedback as used in a bipolar transistor circuit. The signal voltage at the emitter is approximately equal to that at the base. One end of resistor R_Y is connected to the emitter through C (or the voltage that is across the emitter is fed back to this end of

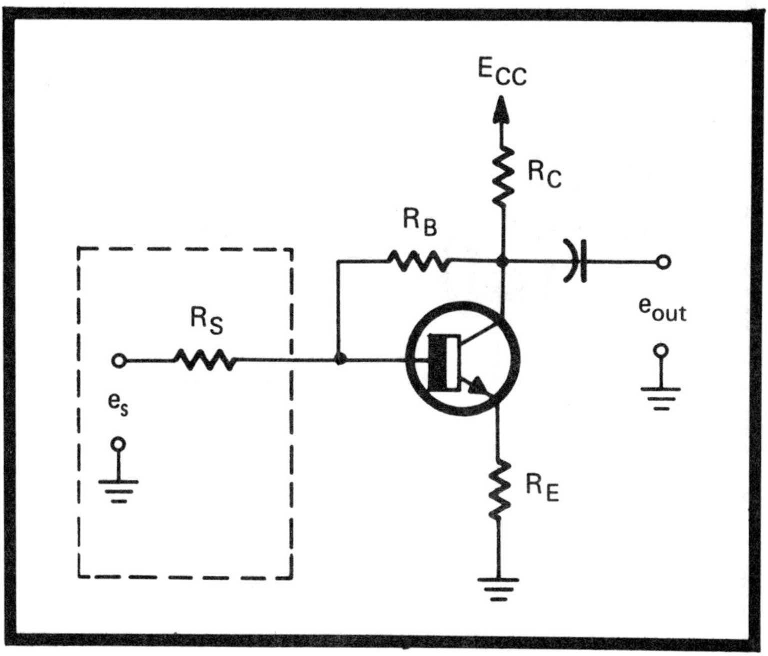

Fig. 10-8. Example of shunt feedback.

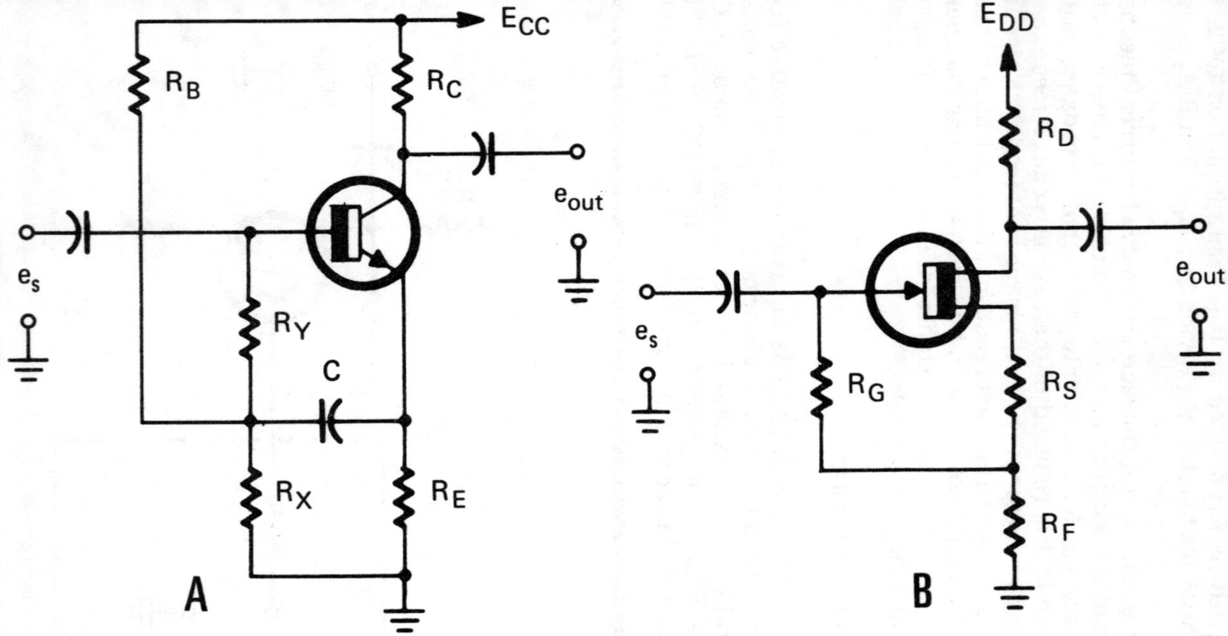

Fig. 10-9. Bootstrapped circuits using positive feedback: Bipolar bootstrap circuit (A) and a JFET bootstrap circuit (B).

R_Y through C) and the other end is connected to the base of the transistor. Since the voltages at both ends of R_Y are equal, zero volts is across R_Y.

Signal generator e_s sees an input impedance equal to Z_i. Impedances seen by signal sources are equal to the input voltage divided by the input current, or e_S divided by i_S, where i_S is the input current. The bulk of the input current must flow through R_Y and is equal to the voltage across R_Y divided by the resistor. Since the voltage across R_Y is zero due to feedback capacitor C, then i_S must be 0 divided by R_Y equals 0. Hence, Z_i equals e_S divided by i_S equals e_S divided by 0 equals ∞. It is, of course, not infinite since the voltages at the two ends of the resistor can never be identical.

A similar circuit can be drawn for the JFET to increase its input impedance. This circuit, in Fig. 10-9B, is a bootstrapped source follower. The bias voltage is developed across R_S and applied through R_G to the gate.

As for the bipolar circuit, the AC portion of Fig. 10-9A can be redrawn as Fig. 10-10. It must be recalled that E_{CC} is at AC

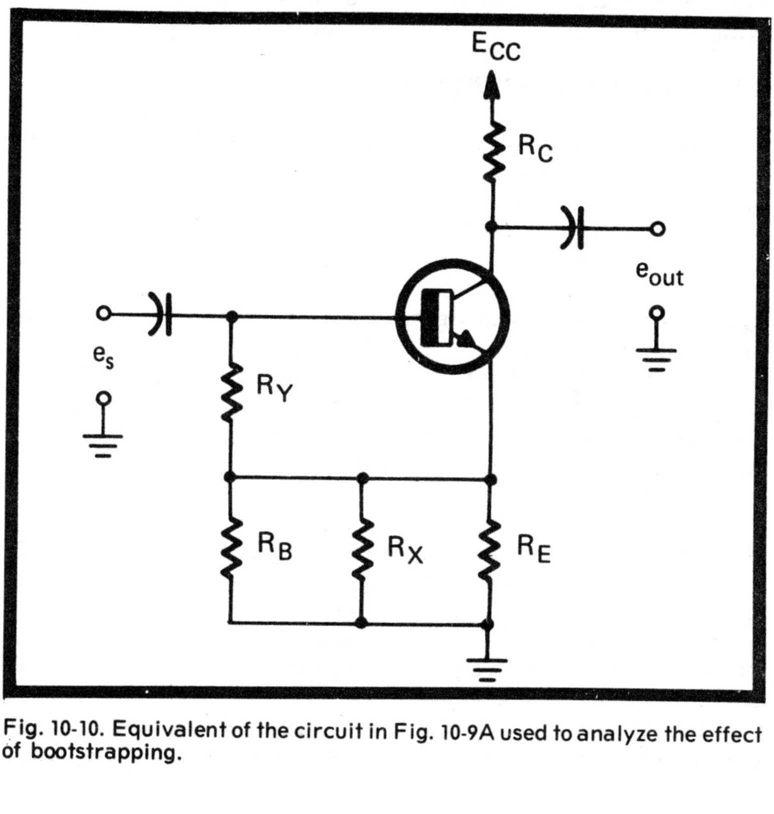

Fig. 10-10. Equivalent of the circuit in Fig. 10-9A used to analyze the effect of bootstrapping.

ground and that C is an AC short circuit. Because E_{CC} is at AC ground, R_B is across R_X. Since C is an AC short, the parallel combination of R_B and R_X is across R_E in the emitter circuit. R_Y is shorted by the base-emitter junction. In the emitter, R_E, R_X and R_B are all in parallel. When reflected into the base circuit, e_{in} sees this parallel combination multiplied by the beta of the transistor. This is the impedance of the circuit, certainly much higher than it would be without the bootstrapped positive feedback components.

The analysis of the JFET circuit is more complex, but it leads to the simple solution that the input impedance is:

$$Z_i = \frac{R_G \left[(1 + g_m (R_S + R_F)\right]}{1 + g_m R_S} \qquad (10\text{-}13)$$

where g_m is the transconductance of the device at the operating point.

This bootstrapping concept is important where extremely high input impedance is required. In the quasi-complementary output circuit to be discussed in the next chapter, an important application of this concept is detailed.

FEEDBACK EQUALIZATION

Records and tapes are not recorded with a flat frequency response characteristic. Not all frequencies are recorded with equal amplitude. The amplitudes at the upper end of the audio band are recorded with a rising characteristic so that the music can override the noise. Playback curves at this end of the spectrum must exhibit rolloff to compensate for the emphasis in the recording process, while further improving the signal-to-noise ratio.

On the other hand, in the recording process the low frequencies are reduced in amplitude with respect to the mid-frequencies so that the width of the record groove can be maintained within reasonable limits. Therefore, the playback curve must emphasize the low frequencies.

The final factor affecting the frequency characteristics of the reproduced record or tape is the transducer. The widely used magnetic type cartridge does not have an output with a linear relationship to the amplitude being reproduced. It is a velocity-sensitive device in which the output voltage is proportional to the frequency of the signal when identical amplitudes are being reproduced at all frequencies.

Taking all these factors into account, the phonograph preamplifier section of the amplifier must have the frequency characteristic as described by the curve in Fig. 10-11A. The overall amplified output of the complete system from record through playback will be linear only if the response is as shown. A straight line approximation to the curve has been drawn in Fig. 10-11B to facilitate design.

Notice that the curve has three distinct sections—two 6 db per octave rolloffs starting at 50 and 2000 Hz, and a flat response between 500 and 2000 Hz. The latter region can be simulated by a 6 db per octave rise in amplitude beginning at 500 Hz to offset the equivalent rolloff which began at 50 Hz. The circuit in Fig. 10-12 can be used to simulate this curve. It is similar to the circuit in Fig. 10-3A, except that a reactive network is used to replace R_F as the feedback element.

The design procedure can proceed in several logical steps: A minimum of about 40 db of feedback must be available to be applied around the circuit. This allows for 0 db of feedback at 30 Hz and for more than the required 36 db of feedback at 15 kHz with respect to the 30 Hz. Since 40 db is a voltage ratio of 100:1, the gain of the circuit without feedback must be greater than 100. This is easily accomplished with the two transistors in the figure.

Next, design the R-C networks in the feedback loop. As discussed earlier, the voltage gain, A_{vf}, of the circuit in Fig. 10-3A is about R_F divided by R_E, because the forward loop gain is sizable. The impedance network in the feedback circuit of Fig. 10-12 can be substituted for the R_F in the gain ratio. Refer to this impedance as Z_F. As the gain is now Z_F divided by R_{E1}, the circuit in Fig. 10-12 now has a gain approximately equal to:

$$A_{vf} = \frac{Z_F}{R_{E1}} = \frac{1 + j\,6.28f\,R\,(C_1 + C_2)}{j\,6.28f\,C_1 R_{E1}\,[1 + j\,6.28f\,C_2 R]} \qquad (10\text{-}14)$$

This follows from the fact that Z_F is equal to the reactance of C_1 or 1 divided by $j6.28fC_1$ in addition to the impedance of the parallel combination of R and C_2 or R divided by $(1 + j6.28fC_2R)$. In the equation, j indicates a 90-degree phase shift.

Equalities in the form of Equation 10-13 can easily be analyzed to determine corner frequencies. All factors in the form of $(1 + jx)$ are set equal to $(1 + j)$. All other factors incorporating a j term are set equal to zero. Thus, for the numerator, $1 + j6.28fR(C_1 + C_2)$ equals $1 + j$, and then:

$$f_{o1} = 1/6.28R\,(C_1 + C_2) \qquad (10\text{-}15)$$

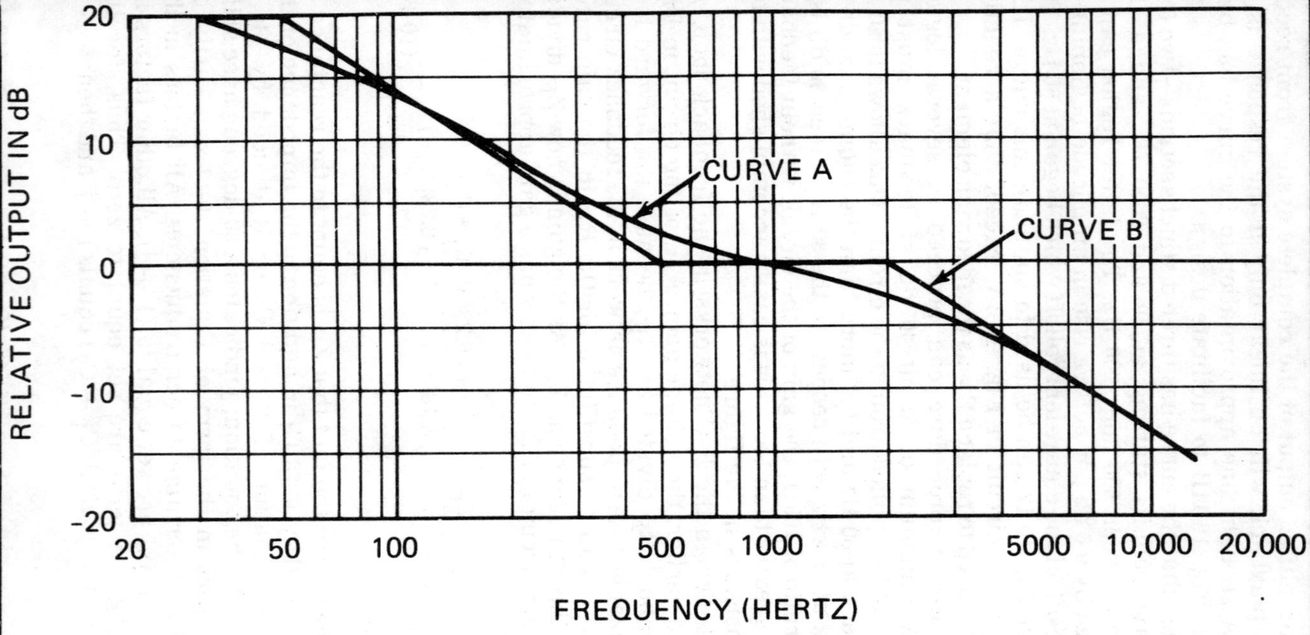

Fig. 10-11. Phonograph playback curves when using magnetic cartridge. The exact frequency response curve is shown at A, and B is a straight-line approximation of the curve at A.

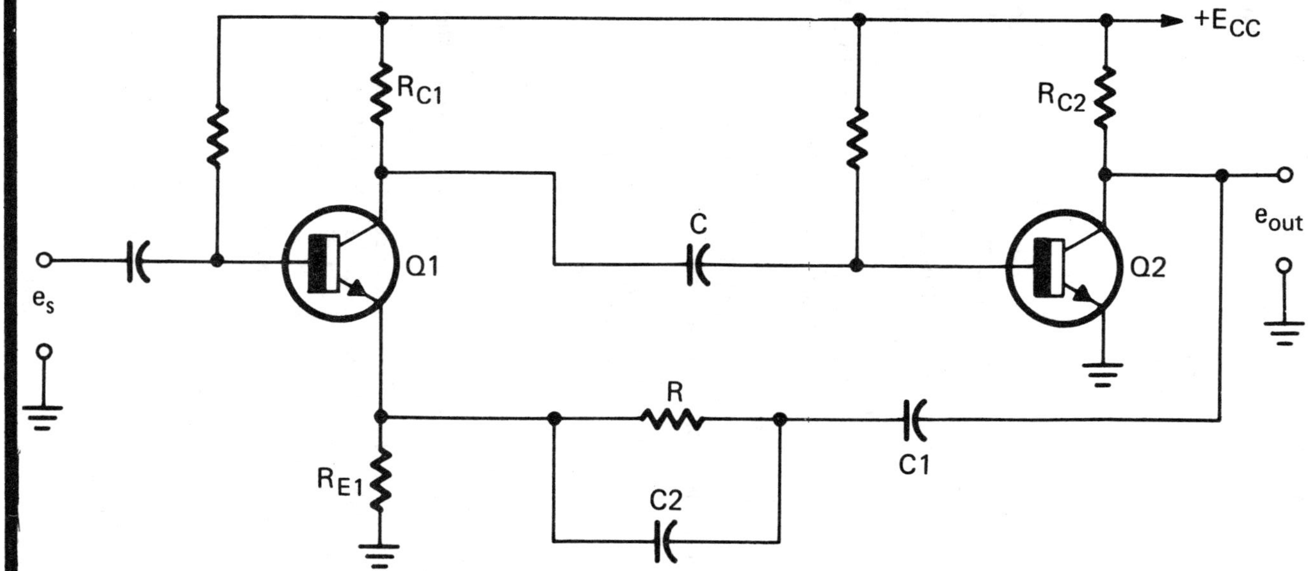

Fig. 10-12. Phonograph preamplifier circuit using feedback for equalization.

f_{o1} is one corner frequency. Similarly, the $(1 + j)$ term in the denominator yields the second corner frequency, f_{o2}:

$$f_{o2} = 1/6.28RC_2 \tag{10-16}$$

The third corner frequency, f_{o3}, is found by setting the $j6.28C_1R_{E1}$ term equal to zero:

$$f_{o3} = 0 \tag{10-17}$$

Now substitute the actual corner frequencies noted in Fig. 10-11A for f_{o1}, f_{o2} and f_{o3}. Curve rolloffs begin at the corner frequencies determined from the factors in the denominator of Equation 10-14, while they rise beginning at frequencies determined from the numerator. Rolloff starts at f_{o3} equals 0 Hz from Equation 10-17 and continues to 500 Hz as determined from f_{o1} in Equation 10-15. It begins to roll off again at f_{o2} equals 2000 Hz, which can be calculated from Equation 10-16. The equations can be solved simultaneously to determine the values of the various components.

You may justifiably ask why the rolloff begins at 0 Hz rather than 50 Hz. Simplification in the calculation is afforded by the expediency of this approximation. Actually, the coupling capacitor between stages in the forward circuit can be adjusted to move the corner frequency from 0 Hz to 50 Hz. A more accurate circuit includes a resistor across C_1 to readjust the corner frequency to its proper location at 50 Hz.

A similar curve may be derived for a tape playback equalized preamplifier. A very rough approximation to the 7½ IPS playback curve is shown in Fig. 10-13. There are two corner frequencies—one at 50 Hz and a second one at 3000 Hz. Once again, a circuit similar to the one in Fig. 10-12 can be used. Just substitute the series R-C circuit in Fig. 10-14 for the network of C_1, C_2 and R in Fig. 10-12. Let Z_F be the impedance of the R-C circuit or Z_F equals R + 1 divided by j6.28fC equals (j6.28fRC + 1) divided by j6.28fC. Since the voltage gain with feedback is approximately Z_F divided by R_{E1}.

$$A_{vf} = (j6.28fRC + 1)/j6.28R_{E1}fC \tag{10-18}$$

The rolloff begins at the frequency where the denominator is equal to zero. This is:

$$f_{o1} = 0\text{ Hz} \tag{10-19}$$

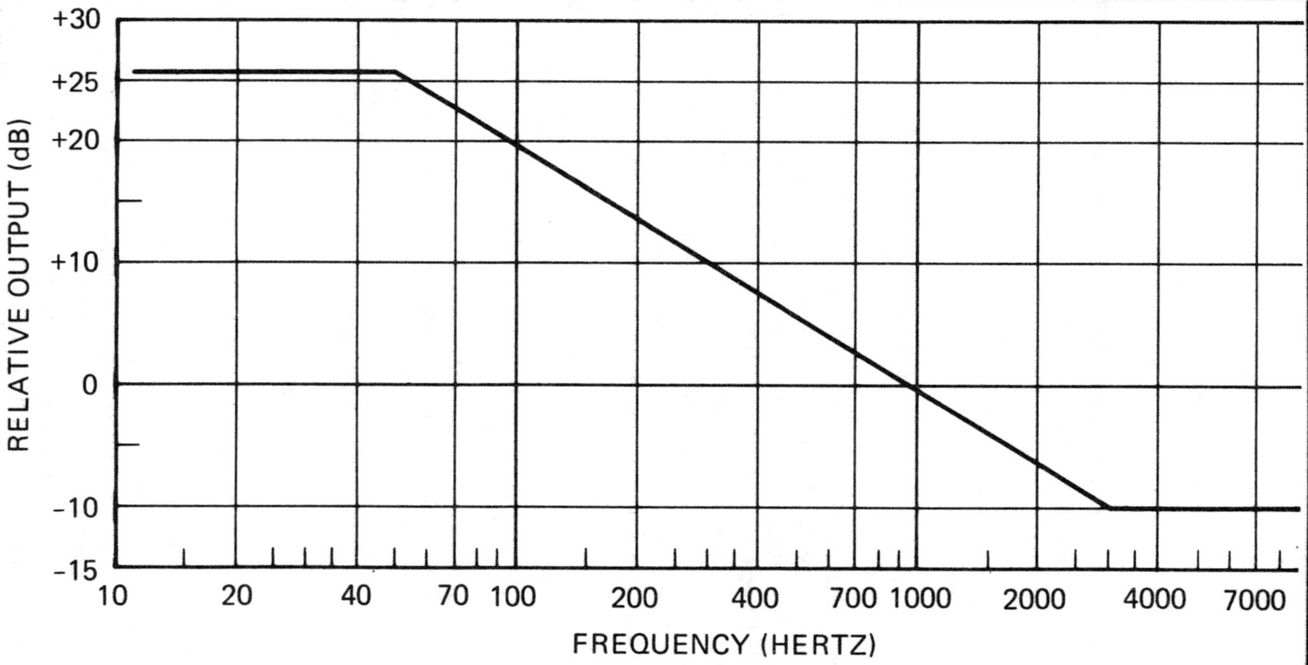

Fig. 10-13. Rough approximation of the 7½ IPS equalization curve for tape reproduction.

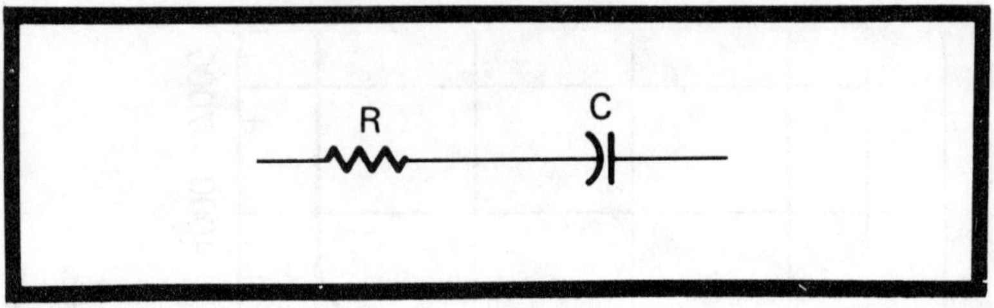

Fig. 10-14. Circuit used in place of R-C1-C2 in Fig. 10-12 to get tape playback equalization.

The rise begins at 3000 Hz or wherever the numerator is equal to j + 1. This frequency is:

$$f_{o2} = 3000 \text{ Hz} = 1/6.28RC \quad (10\text{-}20)$$

Once again, the 50-Hz rolloff point must be treated as in the previous discussion of phono feedback equalization.

Now, the most important step in the design must be pursued. Check the design in the laboratory and adjust the equalization using the actual components. Too many stray factors are usually omitted in the paper design for the calculated components to be sufficiently accurate.

Bipolar devices were used in the examples. JFETs can serve equally well in these applications. In both instances, the first stage must be designed for minimum noise while the second transistor circuit must be adjusted so that there is sufficient swing at its output during peaks in the music. A phonograph amplifier with about 3 or 4 millivolts RMS input sensitivity for an average size signal that will accommodate 60 or 70 millivolts RMS input before the output distorts is indeed satisfactory. A similar ratio of maximum-to-minimum input signal is required for the tape preamplifier, but the minimum input sensitivity in this case should be about 1 millivolt RMS.

OPERATIONAL AMPLIFIER

The operational amplifier, in conjunction with differential amplifier circuitry, is usually associated with computer electronics. Actually, the feedback circuit known as an operational amplifier or anode follower has been in use for many years in tone control circuits in high quality amplifying equipment. Because they are no more expensive or complex than the "lossier" type of base and treble boost and cut controls (based on the passive R-C circuits discussed in Chapter 9), the feedback control is used almost exclusively in all audio equipment. Analysis requires some knowledge of the characteristics of operational amplifiers.

An operational amplifier circuit involving an FET appears in Fig. 10-15. The DC gate bias for this stage is developed across R_S and applied to the gate through R_G. R_G is made as large as practical so as not to affect any other factors in the circuit. It is assumed that **no** AC signal current flows through this resistor.

From the gate through the collector, the stage is an ordinary amplifier where e_{in} will be amplified to produce an output voltage across R_D. However, the actual signal generated, e_s, is applied through R_B to the amplifier. R_F feeds the output signal back to the gate in a feedback circuit. In this circuit, C is considered to be a short circuit for audio signals and is designed into the circuit with the sole purpose of keeping the DC at the collector away from the gate.

The signal current flowing through R_B is $(e_S - e_{in})$ divided by R_B. This current divides between R_G, the gate circuit, and R_F. Since the currents through R_G and the gate circuit are negligible when compared to that flowing through R_F, we can, with reasonable accuracy, conclude that "all" the current flowing through R_B also flows through R_F. The current in R_F

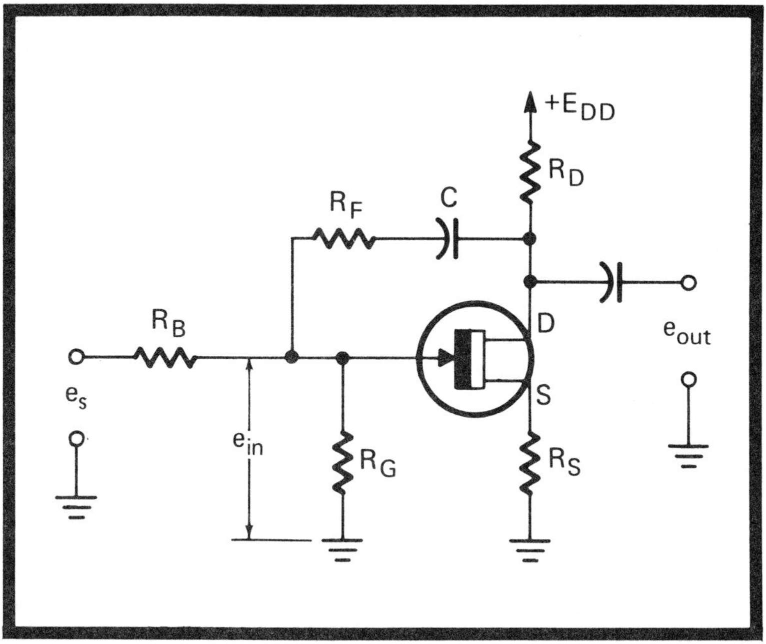

Fig. 10-15. JFET operational amplifier.

is (e_{out} - e_{in}) divided by R_F. Equating the current through R_B with that through R_F, we have:

$$(e_s - e_{in})/R_B = (e_{out} - e_{in})/R_F \qquad (10\text{-}21)$$

We can now write a second equation which considers the gain, A_V, of the amplifier stage itself:

$$e_{out} = e_{in} A_v ; e_{in} = e_{out}/A_v \qquad (10\text{-}22)$$

Since the gain is usually extremely high (assumed infinite), e_{in} approaches zero. Although e_{in} is **practically** zero, the gate is not at ground. This point is referred to as a "virtual ground." Substituting e_{in} equals 0 into Equation 10-21, we get the well known relationship:

$$e_{out}/e_s = R_F/R_B \qquad (10\text{-}23)$$

The ideal operational amplifier has six primary characteristics:

1. Infinite input impedance
2. Zero output impedance
3. Infinite gain
4. Zero offset—zero output signal when the input is zero.
5. Zero response time—instant response at the output when the input signal is applied.
6. Infinite bandwidth

Obviously, no amplifier will fully meet any of these requirements. The closer the circuit approaches the ideal, the more accurate the calculations below will be.

FEEDBACK TONE CONTROLS

Let us now analyze a practical feedback tone control which is, in its complete form, known as the Baxendall tone control circuit. Start with the bass control section in Fig. 10-16. C1 and C4 are short circuits for the audio signals and are used only to prevent DC from entering the gate circuit. R1 is made equal to R3, C2 is equal to C3 and R2 is a linear potentiometer set at the center of rotation.

Compare Fig. 10-16 with Fig. 10-15. R1 plus the parallel combination of C2 and the left-hand half of R2 are the equivalent of the R_B arm in Fig. 10-15, while R3 plus the parallel combination of C3 and the right-hand half of R2 are

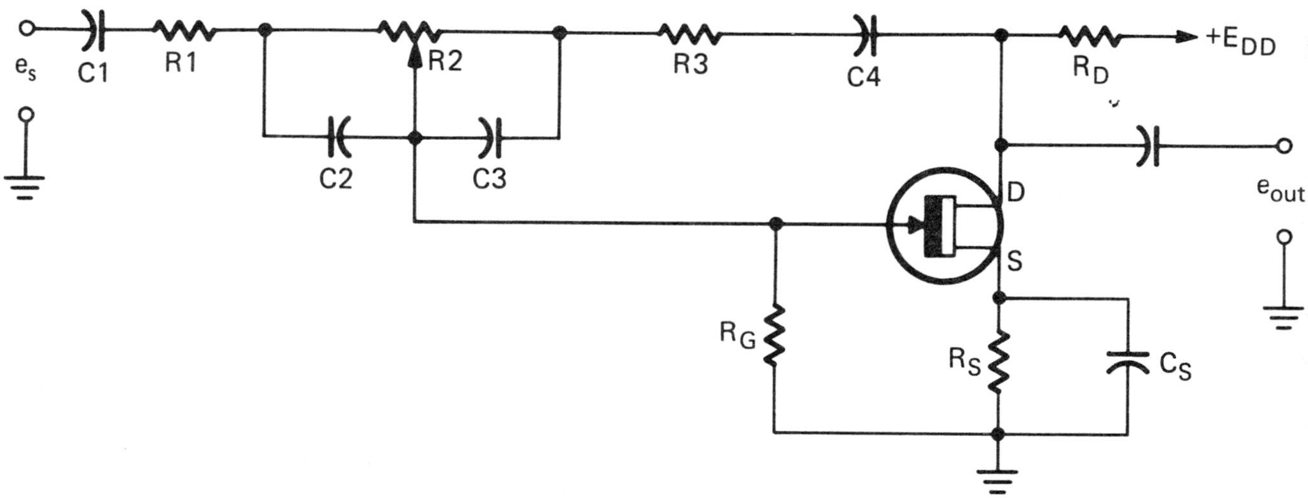

Fig. 10-16. Feedback bass control circuit.

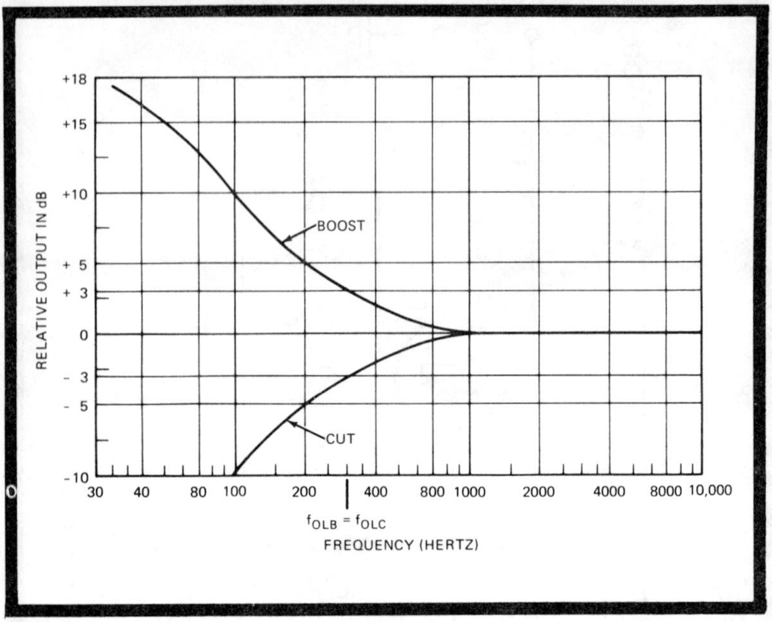

Fig. 10-17. Approximate bass cut and boost curves with the control set at maximum positions.

the equivalent of R_F in Fig. 10-15. Because R_F equals R_B at all frequencies (since it has been specified that R1 equals R3, C2 equals C3 and R2 is linear so that the right-hand half is equal to the left-hand half), the voltage gain of the circuit is e_{out} divided by e_S equals 1 (see Equation 10-23) at all frequencies. The curve relating gain to frequency is theoretically flat from 0 Hz to ∞ Hz.

Now move the wiper arm to the extreme left. C2 is shorted while C3 is a relative short across R2. In this position, the R_B arm of Fig. 10-15 becomes R1 while the R_F arm is effectively R3 in series with C3. The impedance of the R_F arm is R3 + 1 divided by j6.28fC3 equals (j6.28fC3R3 + 1) divided by j6.28fC3. Applying Equation 10-23, the output becomes:

$$e_{out} = e_s \left(\frac{j6.28fC_3R_3 + 1}{j6.28fC_3R_1} \right) \qquad (10\text{-}24)$$

The curve will begin to rise at the frequency f_{oLB} where the numerator is equal to (j + 1) or f_{oLB} equals 1 divided by 6.28C3R3 Hz. It should level off at the frequency where the denominator is zero or f equals 0 Hz. Equation 10-24 defines the maximum bass boost curve shown in Fig. 10-17.

The maximum bass cut occurs when the wiper arm is at the maximum right-hand setting. Here, R_B of Equation 10-22 becomes R1 + 1 divided by j6.28fC2 equals (j6.28fC2R1 + 1) divided by j6.28fC2, while R_F is simply R3. Substituting this into Equation 10-22 yields:

$$e_{out} = e_s \left(\frac{j6.28fC_2R_3}{j6.28fC_2R_1 + 1} \right) \tag{10-25}$$

The curve will begin to roll off at the frequency f_{oLC} where the denominator is equal to j + 1 or f_{oLC} equals 1 divided by 6.28C2R1 Hz. It should level off at f equals 0 Hz, the frequency when the numerator is equal to zero. This bass cut curve is also illustrated in Fig. 10-17. We can now choose the components for this circuit. It is desirable to make the control as large as practical. A 1-megohm potentiometer was the component originally chosen by the inventor.

We can expect 15 db of boost and cut at the extreme settings of the control if we overdesign. For a maximum cut and boost of 20 db. 20 db is an actual voltage ratio of 10:1. Again applying Equation 10-23, e_{out} divided by e_S equals 10 divided by 1 equals R_F divided by R_B. Since R_B equals R_F divided by 10, R1 is made equal to about one-tenth of R2, or about 100,000 ohms. For symmetry of the boost and cut modes, R3 is set equal to R1.

Fifteen db of boost is required at 50 Hz. Applying the curves discussed previously, exhibiting an eventual 6 db per octave boost or rolloff, the corner frequency should be set at about 300 Hz. Substituting 300 Hz for f_{oLB}, C3 is about 5000 pf. For symmetry reasons, f_{oLC} is also 300 Hz and C2 equals C3.

Intermediate settings of this control will give intermediate amounts of boost and cut. 300 Hz will not be the corner frequency at these settings. The corner frequency will shift closer to the low end of the band when less emphasis or attenuation is required. The high and mid-frequencies will not be affected regardless of the setting of the control.

The treble activating circuit is shown in Fig. 10-18. C1, R1, R3 and C4 are from Fig. 10-16A. The potentiometer R2 has been drawn as a short circuit and omitted because at the high frequencies involved C2 and C4 are effective shorts across the bass control. Qualitatively, with R5 at the maximum left-hand setting (maximum treble boost), the control is a shunt across R1 in addition to several other components. Thus, the high frequencies are fed more easily to the gate than are the lower frequencies. This meets the requirements of a treble boost circuit.

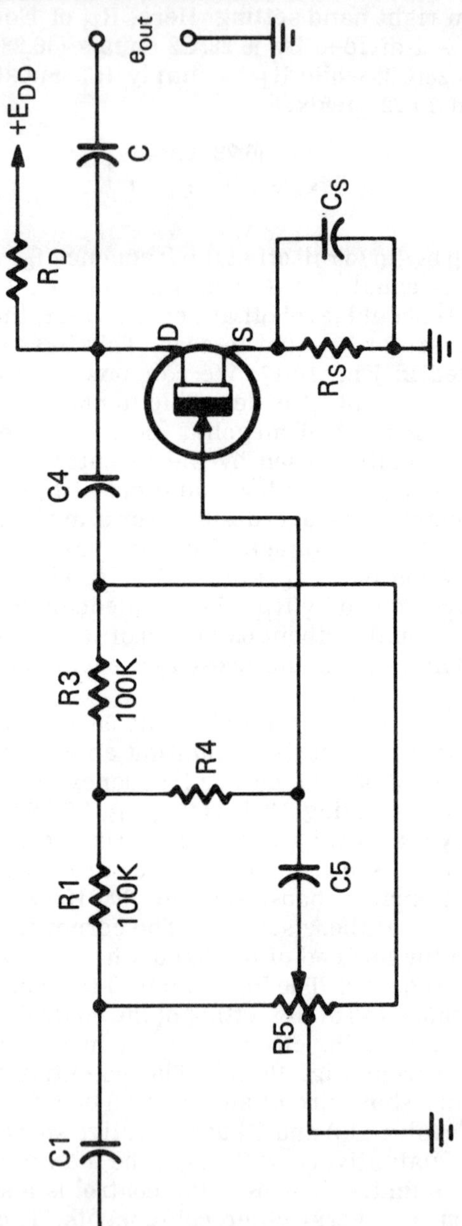

Fig. 10-18. Feedback treble control circuit.

Similarly, at the extreme right-hand setting of R_5, C_5 shunts R_3 as well as several other components in the circuit. It feeds the high frequencies back from the output to the gate more readily than it does the lower frequencies. Hence, there is treble cut.

Both the boost and cut circuits are in the operational amplifier circuit and Equation 10-23 does apply. Converting R_1, R_3 and R_4 mathematically from a "tee" to a "delta" configuration to facilitate analysis will yield a corner boost frequency at f_{oHB} and a corner cut frequency at f_{oHC}. They are both equal to 1 divided by $6.28C_5(R_1 + 2R_4)$.

The intermediate settings of the control will yield intermediate amounts of treble boost and cut. As was the case with the bass control, the corner frequency here too is shifted away from the center frequency when less boost or cut than maximum is required at the upper ends of the bands. The setting of the control will not affect the center- or low-frequency regions of the band.

C_5 was originally chosen at about 100 pf so that it will not load the input circuit excessively and yet be large enough not to be affected by stray capacities in the circuit.

f_{oHB} was chosen so that there will be about 16 db of boost at 10,000 Hz. An approximate curve used to determine the corner frequency is shown in Fig. 10-19. At the maximum setting of the control, f_{oHC} equals f_{oHB} equals 1.5 kHz. Since R_1 and C_5 are already known, R_4 can be calculated to be about 500,000 ohms. R_5 must be made as small as practical when compared to the impedance of C_5 at the highest audio frequency that must be boosted. A 500,000-ohm linear center-tapped control was found to be satisfactory.

A low-gain amplifier or lower impedance bipolar transistor is frequently used in the feedback tone control circuit. Since these components cause the operational amplifier to differ radically from the ideal, the components must change from the calculated values to effect results similar to those outlined above. The circuit should be designed in the laboratory in this case. Since the function of each component has been detailed, the effects of changing a component is known and the design procedure does not have to be haphazard.

The complete tone control circuit in Fig. 10-20 shows the bass and treble controls. The following factors affecting the various functions of the control should be noted. The amount of action of the treble control is affected by R_4 and C_5. Make

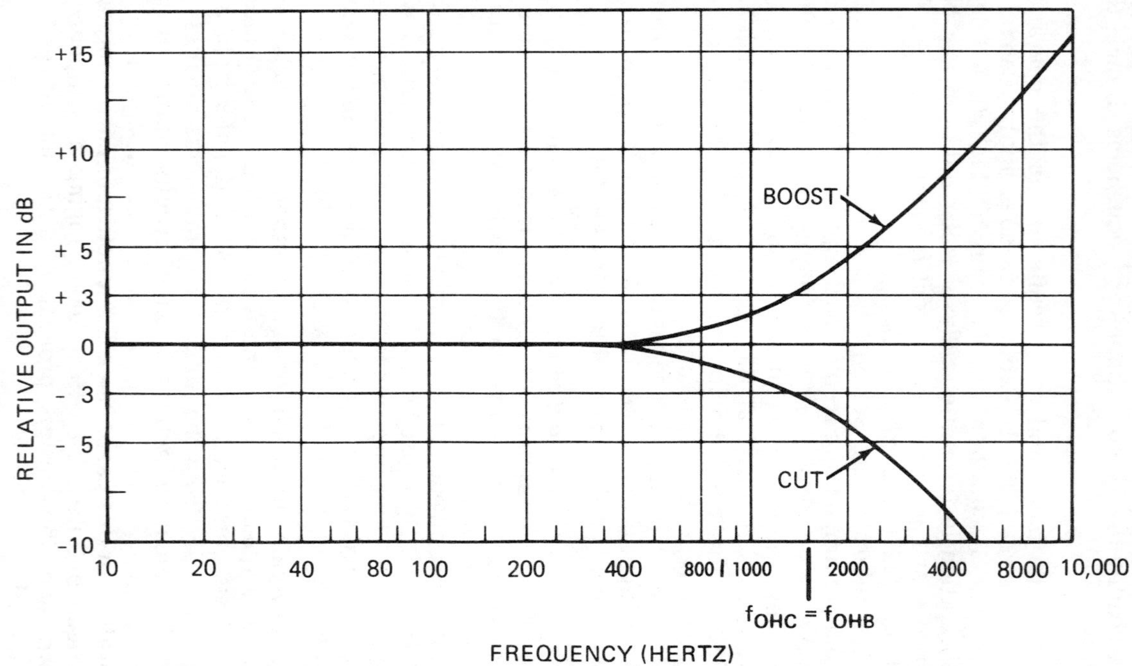

Fig. 10-19. Approximate treble cut and boost curves with the control set at maximum positions.

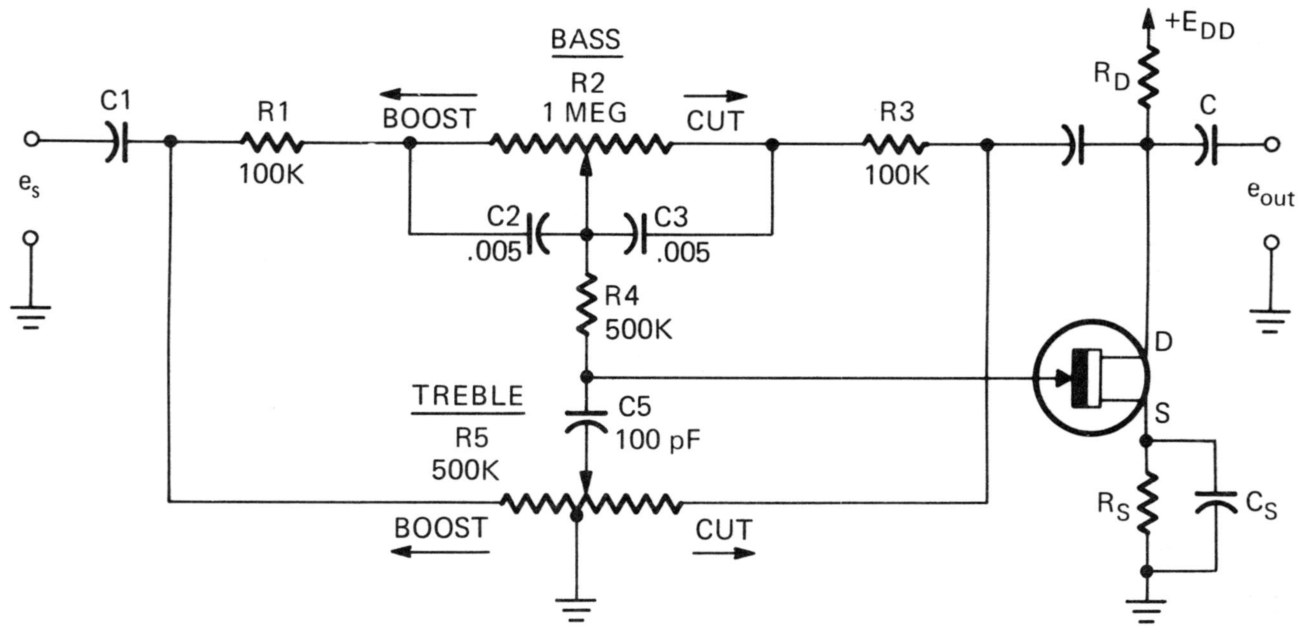

Fig. 10-20. Complete tone control using the operational amplifier feedback circuit.

either component larger if more treble action is required. To a lesser degree, increasing R_1 will increase the amount of treble boost while increasing R_3 will affect the size of the treble cut. As for the bass circuit, C_3 and R_3 must be increased to further emphasize the boost, while C_2 and R_1 must be increased to accentuate the cut.

Output Transformerless Power Amplifiers

Chapter 11

Power amplifiers are discussed in Chapters 7 and 8 on a more or less theoretical and introductory basis. The discussion classifies and details different modes of operation, covers heat sinking considerations, and describes the thermal cycling characteristic. In these circuits output transformers are prescribed as the component used to couple the output transistors to the loudspeaker system.

An output transformer imposes severe limitations on the quality of a design. To overcome the bandwidth restriction, the transformer must be both expensive and bulky. Furthermore, phase shifts at high and low frequencies set boundaries on the amount of feedback that can be placed successfully around a circuit before it becomes unstable.

While practical amplifiers designed for public address applications use output transformers, this component is omitted in just about all pieces of true high fidelity equipment. The loudspeaker is capacitively or direct-coupled to the output devices. The industry uses variations on three basic circuits.

One circuit uses a driver transformer for phase inversion. In this application the transformer is not as taxed as when it is used as an output device. Hence, the size and cost are relatively small. A second arrangement, the quasi-complementary circuit, uses two identical output devices driven by two lower powered complementary transistors arranged to provide the equivalent of phase inversion. The third, the fully complementary amplifier, uses a complementary pair of devices in the output so that phase inversion occurs in the power output transistors themselves, or in the combination of the output transistors and their drivers. The basic characteristics of these three circuits are discussed here.

TRANSFORMER PHASE INVERTERS

A circuit using a transformer as the phase inverter is shown in Fig. 11-1. Q1 and Q2 may be considered as a 2-stage

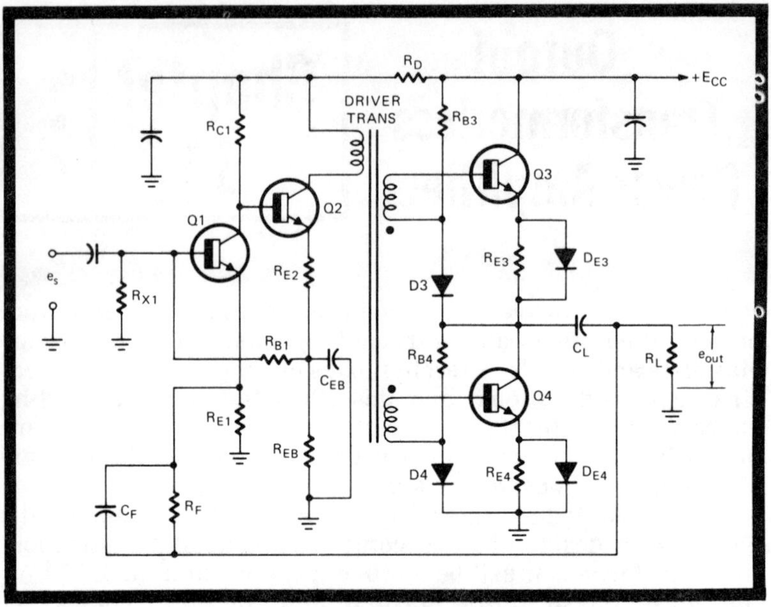

Fig. 11-1. Circuit using a driver transformer for phase inversion.

voltage amplifier driving power transistors Q3 and Q4 through the driver transformer. To be specific, each transistor stage can be thought of as a power amplifier. Thus, Q1 delivers its minute amount of power to drive a somewhat larger device, Q2, which in turn, must deliver enough power to drive the high-power output transistors, Q3 and Q4. Theoretically, Q3 and Q4 are bigger devices than Q2, which is, in turn, bigger than Q1.

The input signal is capacitively coupled to Q1. Direct coupled to Q2, Q1 receives its bias voltage from the emitter circuit of the latter transistor. Since the DC feedback through R_{B1} from the junction of R_{E2} and R_{EB} is substantial, this circuit is extremely temperature stable. R_{EB} in the emitter circuit is bypassed to ground by C_{EB} to prevent any AC from being fed back from this point along with the DC.

The output from Q2 is fed to a transformer with two identical secondary windings—preferably bifilar wound. The phase relationship between the two windings is indicated by the dots. Dots at the ends of two windings indicate that these ends are in phase with respect to the unmarked ends.

Should the portion in a cycle be such that the unmarked ends are positive with respect to the ends with the dot, Q3 is forward biased and will conduct while Q4 is reverse biased. In the next portion of the cycle, the opposite polarity will exist at the bases and Q4 will conduct while Q3 will remain idle. The composite signal is reconstituted across R_L.

The impedance ratio of the transformer is based on the goal of presenting an ideal load to the driver transistor. Conventional designs use an impedance ratio of about 9:1. The ratio should be optimized in the laboratory, specifying this ratio for minimum overall distortion.

Assuming adequate transistor capability and heat sinking, the amount of power the circuit can deliver is based upon the size of the supply voltage, E_{CC}, the collector-to-emitter saturation voltage, and the voltage across emitter resistor R_{E3} or R_{E4}. Power is related to the load at the output by the equations $e_{rms}2$ divided by R_L and $i_{rms}{}^2R_L$. Peak-to-peak voltage for a specific power output is $e_{p\text{-}p}$ equals 2.82 e_{rms}, while peak-to-peak current is $i_{p\text{-}p}$ equals 2.82 i_{rms}. The supply voltage must be capable of swinging the peak-to-peak voltage $e_{p\text{-}p}$ across the load in addition to the peak-to-peak current across one of the emitter resistors, or $i_{p\text{-}p}R_{E3}$.

Collector-to-emitter saturation voltage limits the swing of the voltage across the load. Because two transistors are involved, the sum of both saturation voltages at the peak of the collector current swing must be added to $e_{p\text{-}p} + i_{p\text{-}p}R_{E3}$ to estimate the minimum supply voltage required if the amplifier is to deliver a specified amount of power. Keeping the operation in the linear region requires that the specified saturation voltage be multiplied by a factor of at least three before being added to the other quantities already in the relationship.

Diodes D3 and D4 in the output stage are forward biased and are in the circuit in the interest of stabilizing the quiescent current against variations of V_{BE} with temperature. The actual idling current is established by the voltage developed across the diodes as well as across the other resistors in the DC circuit.

In the emitter circuit, resistors R_{E3} and R_{E4} are used primarily to supply AC and DC feedback and provide some relief for the distortion and DC stability problems inherent in this circuit. As a fringe benefit, it limits somewhat the emitter current to offer some protection to the output devices against overdissipation when the load, R_L, is accidentally shorted. In Class A amplifiers, the design is such that about 0.5 to 1.5 volts is developed across the resistor. Similar voltages are desirable in Class AB or Class B circuits during peak current intervals in the cycle.

Diodes D_{E3} and D_{E4} are not absolutely required in this circuit. Should they be used, the emitter resistors can be increased in size to improve the temperature stabilization characteristics. These diodes will bypass large resistors, thus allowing sufficient collector current swings during peaks in the cycle.

The signal at the output transistors is coupled to the load, R_L, through capacitor C_L. The capacitor limits the gain at low frequencies. Use the smallest C_L that will not limit the performance of the overall circuit.

R_F and C_F, in conjunction with R_{E1}, are the primary components of the feedback circuit. C_F is usually adjusted in the laboratory for the most faithful reproduction of square waves.

The output circuit in Fig. 11-1 can, of course, be simplified by omitting all diodes and replacing those in the base circuit with properly sized resistors. This can be done, but at the expense of performance and temperature stability.

Circuits using driver transformers are necessary primarily when germanium output devices are being used. Due to the relatively large leakage currents, complete isolation of the output transistors by the driver transformer is desirable. Although quasi-complementary circuits without transformers have been used with germanium devices, this circuit emerged as the primary arrangement when silicon transistors became readily and economically available.

QUASI-COMPLEMENTARY POWER AMPLIFIERS

The basic circuit of the quasi-complementary arrangements is shown in Fig. 11-2. It is direct-coupled throughout. The signal is amplified by Q1 and fed to the complementary pair, Q2 and Q3. During the positive portion of the cycle, the bases of the complementary pair are positive with respect to the emitters; the NPN device (Q2) will conduct while the PNP transistor (Q3) is turned off. The reverse is true during the alternate half cycle.

The half cycles are supplied to the output devices, Q4 and Q5, after amplification by the complementary pair. Both portions of the cycle are fed to R_L through C_L and reconstituted across the load resistor. Feedback is accomplished through the parallel combination consisting of C_F and R_F.

The DC quiescent conditions are such that half the supply voltage should be present at the point labeled in the drawing as $E_{CC}/2$. Determined by resistors R_{B1} and R_{X1}, the bias current through Q1 is instrumental in establishing this quiescent condition. The collector load on the transistor consists essentially of $R_{B2} + R_{Y2} + D1 + D2$. The diodes are used to set and maintain the idling current in the output circuit despite temperature fluctuations, and may be replaced by other temperature-sensitive devices. Resistors are often used when compensation for temperature variation is not essential.

There are several inherent problems with this circuit, and all can be solved with the addition of capacitor C2 in a positive feedback bootstrapping arrangement. Notice also that the resistors in the base circuit used to bias Q2 have been split into two components, R_{B2} and R_{Y2}, providing a junction to accept C2. On large signals, the bias on the driver transistors tends to shift the operation to Class B, producing crossover distortion. To compensate for this, large amounts of feedback must be placed around the circuit. Gain must be large to accommodate all the feedback that is required. Positive feedback supplied by capacitor C2 increases the load impedance the complementary pair presents to Q1, with the consequent increase in gain of the circuit.

Large positive peaks in the signal tend to cut off Q2 by placing the base and emitter of the transistor at $+V_{CC}$. However, there is a voltage across C2 due to its being charged while the circuit is idling. This voltage keeps the base at a positive potential with respect to the emitter so that Q2 will continue conducting during all portions of the cycle.

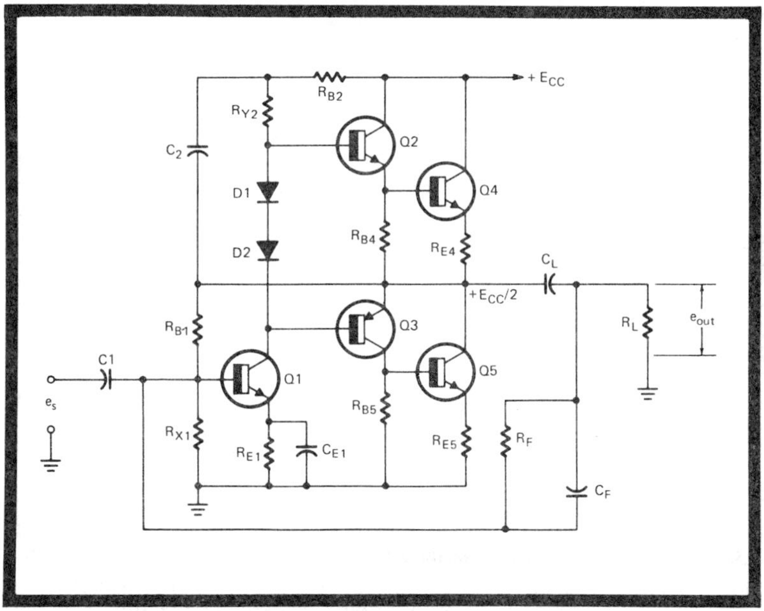

Fig. 11-2. Quasi-complementary circuit.

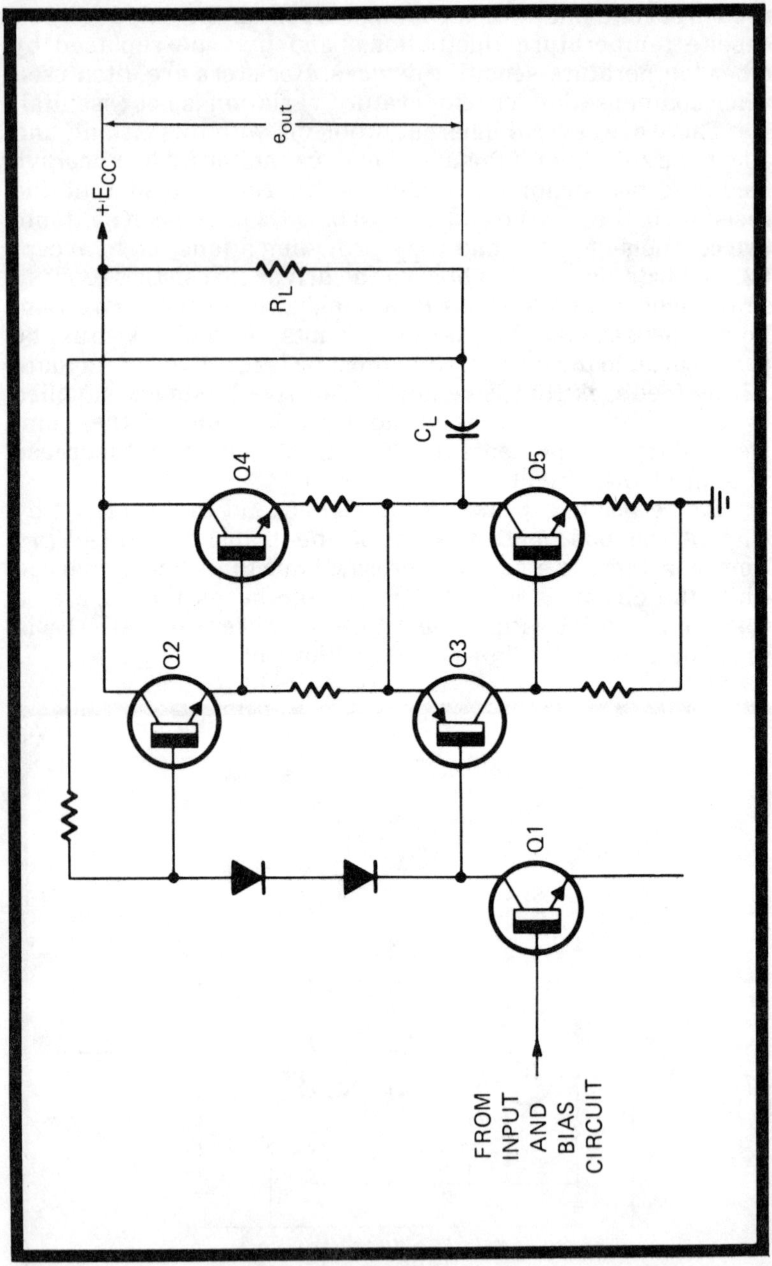

Fig. 11-3. Circuit designed to eliminate the need for a special bootstrap capacitor. C_L doubles as the coupling and bootstrap capacitor. The load is returned to AC ground, $+E_{CC}$.

The design of the bootstrap circuit is quite simple. Since R_{B2} and R_{Y2} are essentially across the load through C2, they should be made as large as possible consistent with the current requirements of the base circuit of Q2. Both resistors are usually specified as equal to each other. Under quiescent conditions, $E_{CC}/2$ appears across the series circuit formed by R_{B2} and R_{Y2} as well as across the series circuit formed by R_{B2} and C2. Since RB2 equals R_{Y2}, the voltage across C2 is one half of $E_{CC}/2$ or is equal to $E_{CC}/4$. Charged to this voltage, C2 maintains a constant current through R_{Y2} and the base-emitter junction of Q2. C2 must be so large as to maintain its charge even when low frequencies are being reproduced.

An alternate bootstrap circuit designed to eliminate the need of capacitor C2 and resistors R_{B2} and R_{Y2} is shown in Fig. 11-3. Instead of these components, a resistor is connected from the junction of C_L and R_L to the base of Q2. C_L doubles as the bootstrap capacitor besides fulfilling its function to couple the signal to the output load resistor or loudspeaker. R_L is connected to $+E_{CC}$, an AC ground. The major drawback of this circuit is that the DC base current for the drivers flows through the load R_L. If this current is very small, it should not affect the operation of the loudspeaker usually used as R_L.

A constant-current source at the bases of the complementary drivers can eliminate the need for the bootstrap capacitor. A drawing of this circuit is in Fig. 11-4. A constant current is supplied to the drivers while a high impedance is presented to the voltage amplifier stage, Q1. The voltage drop between the base of Q6 and $+E_{CC}$ should be as small as practical so as not to limit the output voltage swing. Hence, low forward-voltage-dropping silicon diodes should be used in the constant-current circuit, rather than the higher voltage zener diodes. Main advantages of this circuit include the improved distortion at low frequencies and more symmetrical clipping of the peaks in both halves of the signal.

Direct-Coupled Load

In all quasi-complementary circuits discussed thus far, the load was coupled to the output transistors through a large electrolytic capacitor, C_L. Although frequently used, the capacitor has several drawbacks, not the least of which is the inherent nonlinearity of electrolytic coupling devices. Other reasons for eliminating C_L are the low frequency rolloff due to the R_L-C_L high-pass filter, and the corner frequency created by this rolloff which can contribute to instability when feedback is applied around the circuit. Finally, and perhaps the

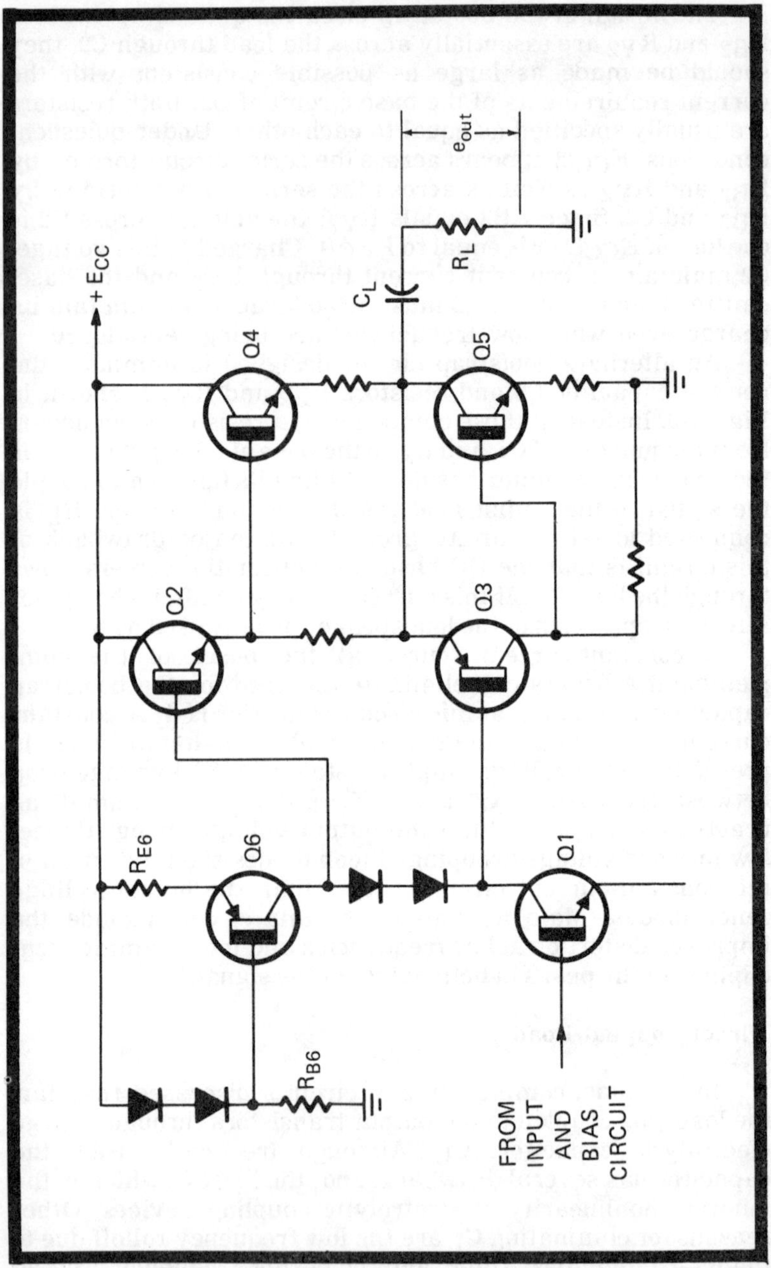

Fig. 11-4. Quasi-complementary circuit using a constant-current source, Q6, to maintain the quiescent current through the output devices at all times.

most important drawback is that this capacitor must be charged through the output transistors. If, in the process, the transistor handles more energy (power x time) than it can dissipate, it will destroy itself.

In Fig. 11-2, one end of R_L is connected to ground. When idling, the other end of R_L must be at the same ground potential if there is to be no DC flowing through the resistor (or loudspeaker) load. This is easily accomplished when a coupling capacitor is used. In the absence of C_L, the junction of Q4 and Q5 must be placed at a zero potential with respect to ground while the circuit is idling. To accomplish this, a positive voltage with respect to ground, $+E_{CC}$, is placed at the collector of Q4 while an identical negative voltage, $-E_{CC}$, is placed at the emitter of Q5, or more exactly at the lower end of resistor R_{E5}. If both transistors, Q4 and Q5, conduct identical currents during the idling period, there is zero voltage at the junction of the two devices to which the load is connected (or at the junction of the collector of Q5 and the lower end of R_{E4}). With a signal applied, the positive-going portion swings the voltage across R_L from zero towards $+E_{CC}$ while the negative portion of the signal swings it from zero towards $-E_{CC}$.

All would be great if the quiescent current can be maintained constant at all times, so that the voltage across R_L will not shift from zero. Unfortunately, Q1 is subject to drift with temperature, as is the case with all transistors. But any change in the collector current through Q1 will upset the balance at the output more than drifts in transistors further advanced in the chain. Drift due to the drivers and outputs is minimized by maintaining the upper and corresponding lower devices at equal temperatures on heat sinks or in free air. In this manner, drift in one half of the driver and output circuit is balanced by the drift in the other half. Q1 has no corresponding transistor with which to overcome or balance collector current changes due to drift. A partner must be added.

Differential amplifier circuits use two transistors. If one is placed in the proximity of the other, or in the same physical case, temperature variations affect both equally. This is especially true if they are matched pairs. A drawing of this type of arrangement is shown in Fig. 11-5. The differential circuit placed before the + and - drives replaces Q1 in the interest of stability.

Q1 and Q2 in Fig. 11-5 are a differential pair driving a second differential pair, Q3 and Q4. A signal at the input to Q1 appears amplified at the collectors of Q1 and Q2. While the input signal is in phase with the signal appearing at the

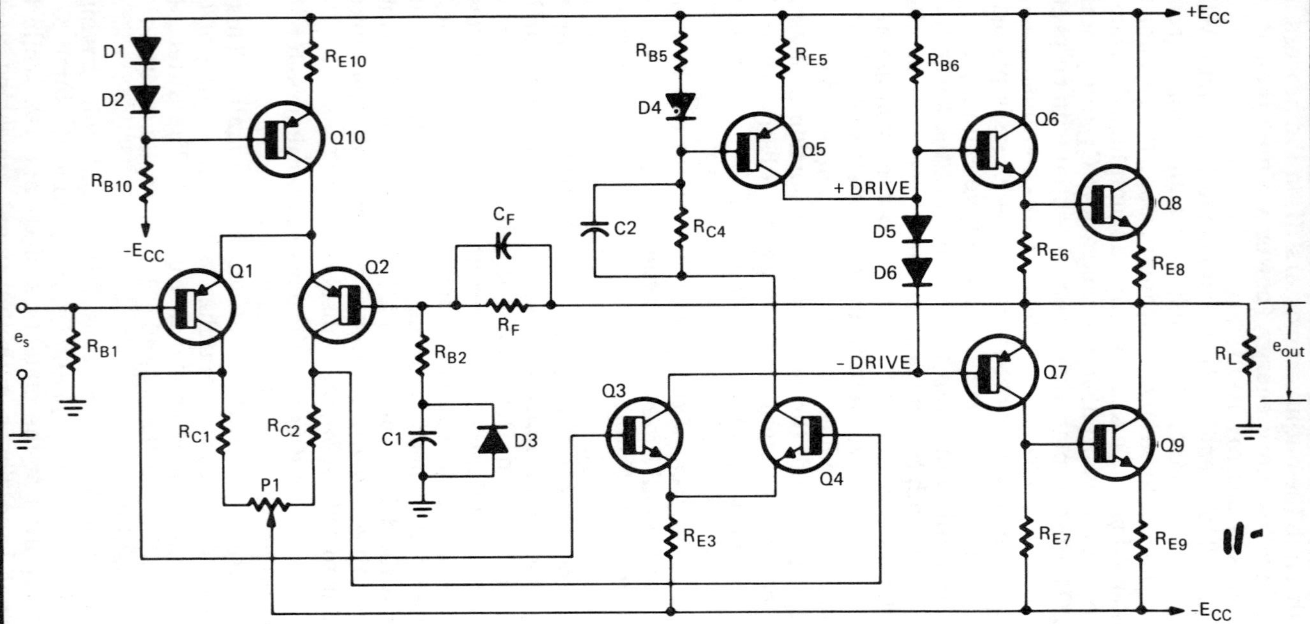

Fig. 11-5. Differential amplifier driving quasi-complementary circuits.

collector of Q2, it is out of phase with the signal at the collector of Q1. When amplified further by Q3 and Q4, the signal undergoes a 180-degree phase reversal between the bases and collectors of the respective transistors. The collectors of both transistors are driven to $-E_{CC}$ by the signal. "-Drive" is provided at the collector of Q3 to the lower pair of output devices, Q7 and Q9.

From Q4, the signal is fed to the base of Q5, which is arranged as a unity gain amplifier. The phase of the signal is once again reversed. The signal drives the collector of Q5 to $+E_{CC}$, providing the "+ drive." The phase reversal in Q5 makes the + drive signal identical in phase with the - drive signal at Q3. Two signals with identical phase characteristics are fed to the quasi-complementary output circuit consisting of Q6 through Q9. Proper phase relationships are achieved here to provide the reconstituted amplified output of the input signal. The + drive causes the output of the upper pair of devices to swing to $+E_{CC}$ while the output of the lower pair of devices swings from $-E_{CC}$ due to the - drive. Equal output excursions of both halves of the output circuit to $+E_{CC}$ and $-E_{CC}$ makes bootstrapping unnecessary.

Feedback is applied from the output through R_F and C_G. AC and DC feedback are developed across R_{B2}. It is connected to ground through C1 as a return for the AC feedback and connected to ground through D3 for the DC feedback return.

The circuit shown is raw. It will not operate properly unless it is designed carefully in each aspect. Instability will be prevalent unless the frequency rolloff is carefully controlled at both ends of the audio band. Let us now carefully analyze various sections of the circuit.

Q10, the constant-current amplifier, is designed to establish and maintain the sum of the idling currents through Q1 and Q2. Assuming all silicon devices are used, the voltage across the base-emitter junction and across D2 are identical. Hence, the voltage across D1 and R_{E10} are identical and equal to about 0.7 volt, the voltage across a silicon diode. If the sum of the idling currents through Q1 and Q2 are adjusted to, let us say, 2 ma, 2 ma must flow through R_{E10}. Then the resistor will be 0.7 volt divided by 0.002 ampere, or 350 ohms. R_{B10} is used to establish a current through, and consequently fix the voltage across, D1 and D2. The required idling current for Q1 and Q2 can be determined by first noting the maximum current required across the output load at the peak of the signal. It is divided by the product of the current gains of all stages excluding the first. This is the minimum collector current required from each of the two input devices.

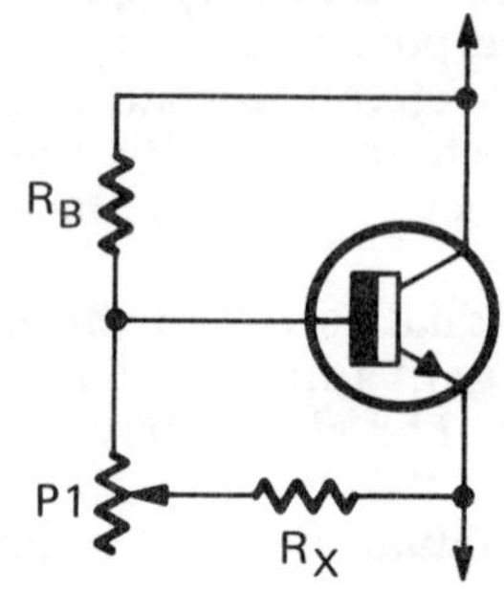

Fig. 11-6. This circuit contains a transistor that enables you to adjust the amplifier's bias while retaining temperature compensation in the unit.

If Q3 and Q4 are to be capable of swinging almost to $-E_{CC}$, the voltage across R_{E3} must be small; let us make it 1.5 to 2 volts. This voltage added to V_{BE} of Q3 or Q4 is present at the collectors of Q1 and Q2. Since, during idling, 1 ma is to flow through each collector, a simple Ohm's Law calculation will help determine the size of the resistances in the collector circuits. A portion of each collector resistor is taken from R_{C1} and R_{C2} to compose potentiometer P1. The control is used to balance the relative quiescent current through each transistor and is adjusted for zero volts across R_L when the circuit is idling. The DC feedback is expected to maintain this idling condition.

The + drive must be identical in amplitude to the - drive. To accomplish this, R_{B5} is made equal to R_{E5}. Voltages across D4 and the base-emitter junction are equal. Hence, the currents through R_{B5} and R_{E5} are equal, producing a - drive current in the collector circuit equal to the current flowing through R_{B5}. Since the current through R_{B5} is almost identical to the collector current flowing through Q3 and Q4, the - drive is equal to the + drive. R_{C4} dissipates power available at the collector of Q4 and not required at the base of Q5.

D5 and D6 set the quiescent idling current through the output devices. Should adjustment flexibility be desirable, one of the diodes may be replaced by a potentiometer. For better temperature compensation, the diodes can be replaced with the transistor circuit in Fig. 11-6, and the circuit will still retain the ability to stabilize the current in the circuit.

A simplified variation of the circuit of Fig. 11-5 is shown in Fig. 11-7. Here, the differential output is taken from one

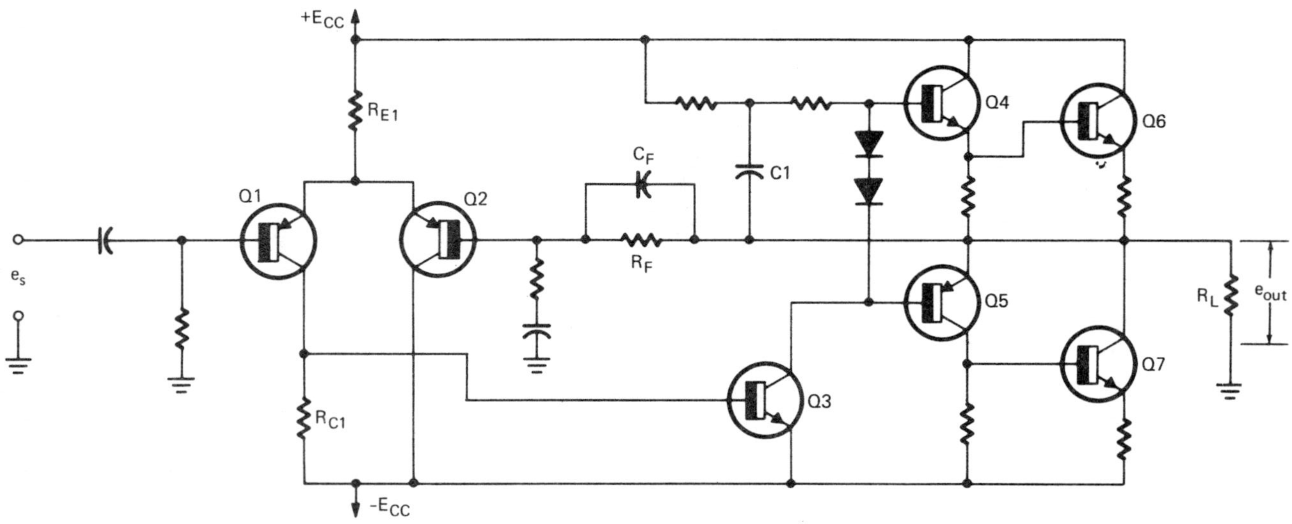

Fig. 11-7. An alternate and economical version of the differential amplifier circuit.

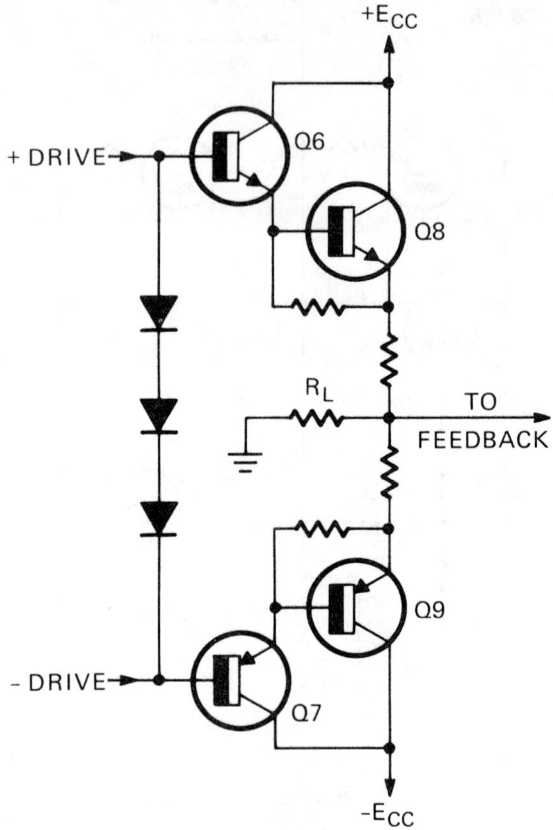

A

transistor of the pair, Q1 and Q2. R_{E1} is made large enough so that a constant-current source is not required, eliminating the cost of one transistor. The signal, fed to Q3, is then passed on to the quasi-complementary circuit consisting of Q4 through Q7. C1 is a bootstrap capacitor required here so that the upper transistor, Q4, can swing to saturation. The independent drivers, capable of signal excursions to the limits of the power supply voltage, make the bootstrapping circuit unnecessary for the arrangement in Fig. 11-5.

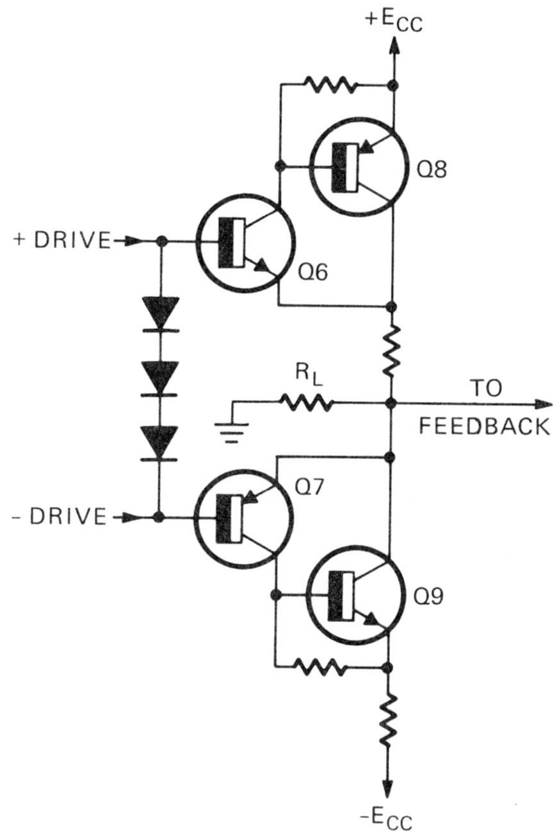

B

Fig. 11-8. Two output circuits used to replace the transistor counterparts in Fig. 11-5—a dual Darlington (A) and a dual beta multiplier (B).

FULLY COMPLEMENTARY OUTPUT

A close look at the quasi-complementary arrangement reveals that the upper two transistors are a Darlington pair while the lower two form a complementary beta-multiplier pair. Although quite similar, the ultimate quality amplifier is designed with identical pairs. That is either a complementary pair or a Darlington arrangement should be used for both halves of the circuit. The two arrangements are shown in Fig. 11-8.

Either circuit can be substituted directly for the quasi-complementary arrangements. Either circuit can be used to replace transistors Q6 through Q8 in the arrangement discussed with respect to Fig. 11-5. Design requirements are not unlike those described above.

PROTECTING THE OUTPUT DEVICES

The output transistors are in a vulnerable position in the circuit. The load plays havoc with the amount of power the device is forced to dissipate. Under normal load conditions, the chances of destroying a transistor are negligible. But consider what happens when the load is accidentally reduced in size from the norm, or shorted. More current can flow through the output devices than is normally possible or required with proper size load resistors.

Should a capacitive load be placed across the nominal rated load or voice coil of a loudspeaker, the amplifier may become unstable and break into oscillation. Transistors can break down as a result of overdissipation when the circuit is oscillating. A situation of this type may occur when an electrostatic tweeter is placed at the output of the power amplifier.

Capacitor Loads

Electrostatic tweeters are not simple capacitors. Equivalent circuit diagrams involve more than a capacitor. However, a capacitor of 2000 to 3000 pf may be considered as an adequate representation of this type of high-frequency speaker for most experimental work.

The amplifier can be tested for stability or marginal stability by simply placing a capacitor across the normal load resistor at the output. The circuit should be checked with capacitors ranging from 500 pf to 1 mfd. Any instability or signs of instability indicates the design requires some additional experimental work to correct the condition.

The test for marginal instability involves feeding a 10-kHz or 20-kHz square wave through the amplifier. Next, place various values of capacitors across the load resistor. This may, or probably will, distort the square wave by rounding the leading edge. Ringing may also be observed on the horizontal portion of the square wave when some sizes of capacitors are placed across the output load resistor. The rounding of the leading edge can be ignored, but the ringing usually cannot be neglected. If there is ringing, the amplifier is marginally unstable. However, if the ringing decays rather than increases as the square wave progresses through the cycle, it need be of no concern to the designer.

A small capacitor placed across a base-collector junction at the output, or between some other judiciously and experimentally chosen points in the amplifier, may remove any signs of instability. A series R-C network across the output load or loudspeaker may at times also be required. A small 2 or 3 microhenry RF choke made from 18 to 20 gauge wire, placed in series with the load, will frequently negate any tendency towards oscillation. The choke is usually shunted by a small resistor; ten ohms or somewhat less will do. Complete elimination of the undesirable oscillation requires a considerable amount of trial-and-error type of experimentation on the wired circuit.

Current Limiting

By far, the primary cause of transistor failure is excess emitter and collector current due to a short circuit across the output load or loudspeaker. Circuits have been devised to sense the output current. At some predetermined output current, a device of some sort is turned on to shunt drive signal away from the power amplifiers. Also, limiting the drive will curtail the amount of current flowing through the output devices.

The most straight-forward circuit uses a small resistor, R_S, in series with the load, as shown in Figs. 11-9 and 11-10.

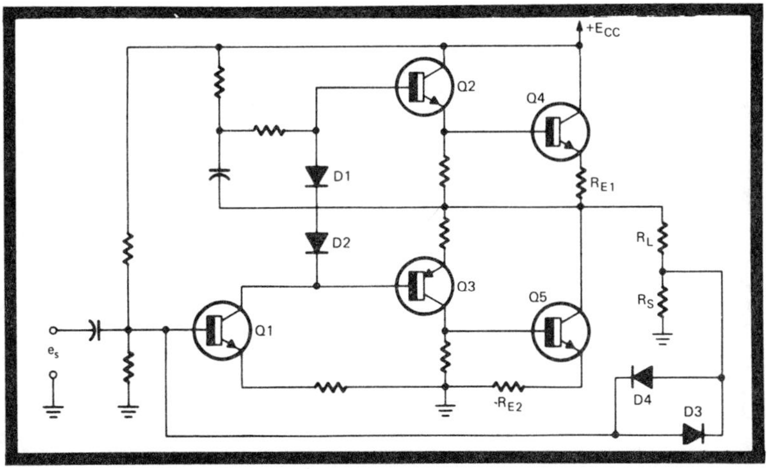

Fig. 11-9. Quasi-complementary circuit in which the output is sensed across R_S and fed back to the input through D3 and D4 when the input voltage is larger than the threshold voltage of the diodes.

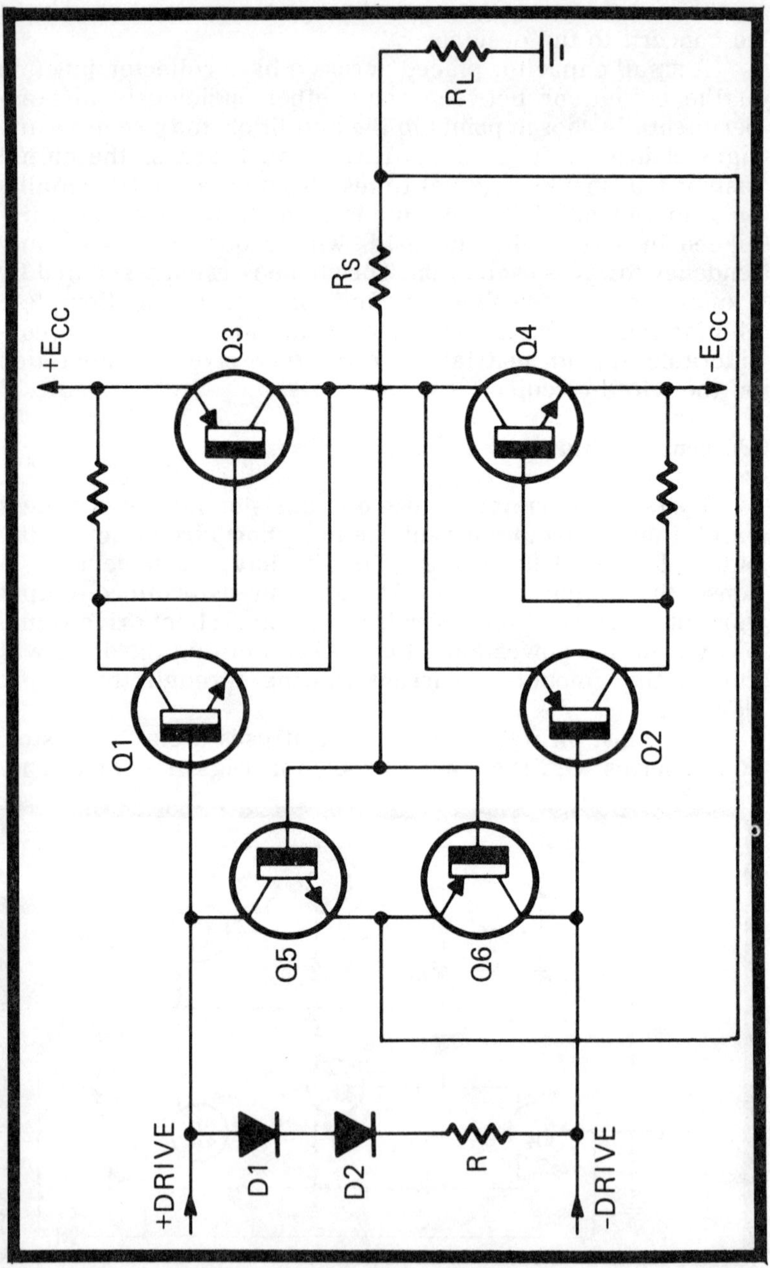

Fig. 11-10. Fully complementary circuit in which the output is sensed across R_S. Here, the signal is used to turn on Q5 and Q6 which shunt the drive signal away from the output transistors.

This resistor should be made as small as practical so as not to substantially affect the damping of the circuit. Never make it larger than one-fourth the impedance of the speaker so that the damping factor will never be more than its minimum acceptable value of 4.

In Fig. 11-9, the output current flows through R_S as well as through the output load resistor, R_L. A voltage is developed across R_S by the output current. The resistor value is chosen so that the portion of the output voltage developed across it will bias diodes D3 and D4 in a conducting mode when more than the permissible amount of current flows through the output circuit. When the peaks of this voltage are sufficiently large so as to turn on either or both diodes, D3 and-or D4, a negative feedback loop is completed through the diodes to the input circuit. Consequently, the output is limited by a reduced drive until the cause of the excess current is eliminated.

The output current in Fig. 11-10 is likewise sensed as a voltage across resistor R_S, this time placed between the load, R_L, and the amplifier. R_S is adjusted so that the voltage developed across it when excess current flows through the output circuit is sufficient to turn on transistors Q5 and Q6. In turn, these transistors shunt signal or drive current away from the output circuit, limiting the output current to predetermined safe levels.

A similar circuit is used in the fully complementary arrangement in Fig. 11-11. Here, the current flowing through the output devices develops voltages across the individual emitter resistors, R_{E1} and R_{E2}. As in Fig. 11-10, this voltage is applied to transistors Q5 and Q6. When the voltage is high due to excessive emitter current, Q5 and Q6 will shunt signal away from the output circuit. Diodes D3 and D4 are provided so that the signal will not forward bias the collector-base junction of transistors Q1 and Q2. Forward bias on these junctions will cause the output signal to distort.

There is nothing tricky and unusual in the circuits described thus far. The output is sensed. Information is then fed back to an earlier stage which limits the drive and consequently controls the amount of output current that can flow.

Simpler methods have been devised to do the job; some are more effective while others are not quite as good. The simplest method is to use excessively large output devices which can easily dissipate and deliver all the power they must, even when the output load is shorted. A second circuit limits the supply voltage on the drivers so that the output transistors cannot be overdriven. A variation of this is the built-in poor

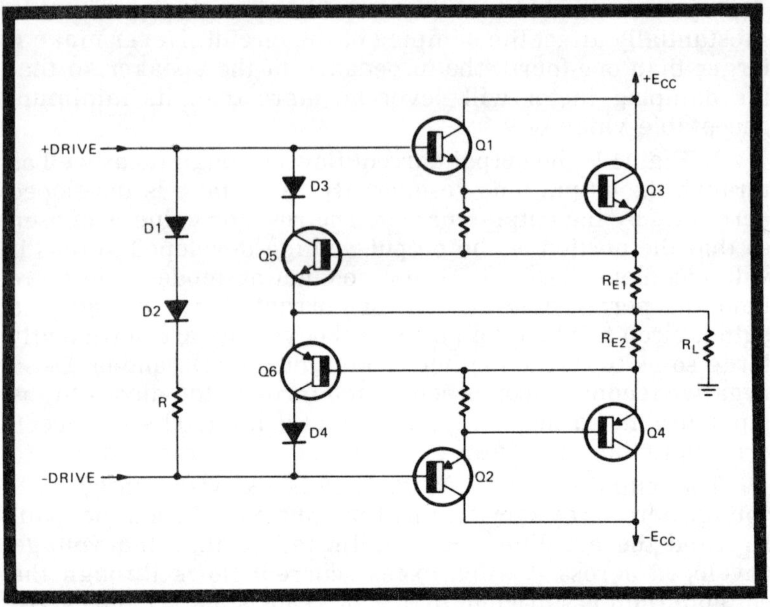

Fig. 11-11. Fully complementary circuit with protection devices similar to the circuit in Fig. 11-10. Here, however, the current is sensed across individual emitter resistors in each output pair of transistors.

regulation in the power supply of the amplifier in order to limit the maximum power available to the output devices.

A third method incorporates fuses in series with the load or in series with each emitter of the output devices (replacing R_{E1} and R_{E2} in Fig. 11-3). The fuses are theoretically supposed to blow before the transistors. Actually, they race the transistors to oblivion. Instead of fuses, small bulbs have been used in place of the emitter resistors in the output circuit. Based on the premise that the bulb resistance will increase when the current is large, it is theoretically a practical measure used to limit current peaks. Unfortunately, bulbs have a thermal time lag. The output transistors are thus not protected during instantaneous bursts of output signal current.

The circuits in Figs. 11-12A and 11-13 are quite effective and simpler than those just discussed. However, the capabilities to limit the output current peaks are not as obvious. In both circuits, the bias is maintained by diodes D1 through D5. In the positive portion of the cycle, transistors Q1 and Q3 conduct. A voltage is developed across R_{E1}. If the voltage across R_{E1} is sufficiently large, the diodes connected across the series circuit, consisting of R_{E1} and the Q1 and Q3 base-emitter junctions, also conduct. In Fig. 11-12A, these diodes consist of D7 in series with D1 through D5. As is obvious from Fig. 11-12B, the series combination of diodes is in parallel with the series circuit consisting of R_{E1} and the base-emitter junctions of Q1 and Q3. Consequently, the sum of the base-emitter junction voltages and the voltage across R_{E1} are equal to the voltage across the combination of diodes D1 through D5 and D7.

For this discussion, assume the voltage across each conducting junction and diode is 0.7 volt. When conducting, the maximum voltage across the series combination of diodes D1 through D5 and D7 is 6 x 0.7 volt or 4.2 volts. This is also the maximum voltage across the series combination of R_{E1} and the junctions of Q1 and Q3. When conducting, the voltage across two of the diodes is equal to the voltage across the junctions of Q1 and Q3. Under normal conditions, the maximum voltage across R_{E1} is identical to that across the remaining four diodes in the circuit, or 4 x 0.7 volt or 2.8 volts. The voltage remaining for the junctions of Q1 and Q3 is 4.2 volts - 2.8 volts or 1.4 volts.

Now assume the current through Q3 and, hence, through R_{E1} is so large that more than 2.8 volts is developed across the resistor. Since the voltage at the base of Q1 is maintained at 4.2 volts maximum by the series diode circuit, less than 1.4 volts will then be across the two base-emitter junctions. Therefore, transistor current will tend to be reduced. R_{E1} is selected so that transistor current will be limited to safe values.

In the negative portion of the cycle, Q2 and Q4 conduct. Should the voltage across R_{E2} be sufficiently large, D6 will be turned on and complete a current path through diodes D1 through D5. An equivalent limiting sequence to that detailed above will then apply to the bottom portion of the transistor circuit.

Should it be that large magnitudes of voltage are required across R_{E1} and R_{E2}, diodes D6 and D7 in Fig. 11-12 can be replaced by zener diodes Z1 and Z2, in series with conventional diodes D8 and D9, as shown in Fig. 11-13. The voltage across D8 is identical to that of the base-emitter junction of Q1. Should the sum of the base-emitter voltage at Q3 and the voltage across R_{E1} exceed the breakdown voltage of Z_1, it will conduct. The drive current reaching the base of Q1 will be reduced. A similar course of events will occur in the lower portion of the circuit involving D9, Z2, R_{E2}, Q2 and Q4. R_{E1} and R_{E2} are adjusted so that Z1 and Z2, respectively, will conduct when more than the maximum allowable current

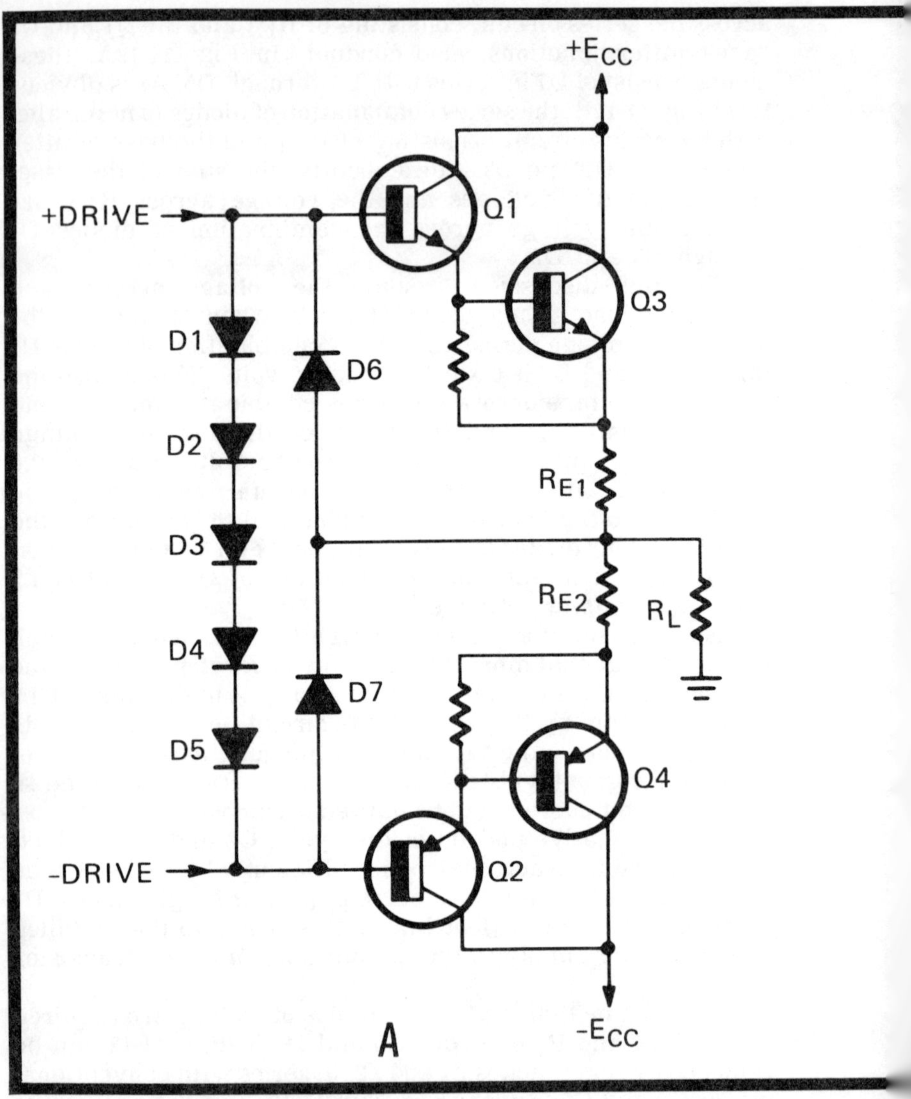

passes through the output device. D8 prevents negative signals from being shunted through Z1, while D9 serves the same function with respect to Z2.

Adding Protection

The current-limiting systems are good. However, more protection can be provided by sensing not only the current

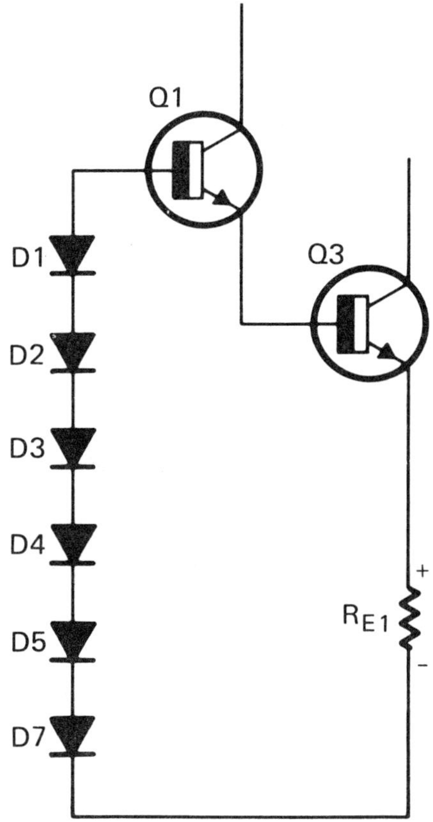

B

Fig. 11-12. (A) Fully complementary circuit in which D6 and D7, in conjunction with R_{E1} and R_{E2}, limit the drive to the output devices. Schematic B is an equivalent circuit of the diodes across R_{E1}.

flowing through a device, but also the power dissipated by the device at every instant in the cycle. Since the power is the product of voltage and current, the input to the power amplifiers can be reduced or completely turned off when the two factors exceed a specific combined level. A circuit of this type is shown in Fig. 11-14.

As far as the upper portion of the pushpull circuit is concerned, the instantaneous current is sensed across R_{E1}

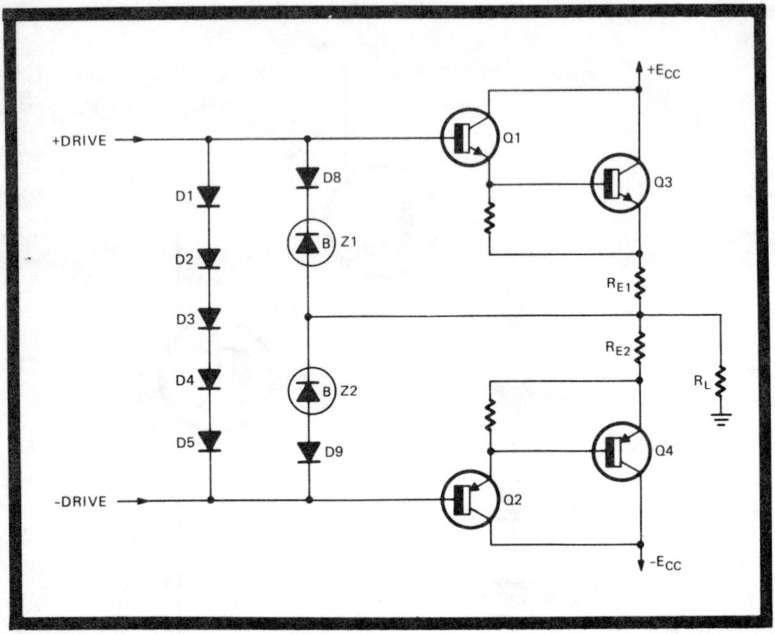

Fig. 11-13. Fully complementary circuit in which zener diodes shunt excess drive away from the respective bases to which they are connected through diodes.

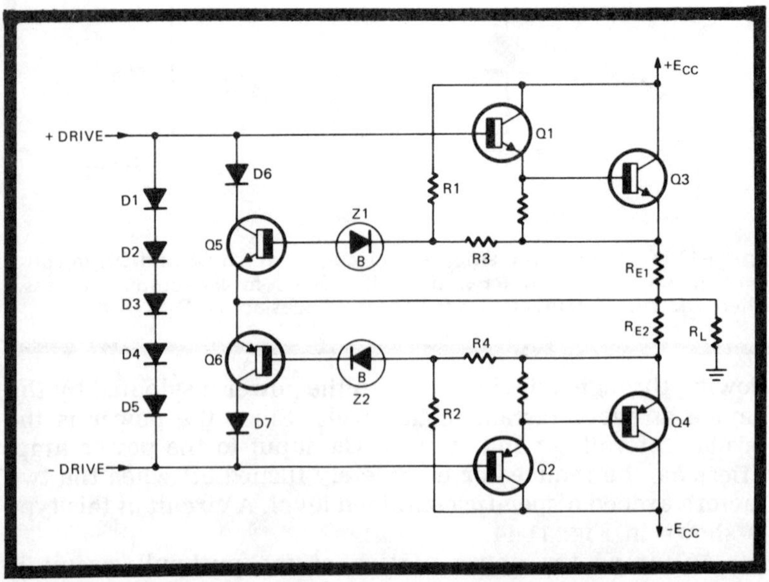

Fig. 11-14. Fully complementary circuit in which dissipation is limited.

and fed through R3 to the cathode end of zener diode, Z1. Meanwhile, resistors R1 and R3 form a voltage divider. The voltage fed to this divider is the instantaneous voltage across transistor Q3. A portion of the sensed voltage is fed to the cathode end of zener diode Z1, along with the sensed current. The information about the current and voltage are added together at the cathode of Z1. If the resulting voltage is large enough, it will cause Z1 to break down and thus turn on Q5. This transistor will shunt excess drive away from the base of Q1 and from Q3. Diode D7 prevents normal signals from being shorted through the collector-base junction of Q5. An identical description can be used to detail the operation of the lower portion of the circuit using Q6 as the key transistor in the protection arrangement.

Circuits can be designed to sense the value of the load resistor. The drive signal can be reduced considerably or entirely eliminated when the output is shorted. A circuit of this type has been drawn in Fig. 11-15. Assume first that the junction of R3 and R4 is connected to the emitter of Q5 and Q6, rather than to ground. The circuit will then revert to that shown in Fig. 11-11 with the exception that the voltage developed across emitter resistor R_{E1} is divided by resistors R1 and R3 before being applied to Q5, and the voltage developed across R_{E2} is divided by R2 and R4 before being applied to Q6. Hence, the circuit will limit the current.

Next, change the connection so that the circuit becomes the one shown in Fig. 11-15. The base-emitter junction of Q5 is connected across the series resistor combination of R3 and R_L. In addition to the voltage across R3, due to the collector current of Q3, the voltage across R_L is a factor in the amount of voltage that appears at the base-emitter junction of Q5. The voltages across R3 and R_L are out of phase and thus buck each other. Obviously, when R_L is large, more signal voltage will be developed across the resistor by the output current than when it is small. Therefore, when R_L is large, more signal voltage is required across resistor R_{E1} (and in turn across R3) to turn on the shunting transistor, Q5, than when R_L is small. This is due to the fact that the large voltage developed across R_L bucks the voltage across R3.

Should R_L be shorted, no voltage is available to buck that developed across R3. A relatively small amount of signal voltage across R_{E1} (and hence across R3) will turn on shunting transistor Q5. Since the voltage across R_{E1} is due to the current flowing through Q3, less current can flow through the output devices under shorted load conditions than under normal load conditions before the shunting transistors are

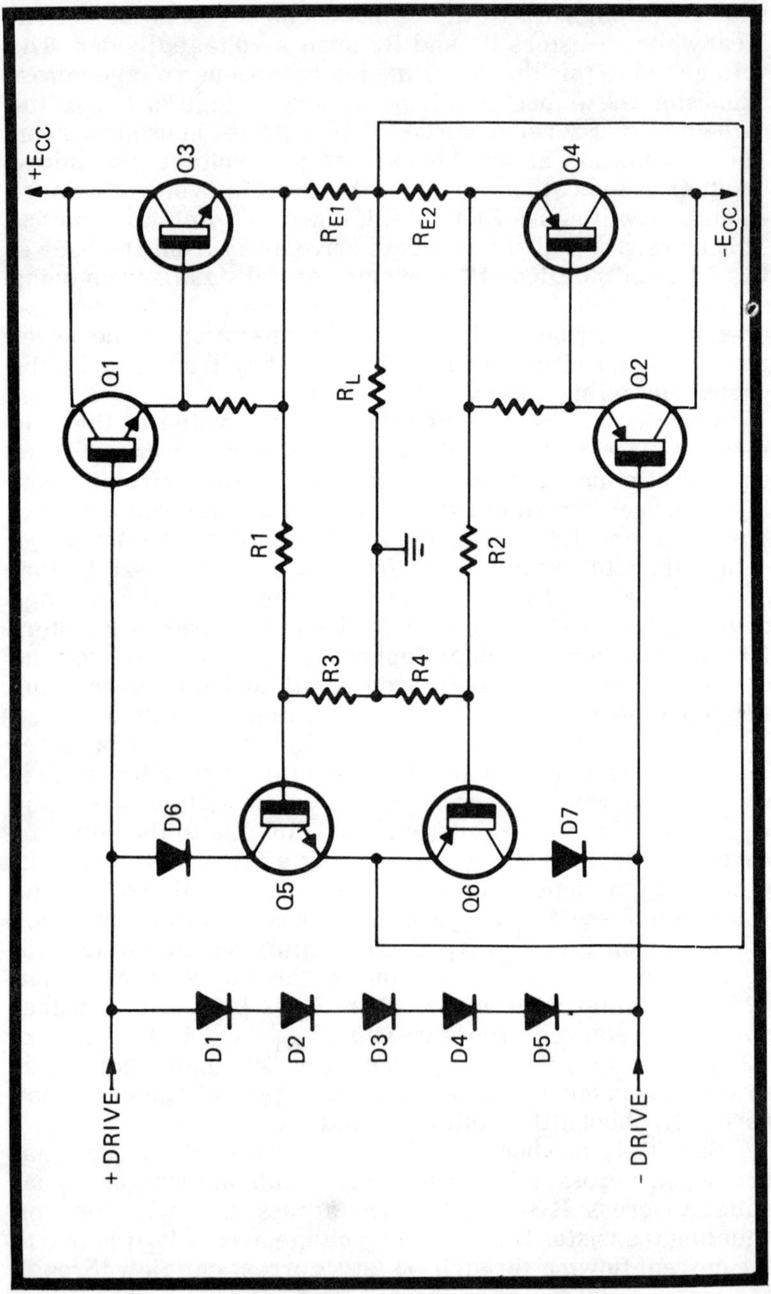

Fig. 11-15. Fully complementary circuit in which the output is a function of the value of load resistor R_L.

activated. Dissipation can be limited to safe values by simply adjusting the size of the resistors in the circuit. A similar description can be evolved for the lower half of the pushpull arrangement.

Unfortunately, high-frequency operation is adversely affected by this circuit. To overcome this drawback, resistors R3 and R4 are shunted to ground through diodes D8 and D9, while a 5000-ohm or 10,000-ohm resistor, R5, is placed between the diodes. The modified portion of the circuit is shown in Fig. 11-16.

The final circuit to be discussed here is used in the 700-watt Phase Linear Company amplifier. The protection arrangement is an integrated circuit computer. Unfortunately, the circuit is a closely guarded secret, but the description is nevertheless interesting and informative. The following is an excerpt from a letter from Mr. Robert Carver, with permission to publish the information.

"The operation of this energy-limiting computer, henceforth known as the limiter, is as follows: Information is fed into the input of the limiter, consisting of the instantaneous current, voltage, time rate of voltage change, and time rate of current change supported by the output transistors. In addition, a "clock pulse" is generated at the zero-axis crossing which serves as a pulse to reset the **integrator**, a circuit internal to the limiter. The output of the integrator, which is an analog voltage proportional to the energy absorbed by the transistor silicon chip, is applied to a summing junction.

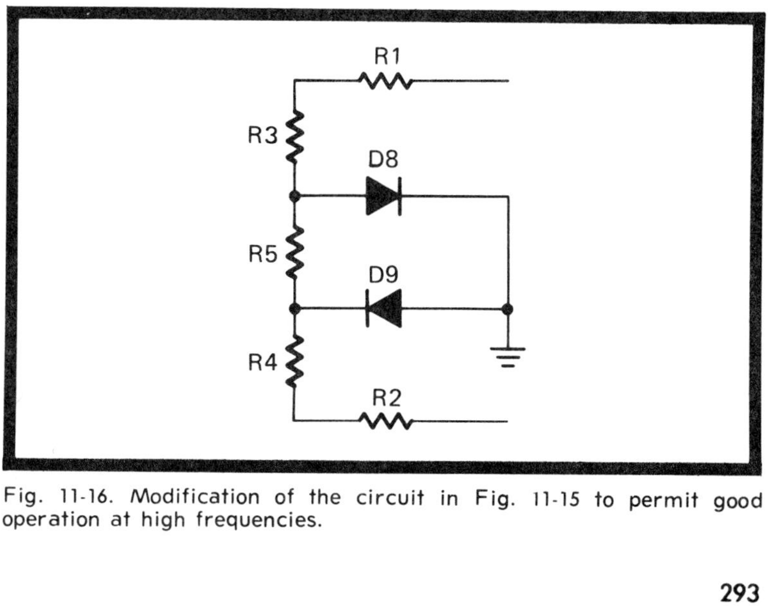

Fig. 11-16. Modification of the circuit in Fig. 11-15 to permit good operation at high frequencies.

"Another section of the limiter, the **common-mode detector**, generates a voltage which is again an analog voltage and which represents the amount of current being carried simultaneously by the output transistors. The common-mode detector generates its voltage indirectly by measuring di-dt and dv-dt (author's note: di-dt is a change in current with respect to time and dv-dt is a change in voltage with respect to time) and performing an appropriate computation. It is not possible to simply act on common-mode current directly unless some sort of frequency-selective process is used, because the ability of the amplifier to drive a low-frequency, high hysteresis load would be limited.

"The **comparator** is self explanatory. The analog outputs are summed at the summing junction, and if this sum (which represents the stress level the output transistors are sustaining) is of sufficient amplitude, the A-D converter, normally in a 0 state, switches to a 1 state, and turns on the **enable**. This causes the disconnect diode to be forward biased and pulls down the drive current at the base of the driver transistors.

"Notice that the protective circuit is either **on** or **off**. This insures that right up to the threshold of limiting, normal operation of the amplifier is not in any way affected."

Power Supplies

Chapter 12

The forward and reverse characteristics of the diode are discussed in Chapter 1, where it is indicated that the diode will conduct when the anode or P-type semiconductor material is made positive with respect to the N slab, the cathode. The diode will not conduct when the voltages are applied in the opposite direction. Due to these unique characteristics, the diode can be effectively used in power supplies.

BASIC POWER SUPPLIES

An audio amplifier uses many different types of power supplies to convert the AC line supply into useful DC power at the required voltage and current levels. They are all based upon the simple half-wave supply shown in Fig. 12-1.

The 120 volts derived from the house power is reduced to usable levels by means of power transformer, T. The voltage ratio of the primary to the secondary is, of course, proportional to the turns ratio of the two windings of the transformer.

The secondary voltage is applied across the load, R_L, through diode D. There is current flowing through the circuit

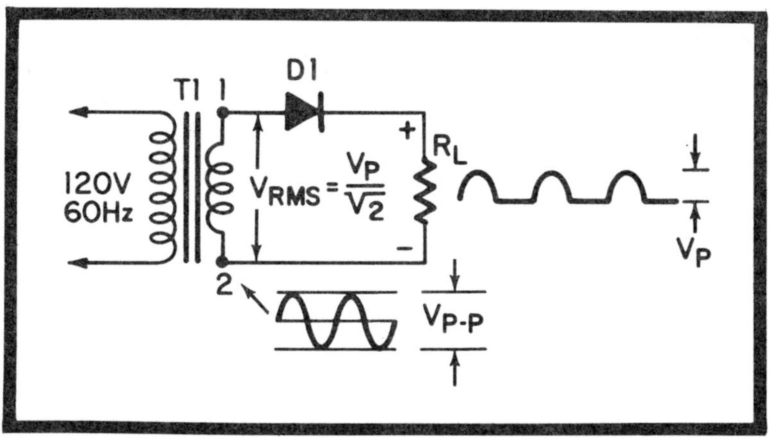

Fig. 12-1. Simple half-wave power supply circuit.

when terminal 1 of the transformer secondary is positive with respect to terminal 2. Only then is the anode of the diode positive with respect to its cathode, thus biasing the diode in its forward-conducting mode. The output across R_L is in positive half cycle pulses. On alternate portions of the cycle, the diode is reverse biased and it does not conduct.

The average voltage across the load is almost V_p divided by π while the RMS voltage of this half cycle is just about 0.5 V_p. The pulsating DC across the load has a large amount of ripple. As it stands, the circuit has little value as a practical power supply for an audio amplifier. But a large capacitor placed across the load will reduce this ripple so that the supply is usable as a smooth source of DC power. The capacitor is charged through the rectifier during the conducting half of the cycle and is only somewhat discharged through the load during the portion of the cycle when the diode does not conduct. It maintains the voltage across itself and the paralleled load at a relatively clean and constant DC.

The DC voltage across the load depends upon several factors, namely:

1. The line frequency, f.
2. The load resistance, R_L.
3. The sum of the resistance, R_T, of the transformer with all other resistors, R_S, in series with the rectifier, as well as the resistance of the rectifier, r_{ac}. (r_{ac}, the forward AC resistance of the rectifier is 26 divided by I, where I is the average or DC current flowing through the rectifier and expressed in milliamperes.)
4. The size of the filter capacitor.

A series of curves relating these factors is shown in Fig. 12-2.

To use this plot, the initial requirement is to determine the particular curve applicable to the circuit in question. To do this, determine the ratio of ($R_T + R_S$) divided by R_L. Use the curve which most nearly approximates this ratio. If you like, you can interpolate between two curves. Next, calculate the product $6.28fR_LC$ for your circuit. Mark this point on the curve you selected and note the ratio V_{RL} divided by V_p at the vertical axis. V_p is the peak voltage at the secondary of the transformer, or the RMS voltage multiplied by $\sqrt{2}$. Then calculate V_{RL}, the DC voltage across the load, from the information derived.

The amount of ripple that can be allowed to ride on the filtered DC depends upon the section of the audio amplifier to which the rectified and filtered voltage is applied. Increasing

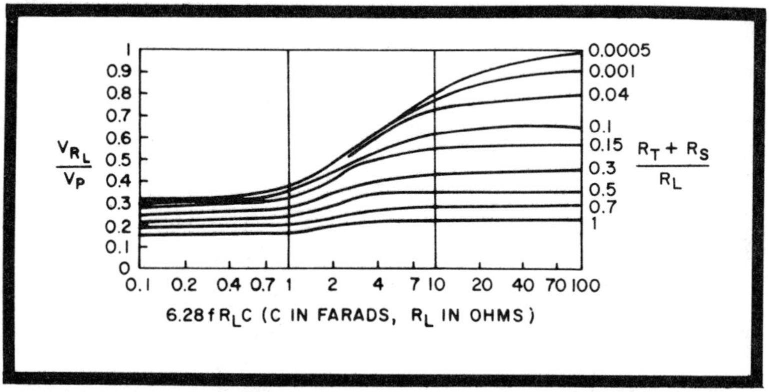

Fig. 12-2. Plot for a half-wave power supply, showing the V_{RL} divided by R_P to $6.28fR_LC$ relationship for various values of R_S divided by R_L.

the size of capacitor C will reduce the ripple. Fig. 12-2 can be used to recalculate the output voltages as related to different Cs. It should be noted that the output remains relatively constant as the product of $6.28fR_LC$ is increased above 100.

A full-wave rectifier system, shown in Fig. 12-3, will also reduce ripple and ease the filtering problem. However, in this circuit the ripple is of the more objectionable 120-Hz type rather than the 60-Hz variety in the half-wave supply. (This, of course, assumes a 60-Hz power supply.)

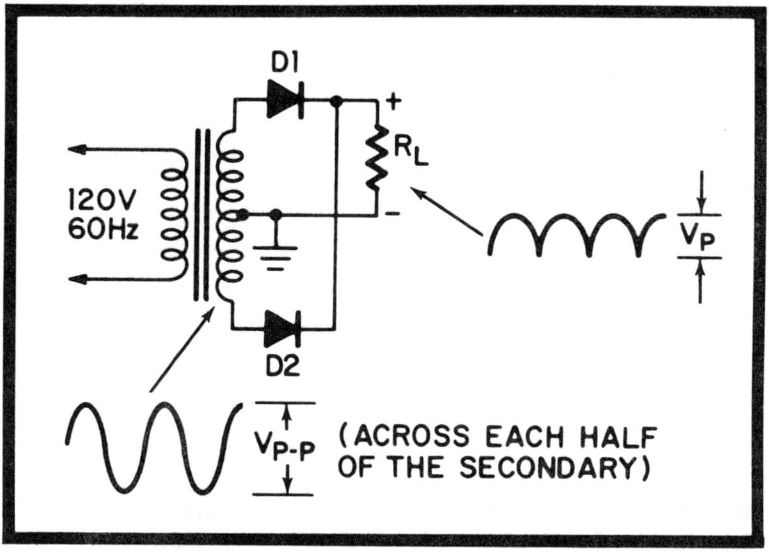

Fig. 12-3. Full-wave power supply circuit.

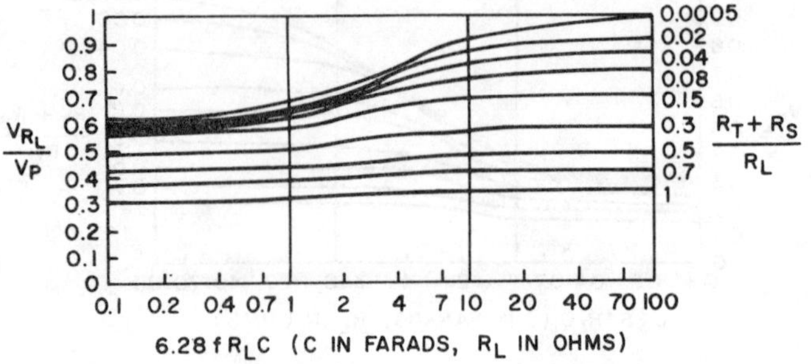

Fig. 12-4. Plot for a full-wave power supply showing the V_{RL} divided by V_p to $6.28fR_LC$ relationship for various values of R_S divided by R_L. (V_p is the peak voltage across half the secondary winding of the power transformer.)

In the full-wave supply, two diodes are used; each conducts on alternate half cycles. The missing half cycle pulse at the output of the half-wave supply in Fig. 12-1 is filled in here. The unfiltered average DC voltage across the load is $2V_p$ divided by π while the RMS voltage is .707 V_p. Should a capacitor be placed across the load, the ripple will be less than in the half-wave supply because the capacitor is not required to hold the charge as long as was previously necessary. The capacitor is recharged each half cycle rather than waiting a full cycle for the next pulse, as is the case with the half-wave supply. The curves in Fig. 12-4 illustrate the various relationships for the full-wave supply. Here, V_p refers to the voltage across half the secondary of the transformer.

PRACTICAL POWER SUPPLIES

The power amplifier using a capacitor to couple the output transistors to the loudspeaker system (see Fig. 11-1) requires a single simple power supply. Only one voltage source is needed with respect to ground. The standard full-wave supply in Fig. 12-3, using a centertapped secondary winding, is commonly used in such cases.

The bridge circuit shown in Fig. 12-5 can also be used. It is not as costly to build as the one with the centertapped secondary on the transformer. For the identical amount of DC output current, less RMS current passes through the transformer in the bridge circuit than in the previously encountered arrangement. Hence the transformer used in the bridge circuit is the cheaper of the two. The major drawback with the bridge circuit is that four diodes are required. However, each diode is rated at the peak voltage across the secondary winding or 1.41 multiplied by the RMS voltage. In the center-tapped arrangement, the PIV of each diode is double this, or 2.82 V_{RMS}. Here, each diode will probably be more expensive than those required for the bridge supply.

The operation of the circuit is self-evident. During the half cycle when the upper leg of the winding is positive with respect to the lower end, electron current flows opposite to the arrows in the diode symbol through D4, R_L and D1. The polarity of the voltage across R_L is as shown. In the next half of the cycle, when the polarity of the voltage across the transformer is reversed, the electron current flows through D2, R_L and D3. Once again the polarity of the voltage developed across R_L is as shown. Current flows in the same direction through the resistor during both halves of the cycle. Full-wave rectification has been accomplished. The curves in Fig. 12-4 can be used to determine the output voltages with the particular filter capacitor used. Since the circuit is "floating," either end of the load resistor can be grounded.

Power amplifier circuits which do **not** use a capacitor to couple the output devices to the loudspeaker system (see Fig. 11-5), require two identical supplies. One must be positive with respect to ground and the other must be negative. The positive voltage is applied to one half of the output stage while the negative voltage is applied to the other half. A circuit which can provide this is shown in Fig. 12-6. It resembles the bridge,

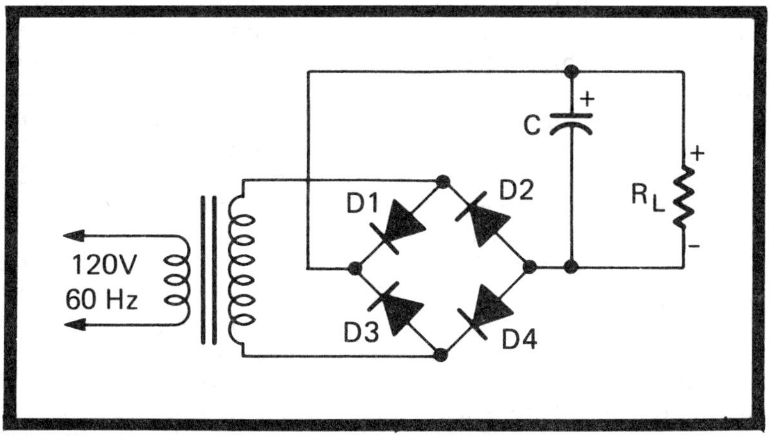

Fig. 12-5. A bridge circuit supplies a full-wave rectified output more economically than does the conventional half-wave center-tapped transformer circuit.

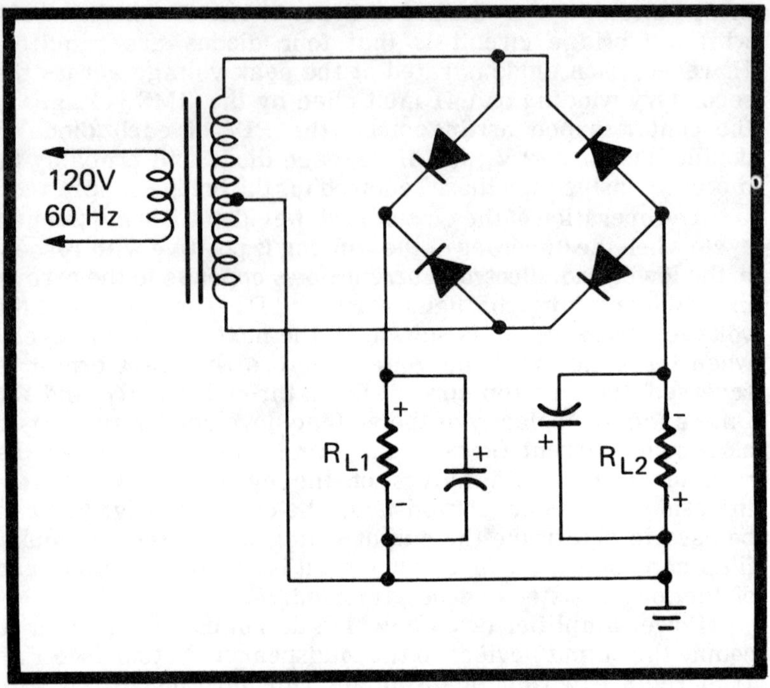

Fig. 12-6. The circuit is designed to supply identical positive and negative voltages to an amplifier.

but is actually analyzed as two full-wave supplies wired across one center-tapped transformer winding.

The DC power supplied to the output stages may contain a considerable amount of ripple, but the hum will nevertheless remain inaudible. Voltage amplifiers preceding the power stage require relatively smooth supplies. Ripple in these portions of the amplifier is magnified by all succeeding stages. A ripple-free supply is most critical when considering phono or tape preamplifiers. Here, the power source must be virtually free of any trace of ripple. One function of the R-C decoupling network in Fig. 12-7 is to accomplish this goal.

Relatively poorly filtered voltage is used as the supply for the output transistors at A. The drivers are supplied power from point B in the decoupling network. The ripple present at A is filtered by the R1-C1 network to provide power with a better ripple factor for the drivers than is supplied to the output devices. Ripple is further reduced before being applied to amplifier stages connected to points C and D by two additional R-C filter sections. The various resistors in the decoupling circuit serve a dual function—as components in the

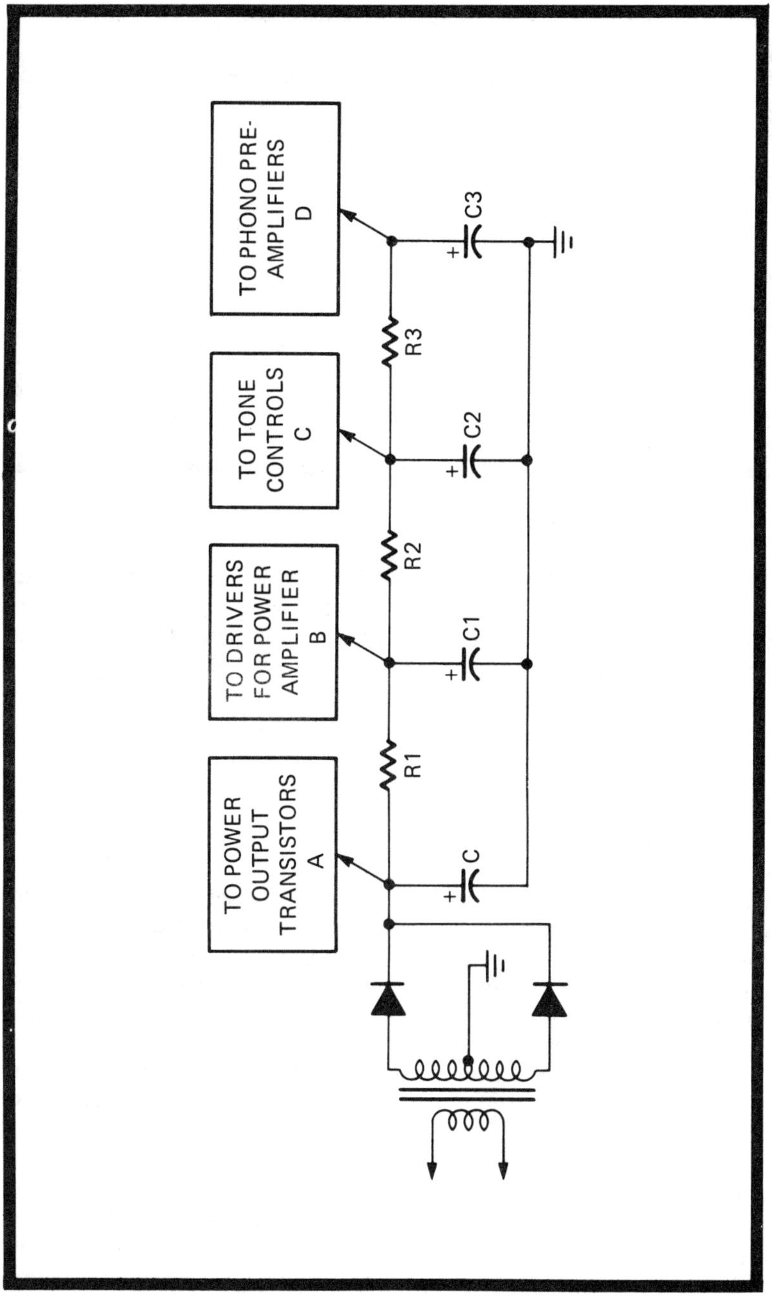

Fig. 12-7. This filter circuit provides the correct operating voltages to each section of the amplifier and prevents unwanted interstage coupling.

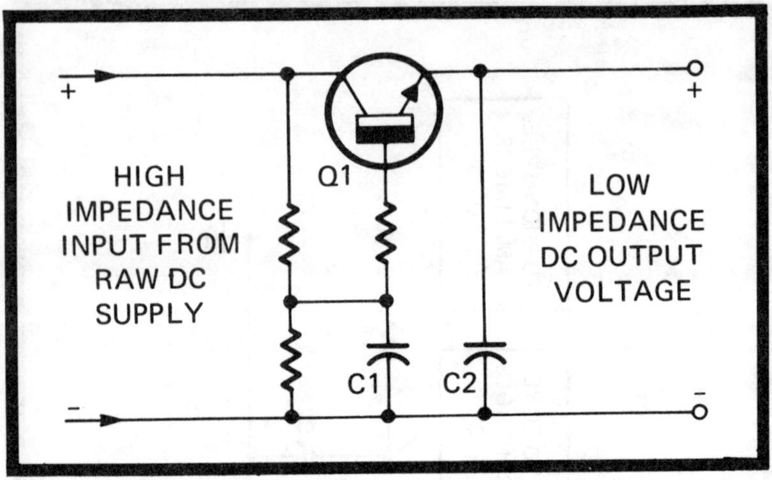

Fig. 12-8. A high input impedance DC voltage is reduced to a low-impedance output voltage by this circuit.

filter network and a voltage divider to establish the proper voltages across the capacitors for the various sections of the amplifier. Once the resistors have been chosen, the capacitors can be determined experimentally. Just observe the contribution of hum by each stage as different capacitors are substituted into the network. Use the smallest capacitor that will do the job.

Decoupling circuits serve another very important function. Each power supply has a specific impedance. Amplifier stages can be inadvertently coupled to each other through the supply. Coupling of this type can cause the amplifier to oscillate at a low frequency. This type of oscillation is known as motorboating. Stages are separated from each other by deriving their required supply voltages across different capacitors of the decoupling circuit. It is best not to connect more than two transistors to any one point in the filter network or decoupling chain.

Some amplifiers are unstable unless the supply voltage presents a low impedance. This characteristic can be achieved by placing a transistor in a power supply circuit shown in Fig. 12-8. The resistors in the circuit are chosen to establish the bias voltage needed to put Q1 in the active region. The capacitors are ripple filters. Voltages at the output across C2 are equal to the input voltage less the few volts dropped across Q1. The stage must be capable of passing all current required at the output. Q1 must not be in saturation. It is not in saturation if the voltage measured between the collector and emitter is greater than the saturation voltage of the device. The power the transistor dissipates is equal to the product of the current through Q1 and the voltage across it.

Should the audio amplifier be combined with a tuner in one circuit, the resulting receiver derives its total power from one supply. The low-impedance power supply is usually not adequate to provide the near perfect DC voltage needed at the tuner section. Voltage supplied to the tuner (and frequently even to the preamplifier section) frequently must be regulated. This is due to the fact that the supply used for the tuner is simply a filtered extension of the supply providing power to the output transistors. Power transistors draw a considerable amount of current. The current varies with the portion of the signal cycle being amplified. Therefore, this current variation produces cyclic voltage fluctuations in the unregulated supply. When used as a power source for a tuner or preamplifier, the supply with large voltage fluctuations can cause the tuner to perform inadequately.

Zener diodes may be used in a circuit to provide the simplest form of voltage regulation. The circuit in Fig. 12-9 can be used as a basic arrangement to help describe a design procedure. Zener diodes are devices in which the reverse breakdown voltage is controlled. Once it breaks down, the voltage across the diode, and across any load in parallel with the diode, is maintained fairly constant despite some changes in the amount of current flowing through the zener device.

In designing the circuit, determine the voltage required across the load, R_L, and the current, I_L, that must flow

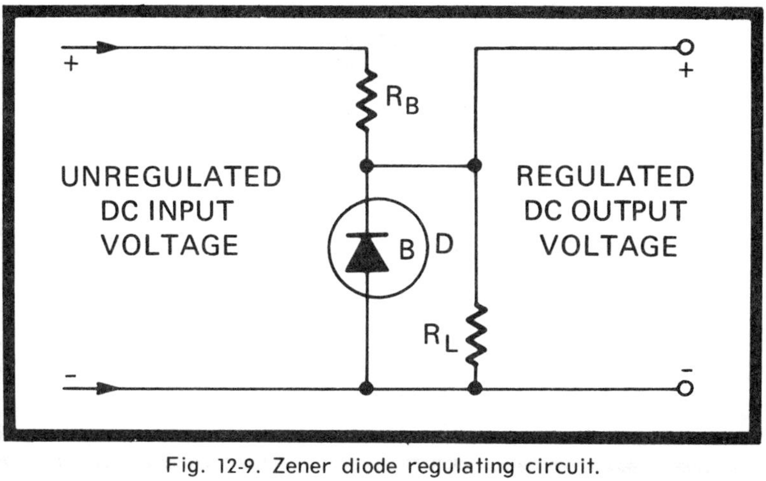

Fig. 12-9. Zener diode regulating circuit.

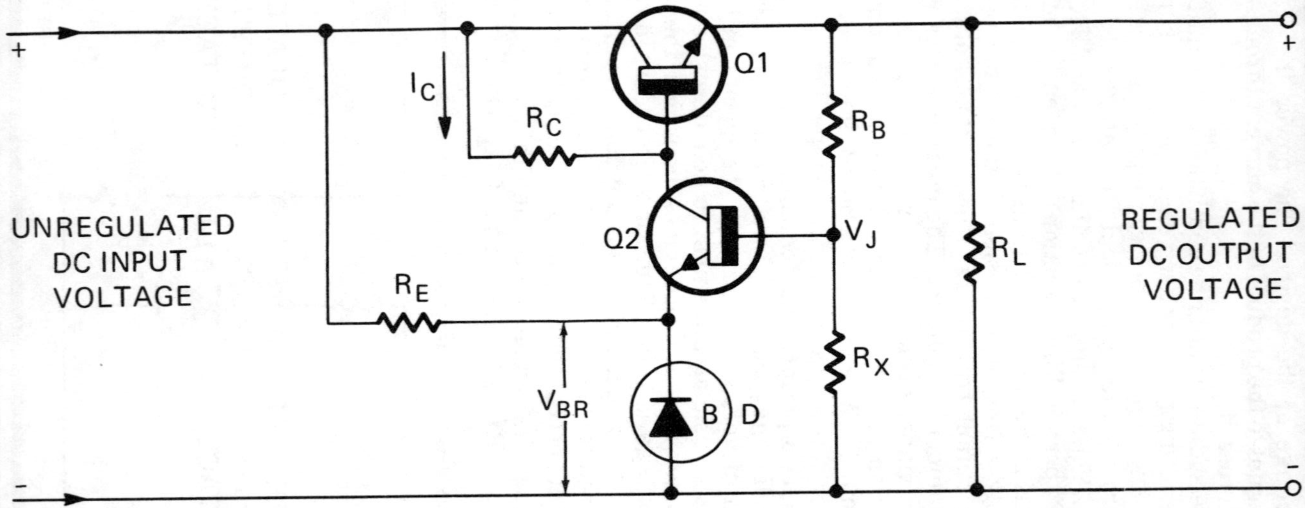

Fig. 12-10. Feedback regulator circuit.

through it. Choose a zener diode with the breakdown voltage, V_{BR}, equal to the voltage required across R_L. The average zener current, I_Z, can be assumed to be 25 percent of that through R_L. The power the zener must be capable of dissipating is $I_Z V_{BR}$ plus a 25-percent safety factor. If the minimum unregulated voltage supplied to the circuit is $V_{sdc}(min)$, the maximum size R_B can be is $(V_{sdc}(min)$ $V_{BR})$ divided by $(I_L + I_Z)$, by Ohm's Law. The power R_B must be capable of dissipating is $(V_{sdc}(max) - V_{BR})^2$ divided by R_B, where $V_{sdc}(max)$ is the maximum unregulated voltage.

Better regulation than can be provided by the zener circuit in Fig. 12-9, is at times required to feed various stages of the amplifier and receiver. A transistor feedback circuit such as the one shown in Fig. 12-10 may be used. Here, the output voltage is sensed and fed back to the base of Q2. This transistor is an important factor in controlling the current flowing into the base and subsequently out of the emitter of Q1. Variations in the current flowing into the base of Q1 corrects the emitter current in such a manner as to maintain it, and the subsequent output voltage developed across the load resistor, at a constant value.

Q1 passes DC current in an emitter-follower circuit. The base current of Q1 is determined by R_C and the collector current of Q2, because R_C limits the total current that can flow from the base of Q1 and collector of Q2. For example, suppose that the DC input voltage increases with a consequent increase in voltage across R_L. The voltage at V_J, the junction of resistor R_B and R_X, is likewise increased. Since the voltage across the zener diode, D, and hence at the emitter of Q2, is constant, the base-emitter voltage is increased. The current through the base-emitter junction of Q2 is likewise increased, as are the emitter and collector currents through this transistor. Q2 now demands a larger portion of the current that can flow through R_C. Less current now remains for the base of Q1, reducing its emitter current as well as the current through and the voltage developed across the load. The voltage across R_L is restored to the level it was before the input voltage shifted. Should the supply voltage be reduced rather than increased, a sequence similar to that outlined above would result. The feedback circuit has been established in a type of operational amplifier arrangement.

In the design procedure, the minimum unregulated DC voltage, $V_{sdc}(min)$, must be determined when a maximum current, $I_{sdc}(max)$, is flowing at the unregulated input of the supply. Likewise, the DC voltage at the input will be at a maximum, V_{sdc}(max), when the minimum current, I_{sdc}(min), is flowing. The input impedance of the unregulated supply is, from Ohm's Law,

$$R_S = \frac{V_{sdc}(max) - V_{sdc}(min)}{I_{sdc}(max) - I_{sdc}(min)} \quad (12\text{-}1)$$

The desired regulated voltage, V_{REG}, must then be specified.

The breakdown voltage of the zener diode in the circuit should be somewhere between V_{REG} divided by 4 and $3V_{REG}$ divided by 4. Diodes with breakdown voltages about midway between these two values are most desirable.

Current through R_c should be about one-twentieth the current through R_L. Regulated current, I_{REG}, is equal to V_{REG} divided by R_L (if R_L is a constant the current as well as the voltage are regulated) so that the current through R_c is approximately V_{REG} divided by $20R_L$ equals I_c.

Ideally, the beta of Q1 should be more than 500. Base current for Q1 will then be about I_{REG} divided by 500. About one-25th of the current that flows through R_c will thus flow into the base of Q1. However, transistors with betas in the range of 500 are hard to get unless they are of the small-signal type. These are seldom useful as the current pass transistor, Q1, in regulated supply circuits due to their limited current and power capabilities. Q1 can be replaced with two or more devices in a Darlington arrangement as shown in Fig. 9-31 or in a complementary circuit arrangement (Fig. 9-34). The product of the betas of all devices used should be greater than 500.

I_{REG} must flow through R_s and develop a voltage across it. $V_{sdc}(min) - I_{REG}R_s$ is the minimum voltage appearing at the junction of R_c, R_E and the collector of Q1. Since the voltage at the base of Q1 is the regulated voltage plus the approximate 0.7 volt between the base and emitter junction of that transistor, the maximum size R_c can be is:

$$R_c = \frac{V_{sdc}(min) - I_{REG}R_S - (V_{REG} + 0.7)}{V_{REG}/20R_L} \quad (12\text{-}2)$$

if it is assumed that I_c equals V_{REG} divided by $20R_L$. Notice that if more than one transistor is used in place of Q1, the 0.7 in the equation must be multiplied by the number of devices involved.

The maximum current, $I_c(max)$, through R_c flows when the unregulated voltage of the supply is at its maximum, $V_{sdc}(max)$:

$$I_c(max) = \frac{V_{sdc}(max) - I_{REG}R_S - (V_{REG} + 0.7)}{R_c} \qquad (12\text{-}3)$$

This is somewhat more than the current flowing through Q2 and D. A decent safety factor dictates that the power Q2 should be capable of dissipating is $I_c(max)(V_{REG} + 0.7 - V_{BR})$.

Current through the zener diode must be sufficient to maintain the device in its constant voltage breakdown region at all times. If it is insufficient, additional diode current can be funneled off the power supply through R_E. The power dissipated by D, the zener diode, is equal to its breakdown voltage multiplied by the sum of $I_c(max)$ and the current flowing through R_E. A substantial safety factor is built into this calculation.

The maximum voltage across Q1 is $V_{sdc}(max) - I_{REG}R_s - V_{REG}$. Multiply this by I_{REG} to determine the power Q1 must be capable of dissipating.

The sum of resistors R_B and R_X is about 100 times the size of R_L. Individual components should be chosen so that V_j is about equal to $V_{BR} + 0.7$. Trimming of the two resistors for the precise regulated voltage desired should be done in the laboratory.

Specifying Components

Three components in any power supply can cause serious problems and must be carefully specified. These are the power transformer, the silicon rectifiers and the input filter capacitor.

Two primary factors which determine the rectifier to be used in any circuit are the DC forward current it must pass and the DC reverse voltage it must block. The current that flows through the load is V_{RL}, the voltage across the load divided by R_L, the load resistance. In the case of the half-wave circuit, the rectifier must carry all this current. In the full-wave circuit, the load current is divided equally between the diodes.

Since a 0.7 forward voltage drop exists across the diode simultaneously with the forward current flow, I_F, the power dissipated by the diode is the product of the RMS current and RMS voltage averaged over the complete cycle. This power generates heat which must frequently be drawn away from the junction by heat sinks. Large rectifiers are mounted in metal cases which can be clamped to a chassis or other mass of metal for this purpose. Small units, rated up to about 3 amperes, use thick leads for conducting heat away from the junction. This latter structure is either a glass or plastic case. The plastic case is more desirable for better heat dissipation, while the glass envelope is used where an absolute hermetic seal is required.

So as a start, specify the average forward DC current. Use maximum figures if the load and diode current vary. Give yourself plenty of leeway; multiply the calculated or measured current by two.

Next, the forward surge current must be considered. Normally, the load current is at its average I_F. However, the filter capacitor across the load is a short circuit at the instant DC is supplied, and it is a partial short until the capacitor is fully charged. During this time, the only resistance in the circuit is the sum of the transformer and diode resistance. The load resistor is shorted by the filter capacitor. A surge current exists initially due to this low circuit resistance. It is approximately the peak voltage V_P (see Figs. 12-1 and 12-3) divided by the sum of all the resistance remaining in the circuit that is not shorted by the filter capacitor.

Diode surge capabilities are denoted by the peak current permitted to flow for one half cycle or for a full cycle. The duration of a cycle of 60-Hz current is one-60th or .0167 second. Half cycle time is obviously half that figure or 0.0083 second. For 50-Hz lines, this becomes 0.02 and 0.01 for full and half cycle time, respectively. It doubles for the 25-Hz line.

The time constant of the power supply circuit is the product of the filter capacitor in microfarads with the sum of the AC resistance of the rectifier (26 divided by I_F with I_F expressed in milliamperes) and the DC resistance of the transformer winding. This product must be less than 0.0167 for a permissible surge current when the diode is rated for a 1-cycle surge and less than 0.0083 for an allowed surge current when the diode is rated for a ½-cycle surge. This, of course, assumes 60 Hz is the base frequency.

Another consideration is the peak inverse voltage rating of the rectifier. The reverse voltage, V_R, the rectifiers must withstand is equal to twice V_P. Frequently, however, there are peaks and pulses in the power line which can exceed the calculated V_R by a large factor. It is wise to use rectifiers with V_R (or V_{BR} or PIV) ratings exceeding the calculated value by two.

The various shapes the reverse breakdown curve can take are shown in Fig. 12-11. Two different types of diode characteristics, A and B, are shown. Both diodes have the same breakdown voltage, V_{BR}. Diode A has a lower AC and DC resistance after breakdown than does diode B. A large current and voltage pulse is more likely to overheat the junction of B than that of A. A diode with the breakdown characteristics of A is more desirable. These are referred to as controlled avalanche breakdown rectifiers.

The power transformer can be specified with the help of the curves in Figs. 12-2 and 12-4. In addition, temperature rise is an important part of the transformer's specifications. Actually, it is desirable to limit the final transformer temperature to 95 degrees C, since the insulation is normally guaranteed to about 105 degrees C. Considering that the transformer should, at worst, operate in a 55-degree C ambient, the temperature rise should be limited to 95 degrees — 55 degrees or 40 degrees Celsius.

The resistance of the copper wire used in the transformer changes with temperature. It is information of this change which is used in a method of determining the maximum temperature at which the transformer is operating. To determine the temperature rise using this change-of-resistance method, the transformer is first permitted to stay in a room with no power being applied to it. The temperature of the room as well as the resistance of the 110-volt or 120-volt primary winding is noted. Refer to this temperature as T_A and to the resistance of the winding at room temperature as R_A.

Now run the equipment with its normal load for about six hours. The best way to do this with a power amplifier is to place an 8-ohm load resistor with an adequate wattage rating

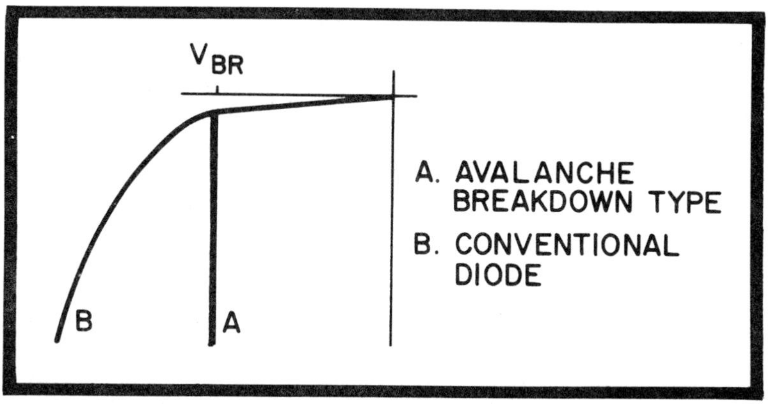

Fig. 12-11. Breakdown characteristic curves for two different types of diodes: avalanche breakdown type (A) and conventional type of diode (B).

at the output and feed a music signal to the amplifier's input. Increase the level of the amplifier so that the peaks in the music just begin to clip, as observed on a scope connected across the 8-ohm load resistor. Keep the amplifier enclosed in its cabinet. After six hours, disconnect the amplifier from the power source and quickly measure the resistance of the primary winding. Refer to this "hot" resistance as R_H. The temperature rise, T_R, above the ambient is close to:

$$T_R = \frac{R_H - R_A}{R_A\,(.00393)} \qquad (12\text{-}4)$$

Add T_R to T_A and you have the total temperature of the windings. The more accurate your readings of R_H and R_A are, the more accurate will be your determination of the temperature rise, T_R.

The third critical component in the power supply is the first or input filter capacitor. Much of the ripple current passes through this capacitor. Ripple current flowing through the capacitor when it is used in a half-wave circuit is:

$$I_R = 6.28fCE_R \qquad (12\text{-}5)$$

where f is the line frequency in Hz, C is the capacity in farads, and E_R is the RMS ripple voltage in volts. For full-wave circuits, I_R is double that noted in Equation 12-5.

Capacitors are designed to pass specific amounts of ripple current. Capacitors in large cans or cases can pass more current before overheating than those in smaller enclosures. For a specific size of case, the ripple capability of a capacitor is proportional to $C_{1/2}$. If the capacitor is subjected to excess current, it will lose a portion of its initial capacity and, consequently, lose its ability to operate properly in the circuit. Once the value of the desired capacitor has been established and the ripple current requirements have been determined, the manufacturer of the capacitor should be consulted to determine if his device will function properly. Due to economic considerations, a capacitor with less than adequate ability to pass ripple current will frequently be used. This will, of course, limit the life of the component.

As an exercise, assume you need a power supply to deliver 30 volts at 1 ampere. You want low ripple voltage so you select the full-wave circuit with a shunt filter capacitor. The load resistance is R_L equals 30 volts divided by 1 ampere or 30 ohms. For reasonable regulation (variation of output voltage with the change in load current) the resistance of the transformer, R_T, should be low. Assume all is satisfactory if the ratio of ($R_S + R_T$) to R_L is 0.08. Since r_{AC} (forward AC resistance of the rectifier) is the only factor comprising R_S (all resistance in series with the rectifier including the forward AC resistance of the rectifier itself) and assuming r_{AC} is negligible compared to R_T, R_T must be $0.08R_L$ or 2.4 ohms.

It can be shown that the ripple voltage is fairly low if $6.28fR_LC$ is made equal to or made greater than 5. Substituting f equals 60 and R_L equals 30 into the equality yields a filter capacitor requirement of at least 443 mfd. Feed this information into the curve in Fig. 12-4.

The ratio of the output voltage, V_{RL}, to the peak supply voltage, V_P, for half the transformer is about 0.73. Since V_{RL} is 30, then V_P equals 30 divided by 0.73 or 43 volts. The RMS voltage across each half of the transformer is V_P divided by $\sqrt{2}$ equals 43 divided by 1.41 or 30.4 volts. From the calculations, you find you need a transformer with a 60.8-volt center-tapped secondary and a maximum effective total winding resistance of 2 x 2.4 ohms or 4.8 ohms.

The continuous DC forward current rating of the silicon rectifiers can be 1 ampere. The surge current in the circuit is V_P divided by R_T equals 43 divided by 2.4 or 18 amperes. The RC time constant is $(2.4)(443 \times 10^{-6})$ equals 1060×10^{-6} equals 0.0016 seconds. Since this is less than the time for one half cycle, a diode with a surge current rating of more than 18 amperes for ½ cycle is satisfactory.

The peak inverse voltage rating for the rectifiers must be greater than $2V_P$ equals 86 volts. A 100-volt rectifier will be satisfactory if there are no instantaneous surge voltage peaks in the line To account for these surges, you can double your voltage specification for the rectifier and-or use a device with a sharp avalanche breakdown characteristic.

Layout and Wiring Considerations

All the derivations and discussions here and above describe decisions essential to design an audio amplifier. Once the paper design has been completed, and even after the basic laboratory experimentation has been done, the amplifier may be a "dud." The actual physical layout and wiring of the various elements of the circuit when built into a unified physical device requires much forethought and consideration.

For example, the power transformer above can wreak havoc with hum. It can radiate so badly as to establish hum current in sensitive circuits and in the chassis proper. Amplify the hum inadvertently and you have a useless audio amplifier. The flux density of a transformer is the determining factor of

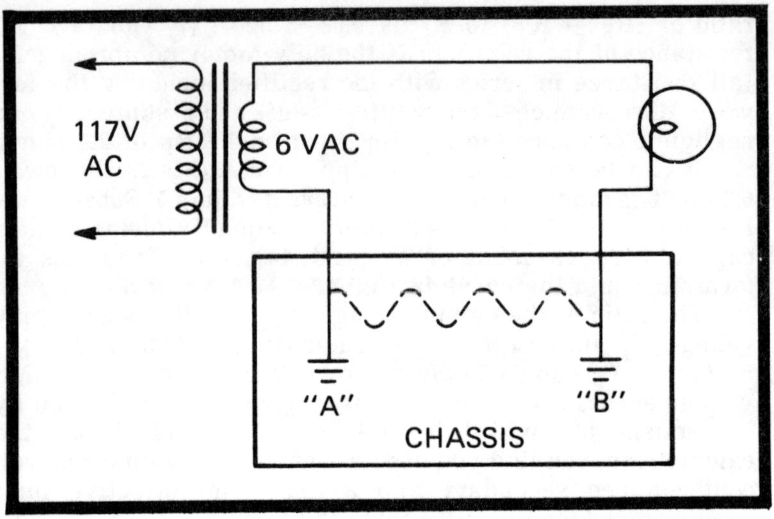

Fig. 12-12. Hum currents are present in a chassis when it is used as part of a pilot-lamp circuit.

just how much of a field it will radiate. You can specify that it be limited to 60,000 or 70,000 gauss, but how are you going to measure it?

In the late 1950s, one transformer manufacturer wound a coil for me consisting of 9200 turns of wire on an American Molded Bobbin 206A form. It may or may not have been a standard throughout the industry. This coil was to be placed against each side of the transformer in turn, when only the primary or 120-volt AC winding was connected. None of the other windings were to be connected and were thus left floating. I did find that if the voltage developed across the leads of the "search coil" was about 1 volt, the radiation of the transformer was within reason for audio applications. Thereafter, I specified that the search coil voltage of all transformers used in audio equipment was to be within 1 volt. If you cannot get the exact coil I have, perhaps you can get another one and calibrate it against the hum voltage induced into it by a known good transformer.

Hum current similar to that generated by the transformer can also be established by improper connections of pilot light bulbs in the amplifier. In Fig. 12-12, the 6-volt winding is grounded to point A on the chassis while the bulb is connected to remote point B on the chassis. AC hum current flows between these two points, establishing a ripple current in this area of the chassis. Voltage is developed across the chassis resistance A — B.

Now suppose that a transistor base circuit is connected to the chassis at points C and D, located somewhere between A and B, as shown in Fig. 12-13. A portion of the hum voltage across A—B is developed across chassis resistor C—D. Being in the base-emitter circuit of the transistor, the hum is amplified by the device along with the input signal, e_{in}. Hum voltage developed across A—C is in series with e_{in} and is also amplified. The voltage developed across D—B at the output may be insufficient to interfere with the much larger output signal.

In practical amplifiers, the power transformer is placed closest to the output stages. Mounted remote from the input circuit, the transformer's contribution to hum is minimized.

The circuit should be wired so that ground loops such as those illustrated in Figs. 12-12 and 12-13 cannot be formed. In Fig. 12-12, connect the 6-volt winding to the bulb with a wire and not through the chassis. The circuit may be grounded at only **one** point on the chassis. In Fig. 12-13, connect A, C and D together with wires and then bring one lead from the floating wires to the chassis.

As a general rule, solder all connections from each individual stage to one floating point or lug. Then connect the individual stages (except the output) together, in sequence from the output to the input. Normally, there will be one chassis ground for the output stages and one for all other

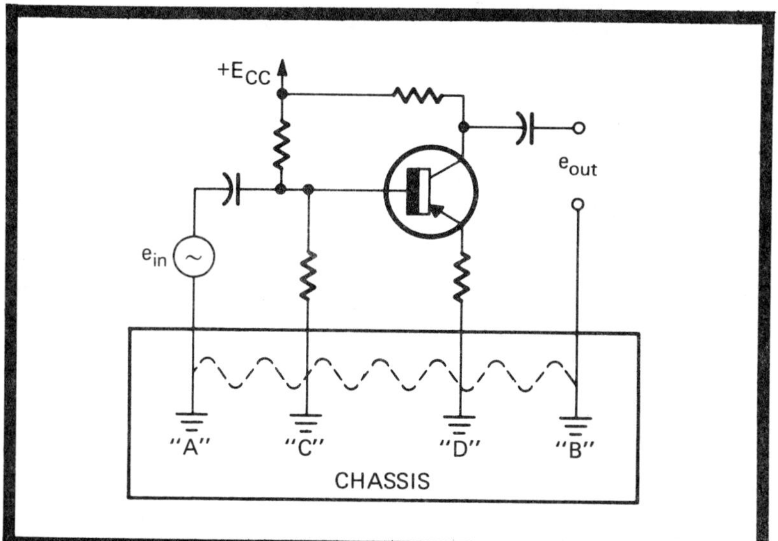

Fig. 12-13. If a transistor circuit is connected with ground returns made to the chassis, the resulting hum currents will be amplified by the transistor.

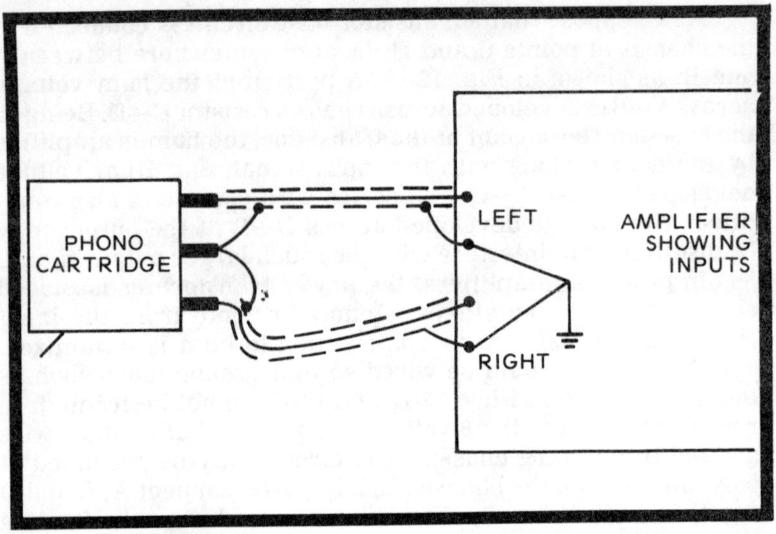

Fig. 12-14. Ground loops can be formed outside the chassis due to an improper method of connecting the signal source to the amplifier. The drawing shows an improper method of connecting an amplifier to a phono cartridge.

stages near the input transistor. In stereo amplifiers, there will be two such sequences of connections.

At times it will be desirable to connect all grounds to one long lug with one end riveted or bolted to the chassis. In this case, the stages should be connected along the length of the lug in the sequence just outlined. The power stage should be connected to a point on the lug closest to the chassis, and all other stages connected in sequence until the input stage is connected to the top of the lug, farthest from the chassis.

Avoid ground loops when connecting phonograph cartridges or tuners to a stereo amplifier. On some tuners and cartridges, one common ground is used at the outputs of both stereo channels. Two shielded leads, grounded at both the source and amplifier, conduct the stereo signals. Because the grounded shields form a complete circuit and can conduct induced current from nearby power lines, hum will appear at the input of the amplifier. This can be illustrated with the help of Fig. 12-14.

A cartridge is shown with one common terminal for both the left and the right channels. Two shielded leads are used to connect the outputs from the cartridge to the inputs on the amplifier. Both shields are grounded at the cartridge as well as at the amplifier, forming a complete circuit. Hum can be readily induced in the shield.

Proper interconnecting wiring practices dictate that one of the shield connections at the cartridge be opened, so that both shields will be grounded only at the amplifier. The closed-circuit loop will no longer exist. Similar tactics can be used when connecting any other equipment to the amplifier, as well as for the connection of stages and wires within the amplifier proper.

Index

A

B

C

D

E

F

G

H

J

L

M

N

O

P

Q